T0141894

Advances in Intelligent Systems and Computing

Volume 747

Series editor

Janusz Kacprzyk, Polish Academy of Sciences, Warsaw, Poland
e-mail: kacprzyk@ibspan.waw.pl

The series "Advances in Intelligent Systems and Computing" contains publications on theory, applications, and design methods of Intelligent Systems and Intelligent Computing. Virtually all disciplines such as engineering, natural sciences, computer and information science, ICT, economics, business, e-commerce, environment, healthcare, life science are covered. The list of topics spans all the areas of modern intelligent systems and computing.

The publications within "Advances in Intelligent Systems and Computing" are primarily textbooks and proceedings of important conferences, symposia and congresses. They cover significant recent developments in the field, both of a foundational and applicable character. An important characteristic feature of the series is the short publication time and world-wide distribution. This permits a rapid and broad dissemination of research results.

Advisory Board

Chairman

Nikhil R. Pal, Indian Statistical Institute, Kolkata, India
e-mail: nikhil@isical.ac.in

Members

Rafael Bello Perez, Universidad Central "Marta Abreu" de Las Villas, Santa Clara, Cuba
e-mail: rbellop@uclv.edu.cu

Emilio S. Corchado, University of Salamanca, Salamanca, Spain
e-mail: escorchado@usal.es

Hani Hagras, University of Essex, Colchester, UK
e-mail: hani@essex.ac.uk

László T. Kóczy, Széchenyi István University, Győr, Hungary
e-mail: koczy@sze.hu

Vladik Kreinovich, University of Texas at El Paso, El Paso, USA
e-mail: vladik@utep.edu

Chin-Teng Lin, National Chiao Tung University, Hsinchu, Taiwan
e-mail: ctlin@mail.nctu.edu.tw

Jie Lu, University of Technology, Sydney, Australia
e-mail: Jie.Lu@uts.edu.au

Patricia Melin, Tijuana Institute of Technology, Tijuana, Mexico
e-mail: epmelin@hafsamx.org

Nadia Nedjah, State University of Rio de Janeiro, Rio de Janeiro, Brazil
e-mail: nadia@eng.uerj.br

Ngoc Thanh Nguyen, Wroclaw University of Technology, Wroclaw, Poland
e-mail: Ngoc-Thanh.Nguyen@pwr.edu.pl

Jun Wang, The Chinese University of Hong Kong, Shatin, Hong Kong
e-mail: jwang@mae.cuhk.edu.hk

More information about this series at http://www.springer.com/series/11156

Álvaro Rocha · Hojjat Adeli
Luís Paulo Reis · Sandra Costanzo
Editors

Trends and Advances in Information Systems and Technologies

Volume 3

Springer

Editors
Álvaro Rocha
Departamento de Engenharia Informática
Universidade de Coimbra
Coimbra
Portugal

Hojjat Adeli
College of Engineering
The Ohio State University
Columbus, OH
USA

Luís Paulo Reis
DSI/EEUM
Universidade do Minho
Guimarães
Portugal

Sandra Costanzo
DIMES
Universita della Calabria
Arcavacata di Rende
Italy

ISSN 2194-5357   ISSN 2194-5365   (electronic)
Advances in Intelligent Systems and Computing
ISBN 978-3-319-77699-6   ISBN 978-3-319-77700-9   (eBook)
https://doi.org/10.1007/978-3-319-77700-9

Library of Congress Control Number: 2018936909

© Springer International Publishing AG, part of Springer Nature 2018
This work is subject to copyright. All rights are reserved by the Publisher, whether the whole or part of the material is concerned, specifically the rights of translation, reprinting, reuse of illustrations, recitation, broadcasting, reproduction on microfilms or in any other physical way, and transmission or information storage and retrieval, electronic adaptation, computer software, or by similar or dissimilar methodology now known or hereafter developed.
The use of general descriptive names, registered names, trademarks, service marks, etc. in this publication does not imply, even in the absence of a specific statement, that such names are exempt from the relevant protective laws and regulations and therefore free for general use.
The publisher, the authors and the editors are safe to assume that the advice and information in this book are believed to be true and accurate at the date of publication. Neither the publisher nor the authors or the editors give a warranty, express or implied, with respect to the material contained herein or for any errors or omissions that may have been made. The publisher remains neutral with regard to jurisdictional claims in published maps and institutional affiliations.

Printed on acid-free paper

This Springer imprint is published by the registered company Springer International Publishing AG part of Springer Nature
The registered company address is: Gewerbestrasse 11, 6330 Cham, Switzerland

Preface

This book contains a selection of papers accepted for presentation and discussion at The 2018 World Conference on Information Systems and Technologies (WorldCIST'18). This conference had the support of the IEEE Systems, Man, and Cybernetics Society, AISTI (Iberian Association for Information Systems and Technologies/Associação Ibérica de Sistemas e Tecnologias de Informação), University of Calabria, and GIIM (Global Institute for IT Management). It took place at Naples, Italy, on March 27–29, 2018.

The World Conference on Information Systems and Technologies (WorldCIST) is a global forum for researchers and practitioners to present and discuss recent results and innovations, current trends, professional experiences and challenges of modern Information Systems and Technologies research, technological development and applications. One of its main aims is to strengthen the drive toward a holistic symbiosis between academy, society, and industry. WorldCIST'18 built on the successes of WorldCIST'13 held at Olhão, Algarve, Portugal, WorldCIST'14 held at Funchal, Madeira, Portugal, WorldCIST'15 held at São Miguel, Azores, Portugal, WorldCIST'16 held at Recife, Pernambuco, Brazil, and WorldCIST'17 which took place at Porto Santo Island, Madeira, Portugal.

The Program Committee of WorldCIST'18 was composed of a multidisciplinary group of experts and those who are intimately concerned with Information Systems and Technologies. They have had the responsibility for evaluating, in a 'blind review' process, the papers received for each of the main themes proposed for the conference: (A) Information and Knowledge Management; (B) Organizational Models and Information Systems; (C) Software and Systems Modeling; (D) Software Systems, Architectures, Applications, and Tools; (E) Multimedia Systems and Applications; (F) Computer Networks, Mobility and Pervasive Systems; (G) Intelligent and Decision Support Systems; (H) Big Data Analytics and Applications; (I) Human–Computer Interaction; (J) Ethics, Computers & Security; (K) Health Informatics; (L) Information Technologies in Education; (M) Information Technologies in Radiocommunications; (N) Technologies for Biomedical Applications.

The conference also included workshop sessions taking place in parallel with the conference ones. Workshop sessions covered themes such as (i) Applied Statistics and Data Analysis using Computer Science, (ii) Artificial Intelligence in Fashion Industry, (iii) Emerging Trends and Challenges in Business Process Management, (iv) Emerging Trends, Challenges, and Solutions in Infrastructures and Smart Building Management, (v) Healthcare Information Systems Interoperability, Security, and Efficiency, (vi) Intelligent and Collaborative Decision Support Systems for Improving Manufacturing Processes, (vii) New Pedagogical Approaches with Technologies, (viii) Pervasive Information Systems, (ix) Technologies in the workplace—use and impact on workers.

WorldCIST'18 received more than 400 contributions from 66 countries around the world. The papers accepted for presentation and discussion at the conference are published by Springer (this book) and by AISTI (one issue in the Journal of Information Systems Engineering & Management) and will be submitted for indexing by ISI, EI-Compendex, SCOPUS, DBLP, and/or Google Scholar, among others. Extended versions of selected best papers will be published in special or regular issues of relevant journals, mainly SCI/SSCI and Scopus/EI-Compendex indexed journals.

We acknowledge all of those that contributed to the staging of WorldCIST'18 (authors, committees, workshop organizers, and sponsors). We deeply appreciate their involvement and support that was crucial for the success of WorldCIST'18.

March 2018   Álvaro Rocha
                                                                        Hojjat Adeli
                                                                      Luís Paulo Reis
                                                                     Sandra Costanzo

Organization

Conference

General Chair

Álvaro Rocha   University of Coimbra, Portugal

Co-chairs

Hojjat Adeli   The Ohio State University, USA
Luis Paulo Reis   University of Minho, Portugal
Sandra Costanzo   University of Calabria, Italy

Advisory Committee

Ana Maria Correia (Chair)   University of Sheffield, UK
Ben Lev   Drexel University, USA
Chris Kimble   KEDGE Business School & MRM, UM2,
                                   Montpellier, France
David Garson   North Carolina State University, USA
Florin Gheorghe Filip   Romanian Academy, Romania
Gintautas Dzemyda   Vilnius University, Lithuania
Janusz Kacprzyk   Polish Academy of Sciences, Poland
Jeroen van den Hoven   Delft University of Technology, Netherlands
João Tavares   University of Porto, Portugal
Jon Hall   The Open University, UK
Karl Stroetmann   Empirica Communication & Technology
                                   Research, Germany
Kathleen Carley   Carnegie Mellon University, USA

Keng Siau	Missouri University of Science and Technology, USA
Ladislav Hluchy	Slovak Academy of Sciences, Slovakia
Marjan Mernik	University of Maribor, Slovenia
Michael Koenig	Long Island University, USA
Miguel-Angel Sicilia	University of Alcalá, Spain
Péter Kacsuk	University of Westminster, UK
Peter Sloot	University of Amsterdam, Netherland
Robertas Damaševičius	Kaunas University of Technology, Lithuania
Sujeet Shenoi	University of Tulsa, USA
Ted Shortliffe	Arizona State University, USA
Wan Kyun Chung	POSTECH, Korea
Wim Van Grembergen	University of Antwerp, Belgium
Yu-Chuan (Jack) Li	Taipei Medical University, Taiwan

Program Committee

Adnan Mahmood	Waterford Institute of Technology, Ireland
Adriana Fernandes	ISCTE-IUL, Portugal
Adriana Peña Pérez Negrón	Universidad de Guadalajara, Mexico
Adriani Besimi	South East European University, Macedonia
Agostino Forestiero	ICAR-CNR, Italy
Ahmed El Oualkadi	Abdelmalek Essaadi University, Morocco
Alan Ramirez-Noriega	Universidad Autónoma de Baja California, Mexico
Alberto Freitas	FMUP, University of Porto, Portugal
Alcides Fonseca	University of Lisbon, Portugal
Alessio Ferrari	ISTI-CNR, Italy
Alexandru Vulpe	University Politehnica of Bucharest, Romania
Ali Alsoufi	University of Bahrain, Bahrain
Ali Idri	ENSIAS, University Mohamed V, Morocco
Almir Souza Silva Neto	IFMA, Brazil
Alvaro Arenas	IE Business School, Spain
Amit Shelef	Sapir Academic College, Israel
Ana Luis	University of Coimbra, Portugal
Anabela Tereso	University of Minho, Portugal
Anacleto Correia	CINAV, Portugal
Anca Alexandra Purcarea	University Politehnica of Bucharest, Romania
André Marcos Silva	Centro Universitário Adventista de São Paulo (UNASP), Brazil
Aneta Poniszewska-Maranda	Lodz University of Technology, Poland
Angeles Quezada	Instituto Tecnologico de Tijuana, Mexico
Ankur Singh Bist	KIET, India
Antonio Borgia	University of Calabria, Italy

António Gonçalves	University of Lisbon & INESC ID, Portugal
Antonio Jiménez-Martín	Universidad Politécnica de Madrid, Spain
Antonio Pereira	Polytechnic of Leiria, Portugal
Antonio Raffo	University of Calabria, Italy
Armando Mendes	University of Azores, Portugal
Armando Toda	University of São Paulo, Brazil
Arsénio Monteiro Reis	University of Trás-os-Montes e Alto Douro, Portugal
Arslan Enikeev	Kazan Federal University, Russia
Benedita Malheiro	Polytechnic of Porto, ISEP, Portugal
Bing Li	Arizona State University, USA
Borja Bordel	Universidad Politécnica de Madrid, Spain
Branko Perisic	Faculty of Technical Sciences, Serbia
Carla Pinto	Polytechnic of Porto, ISEP, Portugal
Carla Santos Pereira	Universidade Portucalense, Portugal
Carlos Costa	ISCTE-IUL, Portugal
Catarina Reis	Polytechnic of Leiria, Portugal
Cédric Gaspoz	University of Applied Sciences Western Switzerland (HES-SO), Switzerland
Cengiz Acarturk	Middle East Technical University, Turkey
Cesar Collazos	Universidad del Cauca, Colombia
Christophe Feltus	LIST, Luxembourg
Christophe Soares	University Fernando Pessoa, Portugal
Christos Bouras	University of Patras, Greece
Ciro Martins	University of Aveiro, Portugal
Claudio Sapateiro	Polytechnic of Setúbal, Portugal
Cristian García Bauza	PLADEMA-UNICEN-CONICET, Argentina
Cristian Mateus	ISISTAN-CONICET, UNICEN, Argentina
Dagmar Cámská	Czech Technical University, Prague, Czech Republic
Dalila Durães	Technical University of Madrid, Spain
Daniel Castro Silva	University of Porto, Portugal
Dante Carrizo	Universidad de Atacama, Chile
David Cortés-Polo	Fundación COMPUTAEX, Spain
Djamel Kehil	ENSET SKIKDA, Algeria
Dorgival Netto	IFMS - Federal Institute of Mato Grosso do Sul, Brazil
Edita Butrime	Lithuanian University of Health Sciences, Lithuania
Edna Dias Canedo	University of Brasilia, Brazil
Eduardo Santos	Pontifical Catholic University of Paraná, Brazil
Egils Ginters	Riga Technical University, Latvia
Elena Mikhailova	Saint Petersburg State University, Russia
Emiliano Reynares	CONICET - CIDISI UTN FRSF, Argentina
Evandro Costa	Federal University of Alagoas, Brazil

Farhan Siddiqui   USA
Felix Blazquez-Lozano   University of Coruña, Spain
Fernando Bobillo   University of Zaragoza, Spain
Fernando Moreira   Portucalense University, Portugal
Fernando Ribeiro   Polytechnic of Castelo Branco, Portugal
Filipe Portela   University of Minho, Portugal
Filippo Neri   University of Naples, Italy
Fionn Murtagh   University of Huddersfield, UK
Firat Bestepe   Republic of Turkey Ministry of Development,
                                      Turkey
Floriano Scioscia   Polytechnic University of Bari, Italy
Francesca Venneri   University of Calabria, Italy
Francesco Bianconi   Università degli Studi di Perugia, Italy
Francisco García-Peñalvo   University of Salamanca, Spain
Frederico Branco   University of Trás-os-Montes e Alto Douro,
                                      Portugal
Gali Naveh   Shamoon College of Engineering, Israel
Galim Vakhitov   Kazan Federal University, Russia
George Suciu   BEIA, Romania
Ghani Albaali   Princess Sumaya University for Technology,
                                      Jordan
Gian Piero Zarri   University Paris-Sorbonne, France
Giuseppe Di Massa   University of Calabria, Italy
Gonçalo Paiva Dias   University of Aveiro, Portugal
Goreti Marreiros   ISEP/GECAD, Portugal
Graciela Lara López   University of Guadalajara, Mexico
Habiba Drias   University of Science and Technology Houari
                                      Boumediene, Algeria
Hafed Zarzour   University of Souk Ahras, Algeria
Hatem Ben Sta   University of Tunis at El Manar, Tunisia
Hector Fernando Gomez   Universidad Tecnica de Ambato, Ecuador
  Alvarado
Helia Guerra   University of the Azores, Portugal
Henrique da Mota Silveira   University of Campinas (UNICAMP), Brazil
Hing Kai Chan   University of Nottingham Ningbo China, China
Hugo Paredes   INESC TEC and Universidade de
                                      Trás-os-Montes e Alto Douro, Portugal
Igor Aguilar Alonso   Universidad Politécnica de Madrid, Spain
Ilham Slimani   ENSIAS, Morocco
Imen Ben Said   Université de Sfax, Tunisia
Ina Schiering   Ostfalia University of Applied Sciences,
                                      Germany
Inês Domingues   University of Coimbra, Portugal
Isabel Lopes   Instituto Politécnico de Bragança, Portugal
Isabel Pedrosa   Coimbra Business School - ISCAC, Portugal

Ivan Lukovic   University of Novi Sad, Serbia
J. João Almeida   University of Minho, Portugal
Jan Kubicek   Technical University of Ostrava, Czech Republic
Javier Medina   Universidad Distrital Francisco José de Caldas,
                                      Colombia
Jean Robert Kala Kamdjoug   Catholic University of Central Africa, Cameroon
Jezreel Mejia   CIMAT, Unidad Zacatecas, Mexico
Jikai Li   The College of New Jersey, USA
Joao Carlos Silva   IPCA, Portugal
João Manuel R. S. Tavares   University of Porto, FEUP, Portugal
João Rodrigues   University of the Algarve, Portugal
Jorge Esparteiro Garcia   Polytechnic Institute of Viana do Castelo,
                                      Portugal
Jorge Gomes   University of Lisbon, Portugal
Jorge Oliveira e Sá   University of Minho, Portugal
José Braga de Vasconcelos   Universidade New Atlântica, Portugal
Jose Luis Herrero Agustin   University of Extremadura, Spain
José Luís Reis   ISMAI, Portugal
José M. Parente de Oliveira   Aeronautics institute of Technology, Brazil
José Martins   University of Trás-os-Montes e Alto Douro,
                                      Portugal
José Manuel Torres   University Fernando Pessoa, Portugal
José-Luís Pereira   Universidade do Minho, Portugal
Juan Jesus Ojeda-Castelo   University of Almeria, Spain
Juan M. Santos   University of Vigo, Spain
Juan Pablo Damato   UNCPBA-CONICET, Argentina
Julie Dugdale   University Grenoble Alps, France
Juncal Gutiérrez-Artacho   University of Granada, Spain
Justin Dauwels   Nanyang Technological University, Singapore
Justyna Trojanowska   Poznan University of Technology, Poland
Kalai Anand Ratnam   Asia Pacific University of Technology
                                      & Innovation, Malaysia
Katsuyuki Umezawa   Shonan Institute of Technology, Japan
Kevin K. W. Ho   University of Guam, Guam
Khalid Benali   LORIA - University of Lorraine, France
Korhan Gunel   Adnan Menderes University, Turkey
Krzysztof Wolk   Polish-Japanese Academy of Information
                                      Technology, Poland
Kuan Yew Wong   Universiti Teknologi Malaysia (UTM), Malaysia
Laila Cheikhi   University Mohammed V, Rabat, Morocco
Laura Varela-Candamio   Universidade da Coruña, Spain
Laurentiu Boicescu   E.T.T.I. U.P.B., Romania
Leonardo Botega   University Centre Eurípides of Marília
                                      (UNIVEM), Brazil
Leonel Morgado   University Aberta and INESC TEC, Portugal

Leonid Leonidovich   Moscow Aviation Institute (National Research
  Khoroshko   University), Russia
Letícia Helena Januário   Federal University of São João do Rei, Brazil
Lila Rao-Graham   University of the West Indies, Jamaica
Luis Alvarez Sabucedo   University of Vigo, Spain
Luis Gomes   Nova de Lisboa University, Portugal
Luis Gomes   University of the Azores, Portugal
Luis Silva Rodrigues   Polytechnic of Porto, Portugal
Luz Sussy Bayona Oré   Universidad Nacional Mayor de San Marcos,
                                Peru
Magdalena Diering   Poznan University of Technology, Poland
Majida Alasady   Tikrit University, Iraq
Manuel Pérez-Cota   University of Vigo, Spain
Manuel Silva   Polytechnic of Porto and INESC TEC, Portugal
Marcelo Mendonça Teixeira   Federal Rural University of Pernambuco, Brazil
Marco Ronchetti   University of Trento, Italy
Mareca María PIlar   University Politécnica de Madrid, Spain
Marek Kvet   Zilinska Univerzita v Ziline, Slovakia
Maria João Ferreira   University Portucalense, Portugal
Maria João Varanda Pereira   Polytechnic of Bragança, Portugal
Maria José Sousa   University of Coimbra, Portugal
Maria Koziri   UTH, Greece
María Teresa García-Álvarez   University of A Coruna, Spain
Marijana Despotovic-Zrakic   Faculty Organizational Science, Serbia
Marina Ismail   Universiti Teknologi MARA, Malaysia
Mário Antunes   Polytechnic of Leiria & CRACS INESC TEC,
                                Portugal
Marisa Maximiano   Polytechnic Institute of Leiria, Portugal
Marisol Garcia-Valls   Universidad Carlos III de Madrid, Spain
Maristela Holanda   University of Brasilia, Brazil
Marius Vochin   E.T.T.I. U.P.B., Romania
Marlene Goncalves da Silva   Universidad Simón Bolívar, Venezuela
Martin Henkel   Stockholm University, Sweden
Martín López Nores   University of Vigo, Spain
Martin Zelm   INTEROP-VLab, Belgium
Mawloud Mosbah   University 20 Août 1955 of Skikda, Algeria
Meryeme Hadni   FSDM, Morocco
Michal Adamczak   Poznan School of Logistics, Poland
Michal Kvet   University of Zilina, Slovakia
Michele Ruta   Politecnico di Bari, Italy
Miguel António Sovierzoski   Federal University of Technology - Paraná,
                                Brazil
Mihai Lungu   University of Craiova, Romania
Mircea Georgescu   Al. I. Cuza University of Iasi, Romania

Mirna Muñoz	Centro de Investigación en Matemáticas A.C., Mexico
Miroslav Bures	Czech Technical University in Prague, Czech Republic
Mohamed Abouzeid	Innovations for High Performance Microelectronics IHP, Germany
Mohamed Serrhini	University Mohamed First Oujda, Morocco
Mokhtar Amami	Royal Military College of Canada, Canada
Monica Leba	University of Petrosani, Romania
Mu-Song Chen	Da-Yeh University, China
Natalia Grafeeva	Saint Petersburg University, Russia
Natalia Miloslavskaya	National Research Nuclear University MEPhI, Russia
Naveed Ahmed	University of Sharjah, United Arab Emirates
Nelson Rocha	University of Aveiro, Portugal
Nikolai Prokopyev	Kazan Federal University, Russia
Noemi Emanuela Cazzaniga	Politecnico di Milano, Italy
Nuno Melão	Polytechnic of Viseu, Portugal
Nuno Octávio Fernandes	Polytechnic of Castelo Branco, Portugal
Patricia Zachman	Universidad Nacional del Chaco Austral, Argentina
Paula Alexandra Rego	Polytechnic of Viana do Castelo & LIACC, Portugal
Paula Viana	Polytechnic of Porto & INESC TEC, Portugal
Paulo Maio	Polytechnic of Porto, ISEP, Portugal
Paulo Novais	University of Minho, Portugal
Paweł Karczmarek	The John Paul II Catholic University of Lublin, Poland
Pedro Henriques Abreu	University of Coimbra, Portugal
Pedro Rangel Henriques	University of Minho, Portugal
Pedro Sobral	University Fernando Pessoa, Portugal
Pedro Sousa	University of Minho, Portugal
Philipp Brune	Neu-Ulm University of Applied Sciences, Germany
Radu-Emil Precup	Politehnica University of Timisoara, Romania
Rafael M. Luque Baena	University of Malaga, Spain
Rahim Rahmani	University Stockholm, Sweden
Ramayah T.	Universiti Sains Malaysia, Malaysia
Ramiro Delgado	Universidad de las Fuerzas Armadas ESPE, Ecuador
Ramiro Gonçalves	University of Trás-os-Montes e Alto Douro & INESC TEC, Portugal
Ramon Alcarria	Universidad Politécnica de Madrid, Spain
Ramon Fabregat Gesa	University of Girona, Spain
Refet Polat	Yasar University, Turkey

Reyes Juárez Ramírez	Universidad Autonoma de Baja California, Mexico
Rui Jose	University of Minho, Portugal
Rui Pitarma	Polytechnic Institute of Guarda, Portugal
Rui S. Moreira	UFP & INESC TEC & LIACC, Portugal
Rustam Burnashev	Kazan Federal University, Russia
Saeed Salah	Al-Quds University, Palestine
Said Achchab	Mohammed V University in Rabat, Morocco
Saide Saide	State Islamic University of Sultan Syarif Kasim Riau, Indonesia
Sajid Anwar	Institute of Management Sciences Peshawar, Pakistan
Salama Mostafa	Universiti Tun Hussein Onn Malaysia, Malaysia
Samantha Jiménez	Universidad Autónoma de Baja California, Mexico
Sami Habib	Kuwait University, Kuwait
Samuel Ekundayo	Eastern Institute of Technology, New Zealand
Samuel Fosso Wamba	Toulouse Business School, France
Sandra Costanzo	University of Calabria, Italy
Sandra Patricia Cano Mazuera	University of San Buenaventura Cali, Colombia
Sergio Albiol-Pérez	University of Zaragoza, Spain
Sergio Inzunza	Universidad Autonoma de Baja California, Mexico
Shahnawaz Talpur	Mehran University of Engineering & Technology Jamshoro, Pakistan
Silviu Vert	Politehnica University of Timisoara, Romania
Simona Mirela	University of Petrosani, Romania
Slawomir Zolkiewski	Silesian University of Technology, Poland
Solange N. Alves-Souza	University of São Paulo, Brazil
Solange Rito Lima	University of Minho, Portugal
Sorin Zoican	Polytechnic University of Bucharest, Romania
Souraya Hamida	Batna 2 University, Algeria
Stanley Lima	Technische Universität Dresden, Germany
Stefan Pickl	UBw München COMTESSA, Germany
Sümeyya Ilkin	Kocaeli University, Turkey
Syed Asim Ali	University of Karachi, Pakistan
Tatiana Antipova	Institute of Certified Specialists, Russia
Thanasis Loukopoulos	University of Thessaly, Greece
The Thanh Van	HCMC University of Food Industry, Vietnam
Thomas Weber	EPFL, Switzerland
Tiago Gonçalves	University of Lisbon, Portugal
Toshihiko Kato	University of Electro-communications, Japan
Tzung-Pei Hong	National University of Kaohsiung, Taiwan
Valentina Colla	Scuola Superiore Sant'Anna, Italy

Victor Alves	University of Minho, Portugal
Victor Georgiev	Kazan Federal University, Russia
Victor Hugo Medina Garcia	Universidad Distrital Francisco José de Caldas, Colombia
Vincenza Carchiolo	University of Catania, Italy
Visar Shehu	South East European University, Macedonia
Vitalyi Igorevich Talanin	Zaporozhye Institute of Economics & Information Technologies, Ukraine
Wolf Zimmermann	Martin Luther University Halle-Wittenberg, Germany
Yair Wiseman	Bar Ilan University, Israel
Yuhua Li	University of Salford, UK
Yuwei Lin	University of Stirling, UK
Yves Rybarczyk	Universidad de Las Américas, Ecuador
Zorica Bogdanovic	University of Belgrade, Serbia

Workshops

Applied Statistics and Data Analysis Using Computer Science

Organizing Committee

Brígida Mónica Faria	Higher School of Health/Polytechnic of Porto (ESS/P.Porto), Portugal
Joaquim Gonçalves	Polytechnic Institute of Cávado e Ave, Portugal
João Mendes Moreira	FEUP - University of Porto, Portugal

Program Committee

Ana Maria Maqueda	University of Valencia, Spain
Armando Sousa	INESC TEC, FEUP, University of Porto, Portugal
Inês Domingues	IPO Porto & University of Coimbra, Portugal
Joao Fabro	UTFPR - Federal University of Technology-Parana, Brazil
Luis Paulo Reis	University of Minho, Portugal
Marcelo Petry	Federal University of Santa Catarina, INESC P&D, Brazil
Nuno Lau	University of Aveiro, Portugal
Pedro Henriques Abreu	FCTUC-DEI/CISUC, University of Coimbra, Portugal

Artificial Intelligence in Fashion Industry

Organizing Committee

Evandro Costa	Federal University of Alagoas, Brazil
Olga C. Santos	UNED, Spain

Hemilis Rocha	Federal Institute of Alagoas, Brazil
Thales Vieira	Federal University of Alagoas, Brazil
Sumit Borar	Myntra Designs, India

Program Committee

Evandro Costa	Federal University of Alagoas, Brazil
Hemilis Rocha	Federal Institute of Alagoas, Brazil
Hiranmay Ghosh	TCS Innovation Labs, India
Julita Vassileva	University of Saskatchewan, Canada
LongLong Yu	Wide Eyes Technologies, Spain
Olga Santos	UNED, Spain
Ralph Deters	University of Saskatchewan, Canada
Shuhui Jiang	Northeastern University, USA
Sumit Borar	Myntra Designs, India
Susana Zoghbi	KU Leuven, Belgium
Thales Vieira	Federal University of Alagoas, Brazil
Tincuta Heinzel	Cornell University/UAUIM, Bucharest

Emerging Trends and Challenges in Business Process Management

Organizing Committee

José Luis Pereira	University of Minho, Portugal
Rui Dinis Sousa	University of Minho, Portugal
Pascal Ravesteijn	HU University, Netherlands

Program Committee

Ana Almeida	School of Engineering - Polytechnic of Porto, Portugal
Armin Stein	University of Muenster, Germany
Daniel Pacheco Lacerda	UNISINOS University, Brazil
Fernando Belfo	ISCAC Coimbra Business School & ALGORITMI Center, Portugal
Frederico Branco	UTAD, Portugal
João Varajão	University of Minho, Portugal
Jorge Oliveira Sá	University of Minho, Portugal
José Martins	UTAD, Portugal
Luis Miguel Ferreira	University of Aveiro, Portugal
Marie-Claude (Maric) Boudreau	University of Georgia, USA

Manoel Veras	Federal University of Rio Grande do Norte, Brazil
Pedro Malta	University of Minho, Portugal
Rafael Paim	Federal Center for Technical Education (Cefet-RJ), Brazil
Sílvia Inês Dallavalle de Pádua	University of São Paulo, Brazil
Vinícius Carvalho Cardoso	Federal University of Rio de Janeiro, Brazil
Vitor Santos	NOVA Information Management School, Portugal

Healthcare Information Systems: Interoperability, Security and Efficiency

Organizing Committee

José Machado	University of Minho, Portugal
António Abelha	University of Minho, Portugal
Anastasius Mooumtzoglou	European Society for Quality in Healthcare, Greece

Program Committee

Alberto Freitas	University of Porto, Portugal
Ana Azevedo	ISCAP/IPP, Portugal
Ângelo Costa	University of Minho, Portugal
Armando B. Mendes	University of Azores, Portugal
Badra Khellat Kihel	University of Oran, Algeria
Cesar Analide	University of Minho, Portugal
Chang Choi	Chosun University, Korea
Chun-Wei Tsai	National Chung-Hsing University, Taiwan
Davide Carneiro	University of Minho, Portugal
Filipe Portela	University of Minho, Portugal
Helia Guerra	University of Azores, Portugal
Henrique Vicente	University of Évora, Portugal
Hoon Ko	J. E. Purkyně University, Korea
Hugo Peixoto	University of Minho, Portugal
Jason Jung	Chung-Ang University, Korea
Joao Ramos	University of Minho, Portugal
Jorge Ribeiro	IPVC, Portugal
Jose Cascalho	Universidade dos Azores, Portugal
José Martins	UTAD, Portugal
Jose Neves	University of Minho, Portugal
Luis Mendes Gomes	University of Azores, Portugal

Mas Sahidayana Mohktar	University of Malaya, Malaysia
Paulo Moura Oliveira	UTAD, Portugal
Paulo Novais	University of Minho, Portugal
Renata Baracho	Universidade Federal de Minas Gerais, Brazil
Tiago Oliveira	National Institute of Informatics, Japan
Victor Alves	University of Minho, Portugal

Intelligent and Collaborative Decision Support Systems for Improving Manufacturing Processes

Organizing Committee

Justyna Trojanowska	Poznan University of Technology, Poland
Magdalena Diering	Poznan University of Technology, Poland
José Machado	Department of Mechanical Engineering, University of Minho, Portugal
Leonilde Varela	University of Minho, Portugal

Program Committee

Agnieszka Kujawińska	Poznan University of Technology, Poland
Boris Delibašić	University of Belgrade, Serbia
Damian Krenczyk	Silesian University of Technology, Poland
Dariusz Sędziak	Poznan University of Technology, Poland
Fatima Dargam	SimTech Simulation Technology, Austria
Filip Górski	Poznan University of Technology, Poland
Grzegorz Królczyk	Opole University of Technology, Poland
Jason Papathanasiou	University of Macedonia, Greece
Krzysztof Żywicki	Poznan University of Technology, Poland
Michał Rogalewicz	Poznan University of Technology, Poland
Sachin Waigaonkar	Birla Institute of Technology & Science, India
Shaofeng Liu	Plymouth University, UK
Varinder Singh	BITS Pilani KK Birla Goa Campus, India
Vijaya Kumar	VIT University, India

New Pedagogical Approaches with Technologies

Organizing Committee

| Anabela Mesquita | CICE- ISCAP/IPP and Algoritmi Centre, Portugal |

Paula Peres   CICE- ISCAP/e-IPP, Politécnico do Porto,
                                        Portugal
Fernando Moreira   IJP - Universidade Portucalense and IEETA –
                                        UAveiro, Portugal

Program Committee

Alex Gomes   Universidade Federal de Pernambuco, Brazil
Armando Silva   Escola Superior de Educação do IPPorto,
                                        Portugal
Ana R. Luís   Universidade de Coimbra, Portugal
César Collazos   Universidad del Cauca, Colombia
Chia-Wen   Tsai, Ming Chuan University, Taiwan
João Batista   CICE/ISCA, University of Aveiro, Portugal
Lino Oliveira   ESEIG/IPP, Portugal
Luisa Moreno   Universidad de Sevilla, Spain
Manuel Perez Cota   University of Vigo, Spain
Paulino Silva   CEOS.PP - ISCAP/IPP, Portugal
Ramiro Gonçalves   UTAD, Vila Real, Portugal
Rosa Vicari   Universidade Federal do Rio Grande do Sul,
                                        Brazil
Stefania Manca   Instituto per le Tecnologie Didattiche, Italy

Pervasive Information Systems

Organizing Committee

Carlos Filipe Portela   Department of Information Systems, University
                                        of Minho, Portugal
Manuel Filipe Santos   Department of Information Systems, University
                                        of Minho, Portugal
Kostas Kolomvatsos   National and Kapodistrian University of Athens,
                                        Greece

Program Committee

António Abelha   University of Minho, Portugal
Christos Anagnostopoulos   University of Glasgow, UK
Cristina Alcaraz   University of Cagliari, Italy
Daniele Riboni   University of Milano, Italy
Fabio A. Schreiber   Politecnico Milano, Italy
Frederique Laforest   Laboratoire Hubert Curien, Univ. Saint Etienne,
                                        France
Hugo Peixoto   University of Minho, Portugal

Jarosław Jankowski	West Pomeranian University of Technology Szczecin, Poland
José Machado	University of Minho, Portugal
Júlio Duarte	University of Minho, Portugal
Karolina Baras	University of Madeira, Portugal
Nuno Marques	New University of Lisboa, Portugal
Ricardo Queiroz	ESMAD- P.Porto & CRACS - INESC TEC, Portugal
Sergio Ilarri	University of Zaragoza, Spain
Spyros Panagiotakis	Technological Educational Institution of Crete, Greece
Teresa Guarda	Universidad Estatal da Peninsula de Santa Elena - UPSE, Portugal
Vassilis Papataxiarhis	University of Athens, Greece

Technologies in the Workplace - Use and Impact on Workers

Organizing Committee

| Ana Veloso | Escola de Psicologia, Universidade do Minho, Portugal |
| Catarina Brandão | Faculdade de Psicologia e de Ciências da Educação, Universidade do Porto, Portugal |

Program Committee

Ana Teresa Ferreira-Oliveira	Technology and Management School, Viana do Castelo Polytechnic Institute, Portugal
Ana Veloso	Escola de Psicologia, Universidade do Minho, Portugal
Catarina Brandão	Faculdade de Psicologia e de Ciências da Educação, Universidade do Porto, Portugal
Esther Gracia	Universidad de Valencia, Spain
Guy Enosh	School of Social Work, Faculty of Welfare and Health Sciences, University of Haifa, Israel
Hatem Ocel	Karabuk University, Faculty of Art, Psychology Department, Turkey
Isabel Silva	Escola de Psicologia da Universidade do Minho, Portugal
Joana Santos	Universidade do Algarve, Portugal
Karin Sanders	UNSW Australia Business School, Australia

Mary Sandra Carlotto	Universidade do Vale do Rio dos Sinos, Brazil
Shay Tzafrir	Faculty of Management, University of Haifa, Israel
Snezhana Ilieva	Sofia University St. Kliment Ohridski, Bulgaria
Vicente Martinez Tur	University of Valencia, Faculty of Psychology, Spain

Contents

**Intelligent and Collaborative Decision Support Systems
for Improving Manufacturing Processes**

Applied Statistics and Data Analysis Using Computer Science

Agribusiness Intelligence: Grape Production Forecast Using Data Mining Techniques

Rosana Cavalcante de Oliveira[✉], João Mendes-Moreira, and Carlos Abreu Ferreira

LIAAD-INESC TEC, Campus da FEUP, Rua Dr. Roberto Frias, 4200-465 Porto, Portugal
{rosana.c.oliveira,joao.mendes.moreira,
carlos.ferreira}@inesctec.pt

Abstract. The agribusiness volatility is related to the uncertainty of the environment, rising demand, falling prices and new technologies. However, generation of agriculture data has increased over past years and can be used for a growing number of applications of data mining techniques in agriculture. The multidisciplinary approach of integrating computer science with agriculture will support the necessary decisions to be taken in order to mitigate risks and maximize profits. The present study analyzes different methods of regression applied in the study case of grapes production forecast. The selected methods were multivariate linear regression, regression trees, lasso and random forest. Their performance were compared against the predictions obtained by the company through the mean squared error and the coefficient of variation. The four regression methods used obtained better predictive results than the method used by the company with statistical significance < 0.5%.

Keywords: Production forecast · Data mining · Agribusiness · Grapes

1 Introduction

Vineyards in Portugal represent 9% of the European Union (EU), the fourth largest after Spain (30%), France (25%) and Italy (19%). In Portugal, the production of quality wines reaches 87.8% of the total, above the EU average (78.2%) [1]. Alentejo is the leading region in the Portuguese market, both in the market share in volume (47%) and in value (46%). The Alentejo wines bring together 1,900 grape growers and 235 companies that sell wines with the guarantee of origin and quality certified by the Alentejo Regional Wine Commission [2].

Agriculture data is highly diversified in terms of nature, interdependencies, and resources. For the balanced and sustainable development of agriculture, these resources need to be evaluated, monitored and analyzed. Sharma and Mehta (2012) [3] analyse the characteristics of agriculture data which is essentially seasonal and uncertain. They suggest the use of data mining techniques as a tool for knowledge management in agriculture. The most used data mining techniques in the field of agriculture were presented in [4–10]. Some of these techniques, such as the k-means, the k-nearest neighbor, artificial neural networks and support vector machines, are discussed and an application in agriculture for each of

© Springer International Publishing AG, part of Springer Nature 2018
Á. Rocha et al. (Eds.): WorldCIST'18 2018, AISC 747, pp. 3–8, 2018.
https://doi.org/10.1007/978-3-319-77700-9_1

these techniques is presented. Regression decision trees using recursive partitioning were applied to agriculture data with good predictive results [11–14].

Appropriate use of agricultural data often leads to considerable gains in efficiency, leading to an economical advantage. Given the importance of wine market and the uncertainty environment of the sector, production forecasting techniques that assist in the decision-making process become strategic. In this work, besides the type of soils, attributes such as caste type, plant age, fertilization plan, and irrigation strategy were analyzed in order to predict the annual production of the campaign. Based on the characteristics of the analyzed data and theoretical references, the following predictive methods were selected to estimate the grapes production: multivariate linear regression [15], regression trees [12], lasso [16] and random forest [17].

2   Methodology

The process of extracting useful knowledge from a collection of data includes data preparation and selection, data cleansing, incorporating prior knowledge on data sets and interpreting accurate solutions from the observed results. The data mining can be described as the process of building models that can be used to assist in our understanding of the world and to make predictions [18]. In this work, the algorithms were developed using R that is a free software environment for statistical computing and graphics [19]. In the case study, the goal is forecasting production of the campaign. This phase starts with the acquisition of data relevant to our task. After making the identified data available, an initial data examination is performed, leading to verification of the quality of the data. Part of the examination is also the computation of basic statistics of key attributes of the data and their correlations. In this case, the target attributes are production field. The information obtained in the exploratory data analysis can be used in the construction of the predictive models. The description of the predictive attributes is in Table 1.

Figure 1 illustrates the stages of the methodology that starts with analyzing and identifying predictive attributes. Based on the characteristics of the analyzed data, data were selected and the data mining techniques applied. The last stage was to compare the errors of the selected techniques with the prediction error of the company using the Coefficient of Variation (CV).

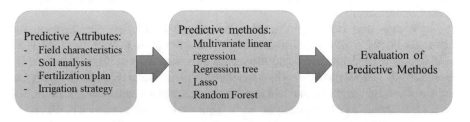

Fig. 1. Stages of the methodology

Table 1. Predictive attributes.

Field characteristics	Description
Caste	A variety that produces grapes with specific characteristics
Number of Plants	Number of Plants for each field
Plant age	Time that was planted in the field
pH water	Acidity of the soil solution
pH KCl	Acidity in the soil solution, plus the reserve acidity in the colloids
Organic matter(OM)	Set of chemical compounds formed by organic molecules found in natural environments
P	Chemical element Phosphorus
K	Chemical element Potassium
Mg	Chemical element Magnesium
PlanFertN	Fertilization plan of Nitrogen
PlanFertP	Fertilization plan of Phosphorus
PlanFertK	Fertilization plan of Potassium
PlanFertMg	Fertilization plan of Magnesium
Irrigation strategy	**Description**
TA	Quality Red Wine
TB	Medium Quality Red Wines
TC	Volume Production Red Wines
BA	Quality White Wine
BB	Medium Quality White Wine
I	New Vineyards

The dataset consists of production data of 2014, 2015 and 2016. The 2014 and 2015 data were used to train the model and the 2016 data was used to test the model. Each training example has 14 attributes (Table 1). Size: 400 training examples, 120 test examples. The target variable is the quantity produced in each field.

The deployment will be the core of the data mining process. Initially occurs the data preparation, the collected data is analyzed using statistical methods. The case study consisted of forecast production in a field of grapes in a farm located in the Alentejo region, Portugal. The methods tested here are computer-intensive algorithms that have been used in data mining and large-scale predictions, particularly in the field of machine learning [20]. Based on the characteristics of the analyzed data, the selected data mining techniques in order to forecast the production were: multivariate linear regression, regression trees, lasso and random forest.

The evaluation measures used to assess the predictive performance of the methods were the Mean Squared Error (MSE) and the Coefficient of Variation (CV). While the root MSE is expressed in the same unit measurement as the target attribute, the CV is independent of the measurement unit used. It is expressed as the ratio between the root MSE and the average value of the target attribute. Finally, the Friedman Test was applied to compare ranked outcomes.

3   Results and Discussion

The multivariate linear regression evaluates the variable-response relationship with each predictor. The learned model is used to make predictions. The residuals analysis is a set of techniques used to investigate the suitability of a regression model based on residues, which is given by the difference between the observed response variable and the estimated response variable.

Regression trees were implemented using the package rpart of R [19] an implementation of classification and regression binary trees. The first node of the tree represents the most important variable among the attributes analyzed, in the case is the type of caste, followed by irrigation strategy and plant age. As the regression tree grows, the sets of examples characterizing each leaf get smaller. Consequently, the estimates of error based on the samples in the leaves are not reliable.

Lasso is a regression method that performs both variable selection and regularization in order to enhance the predictive accuracy and interpretability of the produced model.

The random forest methods can be used to avoid overfitting. In this work after analyzing the problem, 5000 trees were used to build the model. The most important predictive attributes for production forecasting can be visualized in Fig. 2. The variables predicted to be important in the model help us to understand which variables are driving the production forecast. In the Random forest, the measures of variable importance are based on the mean squared error (MSE) and relate to the prediction accuracy of the out-of-bag portion of the data after permuting each predictor variable. The difference between the two MSEs is then averaged over all trees and normalized by the standard error. The higher the %incMSE is, more important the variable [21].

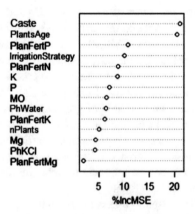

Fig. 2. Variable importance (%MSE)

The method used as baseline is the one used by the company. That is, it predicts for the current year the same production of the previous year. This method is known as persistence. The results obtained can be seen in Table 2.

Table 2. Evaluation of predictive models.

Forecast methods	Coef. variation
Company	0,34
Multivariate linear regression	0,25
Regression trees	0,28
Lasso	0,20
Random Forests	0,18

The Random Forest method presented the best result for this problem followed by the Lasso. Despite the predictive error obtained for multivariate linear regression and regression tree were lower than those predicted by the company, they are worse than the ones obtained by Lasso and Random forest methods, avoiding overfitting and selecting the predictive attributes that most impact the target attribute.

The Friedman Test result, for the production variable, shows a significant difference between the Lasso and Random Forest forecast regarding Company (p-value of 0.0046). These results point out that data mining methods are statistically better to predict the production yield than the company method with p-value $< 0.5\%$.

4 Conclusion

Predictions and trends are a necessity for decision support in an environment with uncertainty. This work applied forecasting techniques that use data mining models that capture the knowledge exhibited by the data. Techniques with greater predictive power are strategic for agribusiness, which despite having an environment of uncertainty, is not able to react quickly to market changes since their production cycles vary according to planted crops. In the case study presented, the production cycle is annual, and as it is verified in Fig. 2, the variables caste and plant age affects directly the production.

The results show that the models Random Forest and Lasso, methods that embed feature selection, have better forecast of grape production. The main advantage of extracting knowledge from a collection of data was the identification of attributes that most affect production to the detriment of the decrease of subjectivity from the model. The company will check the possibility of using the Random Forest method for the next production forecast. As future works, the model can be improved considering environmental variables such as carbon emission and irrigated water volume.

The main difficulty encountered was the small volume of historical data and the lack of a structured database. The next step is the development of an end user interface system that will store the production data and implement the developed methods. In this way, the system will be fed and updated continuously with data generated in the day-to-day business.

Acknowledgements. Rosana Cavalcante acknowledges the grant from project Smartfarming (POCI-01-0247-FEDER-018029), co-financed by the ERDF through COMPETE 2020 under P2020 Partnership Agreement and by the Portuguese Agency ANI. This work is also funded by

8   R. C. de Oliveira et al.

the ERDF through COMPETE 2020 within project POCI-01-0145-FEDER-006961, and by National Funds through the FCT as part of project UID/EEA/50014/2013.

References

1. Eurostat: Statistical office of the European Union. http://ec.europa.eu/eurostat
2. Agronegocios.eu: Portal de Informação Agroalimentar de Portugal. http://www.agronegocios.eu
3. Sharma, L., Mehta, N.: Data mining techniques: A tool for knowledge management system in agriculture. Int. J. Sci. Technol. Res. **1**(1), 67–73 (2012)
4. Allen, P.G.: Economic forecasting in agriculture. Int. J. Forecast. **10**, 81–135 (1994)
5. Mucherino, A., Papajorgji, P., Pardalos, P.: A survey of data mining techniques applied to agriculture. Oper. Res. **9**(2), 121–140 (2009)
6. Bhargavi, P., Jyothi, S.: Applying naive bayes data mining technique for classification of agricultural land soils. Int. J. Comput. Sci. Netw. Secur. **9**(8), 117–122 (2009)
7. Hira, S., Deshpande, P.: Data analysis using multidimensional modeling, statistical analysis and data mining on agriculture parameters. Procedia Comput. Sci. **54**, 431–439 (2015)
8. Ramya, M., Lokesh, V., Manjunath, T., Hegadi, R.: A predictive model construction for mulberry crop productivity. Procedia Comput. Sci. **45**, 156–165 (2015)
9. Khedr, A., Kadry, M., Walid, G.: Proposed framework for implementing data mining techniques to enhance decisions in agriculture sector applied case on food security information center ministry of agriculture, Egypt. Procedia Comput. Sci. **65**, 633–642 (2015)
10. Wen, Q., Mu, W., Sun, L., Hua, S., Zhou, Z.: Daily sales forecasting for grapes by support vector machine. In: Computer and Computing Technologies in Agriculture VII, vol. 420(3), pp. 351–360 (2014)
11. Breiman, L., Friedman, J., Olshen, R., Stone, C.: Classification and Regression Trees. Wadsworth & Brooks, Monterey (1984)
12. Therneau, T., Atkinson, E.: An introduction to recursive partitioning using the RPART routines. Technical report Mayo Foundation (1997)
13. Yi-Yang, G., Nan-Ping, R.: Data mining and analysis of our agriculture based on the decision tree. In: 2009 IEEE ISECS International Colloquium on Computing, Communication, Control, and Management, CCCM 2009, vol. 2, pp. 134–138 (2009)
14. Ruß, G.: Data mining of agricultural yield data: A comparison of regression models. In: Industrial Conference on Data Mining, pp. 24–37. Springer (2009)
15. Draper, N.R., Smith, H.: Applied Regression Analysis. Wiley, New York (2014)
16. Kukreja, S.L., Löfberg, J., Brenner, M.J.: A least absolute shrinlage and selection operator (LASSO) for nonlinear system identification. IFAC Proc. Volumes **39**, 814–819 (2006)
17. Liaw, A., Wiener, M.: Classification and regression by randomforest. R News **2**(3), 18–22 (2002)
18. Williams, G.: Data Mining with Rattle and R: The Art of Excavating Data for Knowledge Discovery. Springer, New York (2011)
19. R-project: The R Project for Statistical Computing. https://www.r-project.org/
20. Gama, J., Carvalho, A.P., Faceli, K., Lorena, A.C., Oliveira, M.: Extração de Conhecimento de Dados. Silabo, vol. 2, pp. 351–360 (2015)
21. Steinberg, D., Colla, P., Kerry, M.: MARS User Guide. Salford Systems, San Diego (1999)

Real-Time Tool for Human Gait Detection from Lower Trunk Acceleration

Helena R. Gonçalves[1]($\boxtimes$), Rui Moreira[1]($\boxtimes$), Ana Rodrigues[2]($\boxtimes$),
Graça Minas[1]($\boxtimes$), Luís Paulo Reis[3,4]($\boxtimes$), and Cristina P. Santos[1]($\boxtimes$)

[1] Center for MicroElectroMechanical (CMEMS),
University of Minho, Braga, Portugal
{a68371,a71335}@alunos.uminho.pt,
{gminas,cristina}@dei.uminho.pt
[2] Neurology Service, Hospital of Braga, Braga, Portugal
a.margarida.r@gmail.com
[3] Information Systems Department, University of Minho, Braga, Portugal
lpreis@dsi.uminho.pt
[4] LIACC - Artificial Intelligence and Computer Science Laboratory,
Porto, Portugal

Abstract. The continuous monitoring of human gait would allow to more objectively verify the abnormalities that arise from the most common pathologies. Therefore, this manuscript proposes a real-time tool for human gait detection from lower trunk acceleration. The vertical acceleration signal was acquired through an IMU mounted on a waistband, a wearable device. The proposed algorithm was based on a finite state machine (FSM) which includes a set of suitable decision rules and the detection of Heel-Strike (HS), Foot-flat (FF), Toe-off (TO), Mid-Stance (MS) and Heel-strike (HS) events for each leg. Results involved 7 healthy subjects which had to walk 20 m three times with a comfortable speed. The results showed that the proposed algorithm detects in real-time all the mentioned events with a high accuracy and time-effectiveness character. Also, the adaptability of the algorithm has also been verified, being easily adapted to some gait conditions, such as for different speeds and slopes. Further, the developed tool is modular and therefore can easily be integrated in another robotic control system for gait rehabilitation. These findings suggest that the proposed tool is suitable for the real-time gait analysis in real-life activities.

Keywords: Gait detection · Lower trunk · Acceleration · Real-time
Wearable

1 Introduction

Walking is one of the most common human physical activities and plays an important role in our daily tasks. The term "gait" is used to describe the way of walking, consisting in consecutive cycles subdivided in a sequence of events triggering transitions from one gait phase to another [1].

Due to the high number of gait pathologies that currently exist, clinicians need a simple, robust and efficient method to quantify patients' gait abnormalities [2].

© Springer International Publishing AG, part of Springer Nature 2018
Á. Rocha et al. (Eds.): WorldCIST'18 2018, AISC 747, pp. 9–18, 2018.
https://doi.org/10.1007/978-3-319-77700-9_2

These methods should work independently of the intra and inter gait variability of each patient and should allow a continuous evaluation. Accurate and efficient new tools for detecting gait events have been developed [2–4]. In fact, these new tools have proved to be important for the assistance and rehabilitation of the human gait, since they can be incorporated into devices, as orthoses or exoskeletons, which may be fundamental in the recovery of the patients' gait [1–4].

Previously, force platforms, stereo photogrammetric systems, optical bars, or video-analysis have been used to analyze the human gait [5, 6]. However, these devices present limitations making them not feasible for measurements on daily-life situations: they do not allow a complete analysis of the entire gait cycle, demand special environments and require long post-processing, especially when used for subjects with gait abnormalities [5]. Wearable sensors, such as Inertial Measurement Units (IMUs) which generally integrate accelerometers, gyroscopes and magnetometers, are an optimal alternative since they allow to evaluate gait in real-time without these restrictions. Furthermore, with the technological advances, these sensors are lighter and smaller, making them suitable to record gait information and be embedded in wearable devices for outdoor ambulatory applications [6].

The placement of IMUs in the body human for gait events identification, firstly considered the lower body segments (shank and foot). However, such approaches commonly require independent sensors for each lower limb, thus increasing the cost of the solution and interference in the users' daily lives [3]. In addition, the correct gait segmentation depends in many cases on the data of more than one of the sensors embedded in the IMU, which becomes more complex to the signal processing. Through the acquisition of lower trunk acceleration, placing an IMU in the lower vertebral column region (lower thoracic region and lumbar region) it is possible to obtain gait information from both lower limbs and from a single axis. In this way, the measurement of the lower torso acceleration is an efficient solution when it is intended to segment the human gait, using just one inertial sensor and without requiring large processing requirements [2].

In the last years, several systems have been developed to detect gait events, through the lower trunk acceleration based on the use of IMUs [2, 3, 7–14]. The development of these detection systems requires the use of sophisticated algorithms specially for real-time contexts, which are actually very important for gait laboratories and outside of rehabilitation environments towards assisted living environments. The implementation of these algorithms varied greatly from system to system and in general, heuristic rules and wavelet-based approaches were the most used. Further, most of the algorithms were constructed based on the antero posterior and vertical plane. In fact, through the study of the vertical acceleration signal (antero-posterior plane) it is possible to identify the follow gait events: the heel strike, foot-flat, toe-off, mid-stance and heel-off for each limb, as is depicted in Fig. 1 [2]. Also, González et al. [3] and Alvarez et al. [7] developed the only two systems which provided a real-time gait events detection, namely the initial and final contact events. The gait detection was based on heuristic roles and it were used two acceleration axes vertical and antero-posterior signals.

This paper addresses the development and validation of a novel adaptive real-time tool for the gait event detection. The proposed system consists of one inertial sensor, in particular, an accelerometer placed in the lower trunk at the T2-L1 inter-vertebral space.

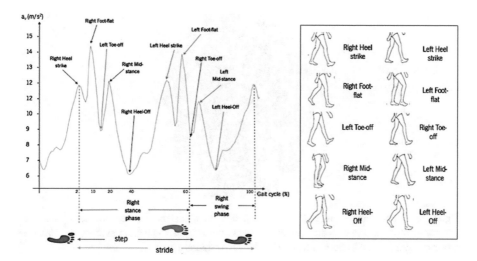

Fig. 1. Vertical acceleration over one stride. Adapted from [2].

Further, it includes a novel tool for gait detection through the lower trunk acceleration signal implemented by a Finite State Machine (FSM) with decision rules and adaptive thresholds. This tool is able to detect five gait events for each leg: Heel Strike (HS), Foot Flat (FF), Mid-Stance (MS), Toe-Off (TO) and Heel-Off (HO). This system follows the idea of implementing a wearable system based on the use of only one IMU, in order to minimize the use of several devices and simplify its use by the user. Lastly, note that the algorithm only uses one acceleration axis - the vertical orientation, aligned with the earth's gravitational axis, in contrast to what was presented in the literature, and especially the two systems developed for real-time segmentation of the gait [3, 7].

2 Methods

2.1 System Overview

In order to achieve all the requirements of portability and ergonomics, the system consisted of a processing unit and an inertial acquisition system embedded in an adjustable waistband for any abdominal diameter, represented in Fig. 2. In addition, it was implemented a data storage system in order to save the inertial gait acquired data in each trial, in a microSD card. These data were storage in the microSD card via SPI protocol.

The processing unit relies on a high-performance microcontroller Atmega 2560 (Arduino Mega) and for the acquisition system it was used an IMU, particularity, the MPU-6050. The MPU-6050, which was the world's first integrated 6-axis motion tracking device, combines 3-axis gyroscope (range: $\pm 2000°/s$) and 3-axis accelerometer (range: ± 16 g) in a small $4 \times 4 \times 0.9$ mm package. In this particular system, data were only recorded from the accelerometer, in particular the vertical acceleration, with a full-scale range of ± 2 g (enough to detect gait events through a lower trunk acquisition) at a sampling frequency of 100 Hz (sufficient to measure one gait cycle).

Fig. 2. Gait detection system developed on a waistband: the bag on the left houses the processing unit and the data storage system, whereas the gait acquisition system, the IMU, has been embedded so as to be placed in T2-L1 inter-vertebral space.

Due the small size, low weight and low admissible current consumption (500 ± μA), this IMU is an optimal solution for the proposed system. The communication between the acquisition system and the processing unit was supported by the I2C protocol.

2.2 Proposed Algorithm

The proposed algorithm consists in five stages: acquisition, calibration, filtering, 1st derivative computation and finite state machine. For the calibration routine, were captured 500 samples which were used to calculate an offset that is withdrawn from each of the samples subsequently acquired.

Then, each acquired sample ($sample_n$), after calibrated, was filtered with an exponential filter, which is ideal for a real-time implementation based on heuristic rules, since it does not cause delays in the signal and smooths the samples. Thus, each sample was filtered based on the following equation:

$$sample_{n_{filtered}} = \alpha.sample_n + (1 - \alpha).sample_{n-1}. \tag{1}$$

Where, α is the smoothing factor ($0 < \alpha < 1$), $sample_n$ corresponds to the current sample and $sample_{n-1}$ to the previous sample. α was set to 0.5 by trial and error.

After filtering the sample ($sample_{n_{filtered}}$), the 1st derivative was determined to detect when the acceleration increases, decreases, or remains constant and, in order to deal with the noise, the derivatives below a threshold (near to zero but empirically set) were assumed as null. This allows to detect only the major variations, that usually are associated with the local peaks. The calculation of the 1st derivative was performed based on the following equation:

$$sample_{n_{diff\_derivated}} = sample_{n_{filtered}} - sample_{n-1_{filtered}}. \tag{2}$$

Once the 1st derivative $\left(sample_{ndiff\_derivated}\right)$ is calculated and the threshold applied, it follows the FSM implemented by means of a switch case statement, which changes the states in accordance with the decision rules.

All these stages, Acquisition, Calibration, Filtering, 1st Derivative and FSM, are presented in the flow chart depicted in Fig. 3. The FSM is constituted by eleven states that correspond to ten gait events and one of reset. Each of these events corresponds to a peak in the filtered acceleration acquired in the lower trunk, as represented in Fig. 1. To detect each of these events, ten decision rules have been implemented that allow to trigger from one state to other which are presented in the Table 1. Also, is indicated the gait event corresponding to the peak in the acceleration signal.

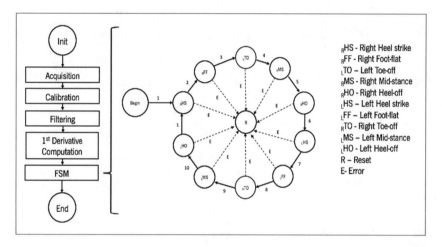

Fig. 3. Flow chart (left) and FSM (right) used to detect the gait events.

To increase the robustness of the algorithm, the thresholds used in the decision rules were adaptively calculated every three gait cycles and the first thresholds were set empirically. Also, after the occurrence of three gait cycles, each of the peaks corresponding to a gait event was detected based on its respective peak of the previous gait cycle, and must belong to a cadence calculated every three gait cycles. In this way, the peaks were only valid if they belonged to this calculated interval.

It is emphasized that the filtering, as well as the calculation of the 1st derivative and the decision rules depend on the previous sample acquired, so this is always stored at the end of each stage (Acquisition, Calibration, Filtering, 1st Derivative and FSM). For the first sample acquired, it is assumed that the previous sample is zero at each of the different stages of the algorithm.

2.3   Validation

The validation of the proposed adaptive tool involved 7 healthy subjects (3 males and 4 females). These subjects present a mean age of 25.13 ± 1.01 years old, mean weight of 71.69 ± 5.84 kg and a mean height of 170.38 ± 3.48 cm. The subjects conducted

Table 1. Gait events detected and corresponding signal acceleration peaks

Gait event	Peak on acceleration signal	Decision rules
Right heel strike	1st Maximum local	$\left(sample_{n_{diff\_derivated}} < 0\right) \& \left(sample_{n-1_{diff\_derivated}} > 0\right)$ $\& \left(sample_{n-1} > th_{1st\,max\,local\,1}\right)$
Right foot-flat	Maximum global	$\left(sample_{n_{diff\_derivated}} < 0\right) \& \left(sample_{n-1_{diff\_derivated}} > 0\right)$ $\& \left(sample_{n-1} > th_{max\,global\,1}\right)$
Left toe-off	Minimum local	$\left(sample_{n_{diff\_derivated}} > 0\right) \& \left(sample_{n-1_{diff\_derivated}} < 0\right)$ $\& \left(sample_{n-1} < th_{min\,local\,1}\right)$
Right mid-stance	2nd Maximum local	$\left(sample_{n_{diff\_derivated}} < 0\right) \& \left(sample_{n-1_{diff\_derivated}} > 0\right)$ $\& \left(sample_{n-1} > th_{2nd\,max\,local\,1}\right)$
Right heel-off	Global minimum	$\left(sample_{n_{diff\_derivated}} > 0\right) \& \left(sample_{n-1_{diff\_derivated}} < 0\right)$ $\& \left(sample_{n-1} < th_{min\,global\,1}\right)$
Left heel strike	1st Maximum local	$\left(sample_{n_{diff\_derivated}} < 0\right) \& \left(sample_{n-1_{diff\_derivated}} > 0\right)$ $\& \left(sample_{n-1} > th_{1st\,max\,local\,2}\right)$
Left foot-flat	Maximum global	$\left(sample_{n_{diff\_derivated}} < 0\right) \& \left(sample_{n-1_{diff\_derivated}} > 0\right)$ $\& \left(sample_{n-1} > th_{max\,global\,2}\right)$
Right toe-off	Minimum local	$\left(sample_{n_{diff\_derivated}} > 0\right) \& \left(sample_{n-1_{diff\_derivated}} < 0\right)$ $\& \left(sample_{n-1} < th_{min\,local\,2}\right)$
Left mid-stance	2nd Maximum local	$\left(sample_{n_{diff\_derivated}} < 0\right) \& \left(sample_{n-1_{diff\_derivated}} > 0\right)$ $\& \left(sample_{n-1} > th_{2nd\,max\,local\,2}\right)$
Left heel-off	Global minimum	$\left(sample_{n_{diff\_derivated}} > 0\right) \& \left(sample_{n-1_{diff\_derivated}} < 0\right)$ $\& \left(sample_{n-1} < th_{min\,global\,2}\right)$

walking experiments on the ground at a comfortable speed, in a distance of 20 m on unobstructed hallway. Each participant performed 3 trials and between each trial repetition, the waistband was removed and then replaced to assess test-retest repeatability.

In order to obtain an effective strategy to determine the performance of the gait identification algorithm implemented, as a ground truth, it was used two FSR sensors (from Interlinks Electronics®) placed on the right heel and toe foot of each subject. The gait events detected were compared with the signals from the FSRs in each gait cycle, more properly the HS and TO. Thus, all participants' steps performed were analyzed

Table 2. Algorithm performance regarding its accuracy, percentage of occurrence and duration of delays (delayed detection) and advances (earlier detection) for HS, FF, TO, MS and HS gait events

Gait event	Accuracy (%)	Delayed (D)		Advanced (A)	
		%	(Mean ± SD) ms	%	(Mean ± SD) ms
HS	98.99	11.1	11.33 ± 2.52	8.33	9.69 ± 7.88
FF	99.98	6.03	2.17 ± 0.67	4.09	1.92 ± 1.24
TO	95.56	12.2	12.2 ± 3.29	5.75	4.81 ± 3.91
MS	93.94	21.8	11.8 ± 4.56	9.09	3.54 ± 1.34
HO	95.04	7.89	10.8 ± 3.55	5.43	9.03 ± 3.78

and HS, FF, TO, MS and HO events were evaluated regarding its accuracy, % of occurrence and duration of earlier and delayed detections.

3   Results and Discussion

The performance of the real-time algorithm is demonstrated in Table 2, where it is possible to analyze the accuracy of correct identification of HS, FF, TO, MS and HS event (considering both feet), in percentage. Besides the accuracy percentage, it is also presented the percentage of delayed and advanced gait events detection and the delay and advance delay times, comparing with the FSR data.

In Table 2, it is verified that the proposed algorithm for gait detection is accurate in the detection of all events, with an accuracy above 93.94%, being the HS and MS events with more accuracy (98.99% and 99.98%, respectively). On the other hand, MS and HS were the events with less accuracy due to changes in cadence and local and global peaks very close. Also, the accuracy is affected by the high susceptibility to noise when using the acceleration signal from the built-in accelerometer of the IMU.

Concerning to the percentage of occurrences of delays and advances, it is observable that the worst results were in the MS event (D: 21.8% and A: 9.09%). The MS corresponds to a maximum local in the trunk vertical acceleration signal, thus this observation is due to the fact that the algorithm detects local peaks that are very close to the local maximum which is supposed to be detected. Also, it was verified that the worst measured delayed (11.8 ms) and advanced (9.98 ms) results to this event does not contribute with significant delay and advanced time in the gait detection, since a normal gait cycle is 1.15 s [15].

It was also found that the delays measured for the initial and final foot-contact (HS - 11.33 ms and TO - 12.2 ms, respectively) were lower than those measured by the González' real-time algorithm proposed (117 and 34 ms, respectively) through the trunk acceleration [3]. These results are probably due to the filter implemented in [3], since it was used a low-order 11-order low-pass filter, which introduces an undesirable delay to the gait events detection, especially when it comes to a real-time implementation. On the other hand, in our tool it was implemented a filter exponential which only smooth the signal, not introducing delays in the real-time detection. Also, it is

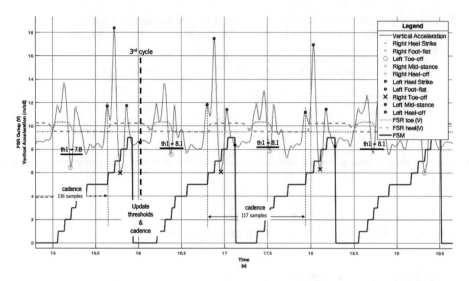

Fig. 4. Representation of gait events detection throughout the vertical acceleration (m/s^2) and FSRs output, in three-steps of a subject. It is pointed out the value of the adaptive thresholds (in this example for the TO detection for the right and left foot, a local minimum) and the value of the cadence (a specific defined range for each gait event) for the HS event.

emphasized that our tool it is based on a set of decision rules implemented by a FSM, depending only on the acquisition of the signal on a single axis. Thereby, the delay is smaller when compared to the algorithm presented in [3], which is based on a set of heuristic rules (zero-crossing) through the analysis of two acceleration axes.

Figure 4 depicts the gait events detection in three-steps of a subject. It is verified that the implemented algorithm detects the right/left heel strike (HS - 1st local maximum), right/left foot-flat (FF - global maximum), right/left toe-off (TO - local minimum), right/left mid-stance (MS - 2nd local maximum), and right/left heel-off (HO - global minimum), for the right and left limb, respectively. These detections provide an innovative character to the proposed tool since all the real-time gait detection algorithms on the literature only detected the initial and final foot-contact on the ground [3, 7].

Note that at the end of a third gait cycle, the threshold and cadence calculation is adapted based on the previous detected values, for each leg. In Fig. 4, it is only represented the threshold for the TO detection for the right limb and the cadence for the HS for the left leg. At the end of the third cycle, the threshold for the TO detection on the left leg (th1) goes from 7.8 to 8.1 and remains with this value during the following three cycles. Also, the calculation of the cadence for HS event was adapted according to the detection of the events during the three previous gait cycles, going from 116 to 117 samples. Thus, we prove the additivity of the implemented tool to detect gait through the vertical acceleration with a single-axis.

Lastly, it can also be verified that the HS and TO events, that the signal from the FSRs detect, are in accordance with the respective events identified through the signal of the trunk acceleration: the event HS corresponds to a rise in the signal of the FSR placed in the heel, which matches with the peak corresponding to the HS in the signal of the acceleration (1st local maximum); and the event TO corresponds to a decrease in the signal of the FSR placed in the toe, which beats with the peak corresponding to the signal of the acceleration (local minimum).

4   Conclusions

A real-time tool for human gait detection from lower trunk acceleration was developed and validated. The vertical acceleration signal (a single axis) was acquired by implementing an IMU on a waistband.

The proposed algorithm was based on a finite state machine using a set of suitable decision rules to detect HS, FF, TO, MS and HS for each leg. The algorithm was validated considering accuracy and time-effectiveness with 7 subjects walking at comfortable walk speeds on the ground. Results allowed to conclude that the projected tool is accurate and time-effective in real-time detection. Moreover, we introduce a new and real-time tool which detects all gait events of the stance-phase, through a single axis by a wearable system with an IMU.

The adaptability of the algorithm has also been verified, indicating the tool will be easily adapted to some gait conditions, such as for different speeds and slopes. Besides that, the implemented algorithm is modular since it can easily be integrated in another robotic control system for gait rehabilitation.

These are the future challenges. In the future, it will be addressed the validation of this algorithm considering different environments and conditions. These validations, besides healthy and (non)-elderly subjects, will include patients with neurological pathologies as Parkinson's Disease or Multiple Sclerosis. Thereby, the algorithm developed will be embedded in a more robust and ergonomic wearable system which will allow the gait monitoring, a special contribution to clinicians in order to facilitate diagnostic techniques but above all, allow to trace paths of motor symptoms' improvements for these devastating diseases.

Acknowledgments. This work is supported by the FCT Fundação para a Ciência e Tecnologia - with the reference project UID/EEA/04436/2013, by FEDER funds through the COMPETE 2020 - Programa Operacional Competitividade e Internacionalização (POCI) - with the reference project POCI-01-0145-FEDER-006941; and the LIACC Project PEst/UID/CEC/00027/2013.

References

1. Rueterbories, J., Spaich, E.G., Larsen, B., Andersen, O.K.: Methods for gait event detection and analysis in ambulatory systems. Med. Eng. Phys. **32**(6), 545–552 (2010)
2. Auvinet, B., Berrut, G., Touzard, C., Moutel, L., Collet, N., Chaleil, D., Barrey, E.: Reference data for normal subjects obtained with an accelerometric device. Gait Posture **16**, 124–134 (2002)
3. González, R.C., López, A.M., Rodriguez-Uría, J., Alvarez, D., Alvarez, J.C.: Real-time gait event detection for normal subjects from lower trunk accelerations. Gait Posture **31**, 322–325 (2010)

4. Félix, P., Figueiredo, J., Santos, C.P., Moreno, J.C.: Adaptive real-time tool for human gait event detection using a wearable gyroscope. In: Human-Centric Robotics, Climbing and Walking Robots and the Support Technologies for Mobile Machines Conference, pp. 1–10 (2015)
5. Muro-de-la-herran, A., Garcia-zapirain, B., Mendez-zorrilla, A.: Gait analysis methods: an overview of wearable and non-wearable systems, highlighting clinical applications. Sensors 14(2), 3362–3394 (2014)
6. Bergamini, E., Picerno, P., Thoreux, P., Camomilla, V.: Estimation of temporal parameters during sprint running using a trunk-mounted inertial measurement unit. J. Biomech. 45, 1123–1126 (2012)
7. Alvarez, J.C., Alvarez, D., López, A., González, R.C.: Pedestrian navigation based on a Waist-Worn inertial sensor. Sensors 12(8), 10536–10549 (2012)
8. Godfrey, A.: Instrumenting gait with an accelerometer: a system and algorithm examination. Med. Eng. Phys. 37(4), 400–407 (2015)
9. Zijlstra, W.: Assessment of spatio-temporal parameters during unconstrained walking. Eur. J. Appl. Physiol. 92(1–2), 39–44 (2014)
10. Köse, A., Cereatti, A., Della Croce, U.: Bilateral step length estimation using a single inertial measurement unit attached to the pelvis. J. Neuroeng. Rehabil. 9(1), 9 (2012)
11. Mansfield, A., Lyons, G.M.: The use of accelerometry to detect heel contact events for use as a sensor in FES assisted walking. Med. Eng. Phys. 25, 879–885 (2003)
12. Menz, H.B., Lord, S.R., Fitzpatrick, R.C.: Acceleration patterns of the head and pelvis when walking on level and irregular surfaces. Gait Posture 18, 35–46 (2003)
13. Storm, F.A., Buckley, C.J., Mazzà, C.: Gait event detection in laboratory and real life settings: accuracy of ankle and waist sensor based methods. Gait Posture 50, 42–46 (2016)
14. Zijlstra, A., Zijlstra, W.: Trunk-acceleration based assessment of gait parameters in older persons: a comparison of reliability and validity of four inverted pendulum based estimations. Gait Posture 38(4), 940–944 (2013)
15. Vaughan, C.L., Davis, B.L., O'Connor, J.C.: Dynamics of Human Gait, 2nd edn. Kiboho Publishers, Cape Town (1999)

Protein Attributes-Based Predictive Tool in a Down Syndrome Mouse Model: A Machine Learning Approach

Cláudia Ribeiro-Machado[1,2,3]($\boxtimes$), Sara Costa Silva[1,2,4]($\boxtimes$),
Sara Aguiar[1]($\boxtimes$), and Brígida Mónica Faria[1,5]

[1] Escola Superior de Saúde, Instituto Politécnico do Porto,
Rua Dr. António Bernardino de Almeida 431, 4249-015 Porto, Portugal
claudia.ar.machado@gmail.com,
sarac.costasilva@gmail.com, sara_aguiiar@hotmail.com,
btf@ess.ipp.pt
[2] i3S - Instituto de Investigação e Inovação em Saúde, Universidade do Porto,
Rua Alfredo Allen, 208, 4200-135 Porto, Portugal
[3] INEB - Instituto de Engenharia Biomédica, Universidade do Porto,
Rua Alfredo Allen, 208, 4200-135 Porto, Portugal
[4] IPATIMUP – Instituto de Patologia e Imunologia Molecular da Universidade
do Porto, Rua Alfredo Allen, 208, 4200-135 Porto, Portugal
[5] LIACC - Lab. Inteligência Artificial e Ciência de Computadores,
Porto, Portugal

Abstract. Down syndrome is a disorder caused by an imbalance in the 21 chromosome, affecting learning and memorizing abilities, which was shown to be improved with memantine administration. In this study we intent to determine the most relevant proteins that could play a role in learning ability, suitable for possible biomarkers and to evaluate the accuracy of several bioinformatic models as a predictive tool. Five different supervised learning models (K-NN, DT, SVM, NB, NN) were applied to the original database and the newly created ones from eight attribute weighting models. Model accuracies were calculated through cross validation. Nine proteins revealed to be strong candidates as future biomarkers and K-NN and Neural Network had the better overall performances and highest accuracies (86.26% ± 0.23%; 81.51% ± 0.48%), which makes them a promising predictive tool to study protein profiles in DS patients' follow-up after treatment with memantine.

Keywords: Down syndrome · Prediction · Learning improvement
Attributes weighting · Data mining · Classification

1 Introduction

Down syndrome (DS) is the most frequent form of mental retardation, with one in each 1000 births worldwide affected by this condition [1] and it significantly impairs health and autonomy of affected individuals [2]. It is known that the mechanism behind this disorder is a triplicate state (trisomy) of all or a critical portion of chromosome 21 and

© Springer International Publishing AG, part of Springer Nature 2018
Á. Rocha et al. (Eds.): WorldCIST'18 2018, AISC 747, pp. 19–28, 2018.
https://doi.org/10.1007/978-3-319-77700-9_3

the subsequent increased level of expression due to gene dosage [1–4]. DS can be coupled with various health issues such as intellectual disability, difficulty in learning and memorizing, congenital heart diseases, Alzheimer's disease (AD), leukemia, cancers and gastrointestinal anomalies [1, 2, 5]. These patients can also present distinctive phenotypic features [2].

Machine learning algorithms are an efficient and inexpensive tool in analysis of biological data since they can accelerate different studies such as biomolecular structure prediction, gene finding, genomics and proteomics. By employing a variety of statistical tools to learn from past data and then use the prior training to classify new data, they are able to identify new patterns or predict novel trends [6]. This approach has been widely used in several studies, namely in DS model, both in diagnosis [7–9] and treatment follow-up [10]. A study by Nguyen, C. et al. hypothesised that locomotor activity was associated with protein level patterns by evaluating the locomotor response MK-801 in a DS mouse model and succeeding in the prediction of this, with accuracies up to 88% [10].

Extended DS research has already been done, namely in possible therapeutics, such as memantine, which has been widely prescribed for the symptomatic treatment of patients with moderate to severe AD and has been shown to improve learning and memory in several animal models [11–14]. Several studies have addressed the memantine effect in the learning/memory abilities in DS individuals [1, 3] using different mouse models [15]. The most complete model is the Ts65Dn mouse [13, 16], since they exhibit significant learning deficits in specific behavioral tasks [13, 16, 17]. Furthermore, a human clinical trial, from University Hospitals Cleveland Medical Center, has been, since October 2014, studying the effect of this drug on DS patients [18]. However, currently the evaluation of success of memantine treatment relies on a behaviour evaluation and, therefore, there is a need to find biomarkers that allow for a continuous follow-up of learning and memory improvement. This deficit represents a significant struggle in the routine life of these patients and an obstacle to their autonomy as well.

In the following study the database from Ahmed et al. (2015), available for public use at the UCI Machine Learning Repository [19] was used, which includes protein expression levels measured in the cerebral cortex in control and Down syndrome mice exposed to context fear conditioning (a task used to assess associative learning) [19]. In addition, in order to assess the effect of memantine in recovering the ability to learn in trisomic mice, some mice have been injected with this drug and others have been injected with saline [19].

Hereby, we intent to determine the most relevant proteins that could play a role in learning ability, suitable for possible biomarkers, in DS patients' follow-up, after memantine administration and apply several bioinformatics models in order to evaluate the accuracy of these as a predictive tool.

2   Methods

2.1   Data Preparation

Mice Protein Expression database was extracted from UCI Machine Learning repository [19]. This data includes the expression levels of 77 proteins that produced detectable signals in the nuclear fraction of cortex. There are 38 control mice and 34 DS mice, for a total of 72. The eight classes are described based on features such as genotype, behavior and treatment. According to genotype, mice can be control or trisomic; according to behavior, some mice have been stimulated to learn (context-shock) and others have not (shock-context) and in order to assess the effect of the drug memantine in recovering the ability to learn some mice have been injected with the drug and others have not. In this work the following nomenclature was applied: c-CS-s: control mice, stimulated to learn, injected with saline; c-CS-m: control mice, stimulated to learn, injected with memantine; c-SC-s: control mice, not stimulated to learn, injected with saline; c-SC-m: control mice, not stimulated to learn, injected with memantine; t-CS-s: trisomy mice, stimulated to learn, injected with saline; t-CS-m: trisomy mice, stimulated to learn, injected with memantine; t-SC-s: trisomy mice, not stimulated to learn, injected with saline; t-SC-m: trisomy mice, not stimulated to learn, injected with memantine. This dataset as imported into Rapid Miner software (Rapid Miner 7.6.001, Dortumund, Germany) and the eight classes of mice were set as label.

2.2 Data Cleaning

Polynominal variables (genotype, behavior and treatment) were omitted, since they were redundant in the information they provided, in comparison to the class. Missing values (1.68% for the total dataset) were replaced with the average expression level of that protein (the higher missing values were 285 out of 1080, for BCL2). Some classification methods require that all input features have a similar range of values so they will have the same influence on the clustering outcome. Otherwise, higher values would have more weight, causing false results. In order to avoid this, a normalization by range (0 to 1) was applied to the data set. After cleaning, the database was labeled as Cleaned database (Cdb).

2.3 Attribute Weighting

Eight different algorithms of attribute weightings were applied to the Cdb in order to identify the most relevant attributes contributing to DS: Information Gain, Chi-square Statistic, Rule, Deviation, Relief, Uncertainty, Support Vector Machine (SVM) and Principal Component Analysis (PCA).

2.4 Attribute Selection

Once the chosen attribute weighting models were applied to the Cdb, each attribute assumed a value between 0 and 1, revealing the importance of said protein. All variables higher than 0.5 were selected and eight new datasets created. This new datasets were named according to the weighting model chosen and, together with the Cdb, were used to perform the upcoming models. Therefore, each dataset was executed 9 times: the original cleaned data set (Cdb) and the 8 new datasets resulting of the attribute weighting.

2.5   Classification and Prediction

Five different learning algorithms were used. The graphics presented were either obtained directly through RapidMiner or obtained using Microsoft Office Excel 2016. The nine datasets were tested on *K-Nearest Neighbour (K-NN)*, with k set to 1; *Decision Tree (DT)*, with pruning and prepruning applied; *Support Vector Machine (SVM)*, *LibSVM* algorithm since it allowed for polynominal label, as it was in this case; *Naïve Bayes (NB)* and, finally, *Neural Network (NN)*, *AutoMLP* algorithm. When no information is given about special parameters, default values were used.

Validation was performed using Cross Validation, with 10 folds since our database had 1080 objects. This validation allows for a division into 10 parts, where 9 are used for training and 1 for test. The final model accuracy was calculated as the average accuracy in all nine datasets and the final weighting attribute accuracy as the average in all five supervised methods. T test between pairs and ANOVA tests were performed as a means to evaluate differences between supervised methods, with a p value of 0.05. Learning curves for each 9 datasets, comparing the 5 supervised learning models, were also generated.

3   Results

The initial dataset had 80 attributes. However, this number decreased to 77 attributes after cleaning.

3.1   Attribute Selection

Eight attribute weighting models were applied to the Cdb and weights were normalized so they would range from 0 to 1. The models corresponded to the operators *Weight by Chi-squared*, *Weight by Deviation*, *Weight by Information Gain*, *Weight by PCA*, *Weight by Relief*, *Weight by Rule*, *Weight by SVM* and *Weight by Uncertainty*. The average of weights obtained by each operator was calculated for every protein expressed (Fig. 1). The proteins with higher average weight value (>0.5) were SOD1_N, CaNA_N, pPKCG_N, pCAMKII_N, pPKCAB_N, PKCA_N, APP_N, Ubiquitin_N, and pERK_N. Variation of selected proteins could be seen between weighting models.

3.2   Classification and Prediction

Five different supervised learning models were applied to the 9 created datasets. Accuracies for each analysis are represented in Table 1. The average time spent executing each model is also represented in Fig. 2A, and learning curves for each dataset comparing the five models is shown in Fig. 2C. SVM's accuracy ranged from 17.06% ± 0.36% (using the *Weight by SVM*) to 27.38% ± 0.6% (*Weight by Uncertainty*) The average accuracy of the model was 24.75% ± 0.13%, and the average time to run the model was 25 s.

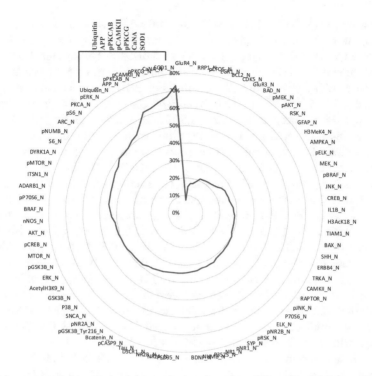

Fig. 1. Average weights obtained by each protein using different weighting models. The proteins with higher average weight values in every dataset tested were SOD1 and CaNA (0.73 and 0.67, respectively).

Table 1. Accuracy of learning models when applied to each datasets after feature selection. All methods had a significance of $p < 0.05$ for each dataset, except for Cdb K-NN/NN ($p = 0.123$) and for Information Gain dataset DT/NN ($p = 0.09$), using t test.

Data set (average ± std)		Supervised methods (average ± std)				
		K-NN	DT	SVM	NN	NB
		86.26% ± 0.23%	54.32% ± 1.92%	24.75% ± 0.13%	81.51% ± 0.48%	61.74% ± 0.35%
	Cdb 70.42% ± 0.91%	99.75% ± 0.07%	49.47% ± 3.86%	25.26% ± 0.15%	99.63% ± 0.1%	78.01% ± 0.39%
	Chi-square 72.08% ± 0.49%	97.53% ± 0.28%	74.66% ± 1.23%	26.84% ± 0.11%	93.21% ± 0.52%	68.15% ± 0.3%
	Information Gain 71.58% ± 0.81%	99.71% ± 0.11%	54.50% ± 3.18%	26.27% ± 0.12%	99.35% ± 0.14%	78.07% ± 0.48%
	Deviation 51.41% ± 0.54%	72.72% ± 0.49	54.54% ± 0.64%	26.69% ± 0.07%	55.61% ± 1.23%	47.48% ± 0.27%
	PCA 66.95%0.57%	98.91% ± 0.15%	60.02% ± 1.84%	21.70% ± 0.14%	98.45% ± 0.26%	55.67% ± 0.45%
	Relief 68.13% ± 0.46%	98.49% ± 0.15%	63.89% ± 0.95%	27.18% ± 0.06%	88.57% ± 0.68%	62.53% ± 0.44%
	Rule 70.19% ± 0.92%	99.77% ± 0.07%	52.04% ± 3.87%	24.34% ± 0.08%	99.32% ± 0.23%	75.49% ± 0.33%
	SVM 20.05% ± 0.33%	18.16% ± 0.39%	14.62% ± 0.15%	17.06% ± 0.36%	24.19% ± 0.58%	26.24% ± 0.17%
	Uncertainty 64.63% ± 0.57%	91.34% ± 0.35%	65.10% ± 1.52%	27.38% ± 0.06%	75.28% ± 0.6%	64.06% ± 0.3%

Using *Weight by Relief, Weight by Chi-square* and *Weight by Information Gain* the accuracies were 27.18% ± 0.06%, 26.84% ± 0.11% and 26.69% ± 0.07% respectively. In all generated learning curves, SVM showed to be the model with lowest performance rate. The DT's accuracy ranged from 14.62% ± 0.15% (*Weight by SVM*) to 74.66% ± 1.23% (*Weight by Chi-square* – Fig. 2B). The average accuracy of this model was 54.32% ± 1.92%; Using *Weight by Uncertainty, Weight by Relief* and *Weight by PCA*, accuracies were 65.10% ± 1.52%, 63.89% ± 0.95% and 60.02% ± 1.84% respectively. The average time spent running this model was 15 s, and it presented a highly irregular learning curve, when inserting new elements. k-NN model's accuracy ranged from 99.77% ± 0.07% (*Weight by Rule*), to 18.16% ± 0.39% (*Weight by SVM* dataset). The average accuracy of this model was 86.26% ± 0.23% and the average time spent running it was 4 s. Using the initial Cdb, *Weight by Deviation* and *Weight by PCA*, accuracies of the model were 99.75% ± 0.07%, 99.71% ± 0.11% and 98.91% ± 0.15% respectively. NN's accuracy ranged from 99.63% ± 0.1% (Cdb) to 24.19% ± 0.58% (*Weight by SVM* dataset). The average accuracy of this model was 81.51% ± 0.48% and the average time spent running it was 18 min and 56 s. Using *Weight by Deviation, Weight by Rule* and *Weight by PCA* the accuracies of the model were 99.35% ± 0.14%, 99.32% ± 0.23% and 98.45% ± 0.26% respectively. Learning curves demonstrated NN and k-NN to be the higher performing models. NB accuracy ranged from 78.07% ± 0.48% (*Weight by Deviation* dataset) to 26.24% ± 0.17% (*Weight by SVM*).

The average accuracy of this model was 61.74% ± 0.35% and the time spent running it was 3 s. Using the Cdb, *Weight by Rule* and *Weight by Chi-square*, the accuracies of the model were 78.01% ± 0.39%, 75.49% ± 0.33% and 68.15% ± 0.3% respectively.

4   Discussion

The chosen database [19] was cleaned by eliminating its polynomial attributes, since they did not bring any new information to the dataset, when considering the class; by replacing missing values by average (taking into account that the missing percentage was 1.68%, there wasn't a significant effect on final results); and by normalizing data, since the values of proteins were of different ranges. After cleaning, the final dataset was composed by 77 attributes, with its objects distributed by eight classes. A previous study using this database reported differences in protein profile between Ts65Dn mice that fail to learn, the control mice that learnt successfully, and in Ts65Dn mice whose learning abilities were rescued by memantine [3]. However, in our study, eight different weighting models were applied to the Cdb in order to select the most relevant proteins (with normalized weight higher than 0.5, as previously stated [20]) that could play a role in learning ability, therefore important in DS follow-up after memantine treatment. These weighting algorithms calculated the relevance of an attribute: by using the value of the chi-squared statistic with respect to the class (*Weight by Chi-squared*); based on the normalized standard deviation of the attributes (*Weight by Deviation*); by computing the information gain in class distribution (*Weight by Information Gain*); using the factors of the first of the principal components as feature weights (*Weight by PCA*);

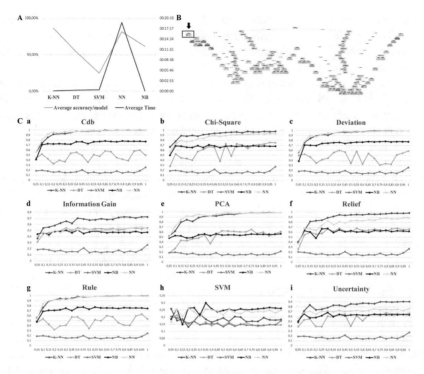

Fig. 2. Evaluation of learning models performance. A. Graphic of average accuracy of each model and time spent executing. **B.** DT produced with the dataset *Weight by Chi-squared* (74.66% ± 1.23%). Arrow represents bar leave with two colors. **C.** Learning curves for the learning models within the nine datasets. K-NN and NN were in the great majority the best methods and SVM the lowest. DT presented a high level of irregularity.

by the quality of features according to how well their values distinguish between the instances of the same and different classes that are near each other (*Weight by Relief*); by constructing a single rule for each attribute and calculating the errors (*Weight by Rule*); using the coefficients of the normal vector of a linear SVM as feature weights (*Weight by SVM*) and by measuring the symmetrical uncertainty with respect to the class (*Weight by Uncertainty*) [6].

By selecting proteins with weight value higher than 0.5, it was observed that from different weighting models, different proteins were considered relevant. For instance, using the *Weight by SVM* only one protein was considered relevant, whereas when using *Weight by Deviation*, 51 proteins had weight values higher than 0.5. In order to obtain which proteins were considered relevant for most of the algorithms, an average of all weighting models was calculated. The results, shown in Fig. 1, revealed SOD1, CaNA, pPKCG, pCAMKII, pPKCAB, PKCA, APP, Ubiquitin, and pERK as the most important. According to literature, an increase of SOD1 can lead to oxidative injury, dysregulation of REDOX signaling and cell death [21]. Furthermore, some studies revealed its presence in degenerating neurons, being related to the increase of fribillogenic Aβ42 peptide production from APP, in adults with DS [22]. Additionally, pCAMKII is related to long-term potentiation and neurotransmitter release, being intensively studied in hippocampal and other neurons as a mediator of plasticity, learning and memory [23].

After weight selection, supervised learning algorithms were applied to all datasets (SVM, DT, k-NN, NN and NB). With the purpose of finding the best learning model and the best weighting algorithm, average accuracy within each model, time spent executing, as well as the learning curve were obtained, as can be seen in Table 1 and Fig. 2. There was not a significant increase in algorithm performances when applied to the created datasets by attribute weighting models in comparison to the cleaned dataset since the highest average accuracy was for *Weight by Chi-Square, Weight by Deviation* and *Weight by Rule* (72.08%, 71.58% and 70.19%, respectively) and the Cdb had an average accuracy of 70.42%. SVM is a non-probabilistic binary linear classifier. Given a set of training examples, each marked as belonging to one of two categories, it builds a model that assigns new examples into one category or the other[1]. LibSVM is a SVM algorithm that allows the existence of a polynomial label, hence it was used in this study. SVM turned out to be the algorithm with lowest performance for every dataset tested (Fig. 2C), with an average accuracy of 24.75% ± 0.13% (Table 1). These results were contrary to the expected, since SVM tends to have a high accuracy, close to NN. Nevertheless, since in this study it was used the default parameters, an optimization of these could lead to an accuracy improvement. Likewise DT, a tree like collection of nodes intended to create a decision[2], showed to have a highly irregular performance when increasing the number of examples (Fig. 2C), being that the best performance occurred when *Weight by Chi-square* dataset was used. The average accuracy of this algorithm was 54.32% ± 1.92% (Table 1). Although in this case, the selected criterion was Ratio Gain, and pre and post pruning were applied to the generated tree, it was still highly complex, leading to a confusing result (Fig. 2B). Moreover, for some branches of the tree, one rule could correspond to more than one class (several color bar leaves – Fig. 2B), demonstrating it was not very precise. NB is a simple probabilistic classifier based on Bayes' theorem with strong (naive) independence assumptions[3]. This algorithm demonstrated an almost linear performance when increasing the number of examples (Fig. 2C), with an average accuracy of 61.74% ± 0.35%. k-NN, an algorithm which compares a given test example with training examples that are similar to it[4], and NN, a model which consists of an interconnected group of artificial neurons that process information[5], showed to be the best performing algorithms, with an accuracy of 86.26% ± 0.23% and 81.51% ± 0.48%, respectively (Fig. 2A). Different k were tested (1 to 5) in k-NN method and k = 1 showed the best result. Accuracies tended to decrease as k increased, although values were not very different. Nevertheless, when analyzing the executing time, it was

---

[1] GmbH R. Support Vector Machine - RapidMiner Documentation 2017.

[2] GmbH R. Decision Tree - RapidMiner Documentation 2017.

[3] GmbH R. Naive Bayes - RapidMiner Documentation 2017.

[4] GmbH R. k-NN - RapidMiner Documentation 2017.

[5] GmbH R. Neural Net - RapidMiner Documentation 2017.

possible to observe that although NN had a high accuracy and an average time of 18 min, for some datasets (as in the case of Cdb) it took 44 min to complete the process. This could potentially represent a problem, when dealing with larger datasets.

5 Conclusion

In this study, several machine learning methods were applied to the dataset in order to find relevant proteins that could play a role in learning ability and, furthermore, evaluate the accuracy of learning models as a predictive tool. Currently there are still no biomarkers capable of predicting the learning ability in DS after treatment [3, 11, 12]. Proteins identified in this work revealed to be strong candidates as possible biomarkers. Regarding the learning models, k-NN and NN had the best overall performances and highest accuracies, which makes them a promising predictive tool to study protein profiles in DS patients' memantine treatment follow-up. However, there is a need for more studies addressing this question, where more proteins can be included and fewer classes considered, as well as a replication of the same analysis in human samples. All data mining methods showed to be an efficient, quick and cheap way of analyzing biological data.

Acknowledgments. This study was funded by QVida+: Estimação Contínua de Qualidade de Vida para Auxílio Eficaz à Decisão Clínica, NORTE-01-0247-FEDER-003446, supported by Norte Portugal Regional Operational Programme (NORTE 2020), under the PORTUGAL 2020 Partnership Agreement, through the European Regional Development Fund (ERDF) and strategic project LIACC (PEst-UID/CEC/00027/2013).

References

1. Higuera, C., Gardiner, K.J., Cios, K.J.: Self-organizing feature maps identify proteins critical to learning in a mouse model of Down syndrome. PLoS One **10**(6), e0129126 (2015)
2. Kazemi, M., Salehi, M., Kheirolahi, M.: Down syndrome: current status, challenges and future perspectives. Int. J. Mol. Cell. Med. (IJMCM) **5**(3), 125–133 (2016)
3. Ahmed, M.M., Dhanasekaran, A.R., Block, A., Tong, S., Costa, A.C., Stasko, M., Gardiner, K.J.: Protein dynamics associated with failed and rescued learning in the Ts65Dn mouse model of Down syndrome. PLoS One **10**(3), e0119491 (2015)
4. Gardiner, K.J.: Molecular basis of pharmacotherapies for cognition in Down syndrome. Trends Pharmacol. Sci. **31**(2), 66–73 (2010)
5. Asim, A., Kumar, A., Muthuswamy, S., Jain, S., Agarwal, S.: Down syndrome: an insight of the disease. J. Biomed. Sci. **22**(1), 41 (2015)
6. Hosseinzadeh, F., KayvanJoo, A.H., Ebrahimi, M., Goliaei, B.: Prediction of lung tumor types based on protein attributes by machine learning algorithms. Springerplus **2**, 238 (2013)
7. Feng, B., Hoskins, W., Zhou, J., Xu, X., Tang, J.: Using supervised machine learning algorithms to screen Down syndrome and identify the critical protein factors. In: Xhafa, F., Patnaik, S., Zomaya, A. (eds.) Advances in Intelligent Systems and Interactive Applications, vol. 686, pp. 302–308. Springer, Cham (2018)
8. Saraydemir, S., Taşpınar, N., Eroğul, O., Kayserili, H., Dinçkan, N.: Down syndrome diagnosis based on gabor wavelet transform. J. Med. Syst. **36**(5), 3205–3213 (2012)

9. Zhao, Q., Rosenbaum, K., Okada, K., Zand, D.J., Sze, R., Summar, M., Linguraru, M.G.: Automated Down syndrome detection using facial photographs. In: 2013 35th Annual International Conference of the IEEE Engineering in Medicine and Biology Society (EMBC) (2013)

10. Nguyen, C.D., Costa, A.C., Cios, K.J., Gardiner, K.J.: Machine learning methods predict locomotor response to MK-801 in mouse models of Down syndrome. J. Neurogenet. **25**(1–2), 40–51 (2011)

11. Rueda, N., Llorens-Martin, M., Florez, J., Valdizan, E., Banerjee, P., Trejo, J.L., Martinez-Cue, C.: Memantine normalizes several phenotypic features in the Ts65Dn mouse model of Down syndrome. J. Alzheimer's Dis. **21**(1), 277–290 (2010)

12. Scott-McKean, J.J., Costa, A.C.: Exaggerated NMDA mediated LTD in a mouse model of Down syndrome and pharmacological rescuing by memantine. Learn Mem. **18**(12), 774–778 (2011)

13. Lockrow, J., Boger, H., Bimonte-Nelson, H., Granholm, A.C.: Effects of long-term memantine on memory and neuropathology in Ts65Dn mice, a model for Down syndrome. Behav. Brain Res. **221**(2), 610–622 (2011)

14. Victorino, D.B., Bederman, I.R., Costa, A.C.S.: Pharmacokinetic properties of memantine after a single intraperitoneal administration and multiple oral doses in euploid mice and in the Ts65Dn mouse model of Down's syndrome. Basic Clin. Pharmacol. Toxicol. **121**(5), 382–389 (2017)

15. Herault, Y., Delabar, J.M., Fisher, E.M.C., Tybulewicz, V.L.J., Yu, E., Brault, V.: Rodent models in Down syndrome research: impact and future opportunities, 1 October 2017

16. Costa, A.C.: On the promise of pharmacotherapies targeted at cognitive and neurodegenerative components of Down syndrome. Dev. Neurosci. **33**(5), 414–427 (2011)

17. Xing, Z., Li, Y., Pao, A., Bennett, A.S., Tycko, B., Mobley, W.C., Yu, Y.E.: Mouse-based genetic modeling and analysis of Down syndrome. Br. Med. Bull. **120**(1), 111–122 (2017)

18. Down Syndrome Memantine Follow-up Study - Full Text View - ClinicalTrials.gov. https://clinicaltrials.gov/ct2/show/NCT02304302

19. Higuera, C., Gardiner, K.J., Cios, K.J.: Mice protein expression data set. UCI MLRep (2015)

20. Hosseinzadeh, F., Ebrahimi, M., Goliaei, B., Shamabadi, N.: Classification of lung cancer tumors based on structural and physicochemical properties of proteins by bioinformatics models. PLoS One **7**(7), e40017 (2012)

21. Cowley, P.M., Nair, D.R., DeRuisseau, L.R., Keslacy, S., Atalay, M., DeRuisseau, K.C.: Oxidant production and SOD1 protein expression in single skeletal myofibers from Down syndrome mice. Redox Biol. **13**, 421–425 (2017)

22. Schupf, N., Lee, A., Park, N., Dang, L.H., et al.: Candidate genes for Alzheimer's disease are associated with individual differences in plasma levels of beta amyloid peptides in adults with Down syndrome. Neuro Aging **36**(10), 2907.e1–2907.e10 (2015)

23. Bustos, F.J., Jury, N., Martinez, P., Ampuero, E., Campos, M., Abarzua, S., Jaramillo, K., Ibing, S., Mardones, M.D., Haensgen, H., Kzhyshkowska, J., Tevy, M.F., Neve, R., Sanhueza, M., Varela-Nallar, L., Montecino, M., van Zundert, B.: NMDA receptor subunit composition controls dendritogenesis of hippocampal neurons through CAMKII, CREB-P, and H3K27ac. J. Cell. Physiol. **232**(12), 3677–3692 (2017)

Artificial Intelligence in Fashion Industry

Enhancing Apparel Data Based on Fashion Theory for Developing a Novel Apparel Style Recommendation System

Congying Guan[1(✉)], Shengfeng Qin[1], Wessie Ling[1], and Yang Long[2]

[1] School of Design, Northumbria University, Newcastle upon Tyne, UK
congying.guan@gmail.com
[2] Open Lab, School of Computing,
Newcastle University, Newcastle upon Tyne, UK

Abstract. Smart apparel recommendation system is a kind of machine learning system applied to clothes online shopping. The performance quality of the system is greatly dependent on apparel data quality as well as the system learning ability. This paper proposes (1) to enhance knowledge-based apparel data based on fashion communication theories and (2) to use deep learning driven methods for apparel data training. The acquisition of new apparel data is supported by apparel visual communication and sign theories. A two-step data training model is proposed. The first step is to predict apparel ATTRIBUTEs from the image data through a multi-task CNN model. The second step is to learn apparel MEAN-INGs from predicted attributes through SVM and LKF classifiers. The testing results show that the prediction rate of eleven predefined MEANING classes can reach the range from 80.1% to 93.5%. The two-step apparel learning model is applicable for novel recommendation system developments.

Keywords: Apparel recommendation · Body · Style
Visual communication system · Apparel data · Deep learning

1 Introduction

In order to improve online shopping experiences and users' satisfactions, apparel recommendation systems that combine online apparel data with intelligent computing technologies have attracted research attentions. Traditionally, obtaining customers information and their purchasing history data is a prerequisite for developing recommendation systems. It suffers from the cold start due to the lack of historical data for new customers. Furthermore, the apparel as a kind of fashion products changes quickly with the latest fashion trend. Recommending clothes that are similar to the previous liked ones may not useful since people tend to seek style changes rather than copy the past. With the growing Artificial Intelligence technologies, recent studies have started to develop more professional apparel recommendation systems. However, the domain knowledge of clothes and dressing is used fragmentally in various systems, such as dressing for bodies in Vuruskan et al.'s [1] recommendation system, dressing for occasions in Liu et al.'s [2] recommendation system and mix-and-match styling recommendation system developed by Vaccaro et al. [3].

© Springer International Publishing AG, part of Springer Nature 2018
Á. Rocha et al. (Eds.): WorldCIST'18 2018, AISC 747, pp. 31–40, 2018.
https://doi.org/10.1007/978-3-319-77700-9_4

This study aims to integrate the profound knowledge and theories of clothes, fashion and dressing into the development of high quality apparel datasets and intelligent learning models to support apparel-oriented smart recommendation systems. We introduce a new apparel data coding system and two datasets: ATTRIBUTE and MEANING, based on clothes communication and semiotic theories. The ATTRIBUTE data captures the visual elements of design such as lines, shapes, colours, patterns, materials, mix and matches etc. and the MEANING data captures the connotation meanings carried by visual clothes and outfits. We also develop a deep learning model that predicts apparel MEANINGs through the apparel images and ATTRIBUTEs. The proposed methodology above can enhance the quality of an apparel recommendation system.

The remaining of this paper is organized as follows: Sect. 2 reviews existing apparel data and recommendation systems. Section 3 investigates the theory of clothes, menswear and fashion. Section 4 introduces the definition, construction and collection of newly developed apparel datasets based on visual communication theories. The following section details the methods of training apparel datasets and testing results. Section 6 draws conclusions and future works.

2 Related Work

Current studies have been focusing on the apparel data revolution in recommendation system development. The earlier systems excavated user's profiles and bought/liked history data and recommend new products based on images similarity. A raising research concern was that apparel recommendations should consider both clothing and dressing issues. Occasion-oriented recommendations extracted clothes image features and labelled it through crowdsourcing method [2]. Customer profile and behavioural data were replaced by occasion data offered by online crowdsourcers. Moreover, some researchers focused on outfit images designed by domain experts for developing mix-and-matches recommendation systems [4, 5]. The newly developed intelligent systems started to integrate with clothing and fashion knowledge such as providing recommendations based on fashion DNA, styling, body dressing [1, 3, 6–8].

Applied methods for developing intelligent recommendation systems can be divided into two parts, feature extractions and classifications. Current research has showed the abilities of using computer to detect texture [9], outline shape [10] and some local features [11] such as the Scale Invariant Feature Transform (SIFT). Automatic extractions could generate a large volume of image-based feature data, but it may be less accurate to represent essential apparel characteristics. It is necessary to bridge the connections between image features and apparel design elements. Regarding to classifications, the conventional machine learning algorithms are the dominant methods such as Neural Networks and Support Vector Machine. This study is going to develop a deep learning model to extract design attributes from raw images and make classifications using separate classifiers.

3   The Theories of Clothes, Menswear and Fashion

The understanding of clothes particular on Menswear fashion involves multiple disciplines such as design, sociology, semiology, economics and cultural studies. According to the fashion theory by Barnard [12] and Men's fashion theories by McNeil and Karaminas [13], we found important theories and knowledge to support apparel data and recommendation system development. The new viewpoint is that clothing is a kind of language for people to communicate meanings. For example, Barnard elaborated the relationship of fashion, clothing, communication and meaning in his book named Fashion as Communication [14]. This theory explained what the apparel might be for the wearer, why people wear clothes and how people use the meanings of clothes for effective communication.

3.1   An Overview of Fashion Theory and Men's Fashion Theory

The context of wearing clothes was discussed in some fashion theories such as Clothes and the Body, Style and Identity, and Clothes and Communication. Firstly, the body could bring meanings to dress through adornment. In dressing male bodies, Odile Blanc stated that dress amplified the body in particular ways and was part of male power strategies [15]. The transposition of the clothes' attributes and the feelings that it invokes is not merely the effect of the overall apparel as a casing, but of the body itself [16]. Dress serves as a visual metaphor for identity.

The message of clothes implies the identity regarding sex and gender, social class, ethnicity and race, as well as culture and subculture [12]. The unique identity in men's style is the Masculinity. It is constructed with certain social background and changing with time, as Peter McNeil indicated that the meanings of men's dress are transformed in different cultural and historical contexts [13]. Tim Edwards addressed men's fashion and masculinity through consumer society [17]. He emphasized the significance of sign value which is the meaning attached to commodities.

Rouse in Understanding Fashion indicated that the important function of clothes is the communication [18]. Barnard in Fashion as Communication stated that clothing is a non-verbal communication medium where one person as the wearer intends to send messages and another person as the spectator/viewer receives the messages [14]. The clothes language and communication theory could fit the men's fashion. These arguments specifically explained the message of men's style and its communication value.

3.2   Embedding the Apparel Communication Theory
into Knowledge-Based Recommendation System Development

Apparel communication is a complex system which occurs at two levels: clothes-body communication and wearers-viewers communication (Fig. 1). In level one, wearers themselves express meanings through the visual perception of clothes on their body, with visual and touch feelings. In level two, viewers perceive meanings through the visual perception of clothes on wearer's body without touch feelings. The overlap area of meanings in both levels evaluates the quality of apparel communication.

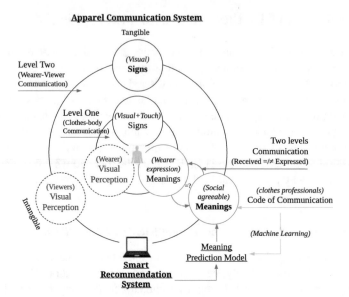

Fig. 1. The relationship of apparel communication system and smart recommendation system

Hence, supporting meanings to be communicated correctly is the target that apparel recommendation system is trying to achieve.

The development of an apparel recommendation system needs to focus on apparel meaning prediction model, aiming to effectively estimate meanings that could match up viewers' perceived with wearer's intended to express. In fact, the code to support apparel communication is better known by clothes professionals since these people have particular knowledge, skill and experiences of clothes and dressing. Thus the method of developing meaning prediction model is using machine learning and deep learning algorithms to train a learning model based on apparel communication theory and practical expert knowledge.

4   Knowledge-Encoded Apparel Data

4.1   Data Definition, Construction and Collection

Figure 2 illustrates an apparel communication network composed of three major components, including apparel signs, denotation meanings and connotation meanings. Signs of apparel refer to visual elements of line, shape, volume, drape and other elements in the listed. The first order of communication is between visual signs and denotation meanings. It commonly indicates obvious or literal meanings of apparel, but this study introduces a second order that indicates the semantics of design. The second order of communication is between denotation and connotation meanings. It generates meanings of mental (or perceptual) feelings that may associate with body, occasions, design genres or personal preferences. The second order of meanings is finally applied for communication.

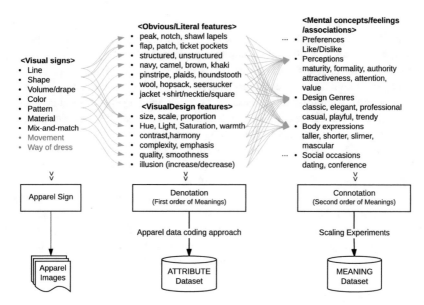

Fig. 2. Apparel communication networks and data definitions

The structure of apparel datasets is connected with the clothes communication networks. Apparel signs are composed by original apparel image data. Denotation meanings indicating apparel itself are belonged to the ATTRIBUTE dataset, and connotation meanings given by people outside of apparel are filled to the MEANING labelling dataset.

ATTRIBUTE Dataset. The ATTRIBTUE data captures the first order of meanings, which is the denotation meanings generated from apparel itself, such as lines, shapes or colours. It nominally denotes the obvious and literal meanings of clothes, such as lapel shape is denoted as peak, notch or shawl. The deeper denotative meaning involves the design of lines, shapes and colours, such as the proportion of lapel shapes. More design features, such as the contrast, harmony, illusion, emphasis and complexity [19] could be adopted to denote deep attributes. An extended semantic attribute data has been defined which denotes both natural and design meaning derived from the internal of apparel.

All of the identified visual features of apparel signs were classified into 25 categorical variables (e.g. proportion) and 20 ordinal variables (e.g. buttons number) based on their properties. For coding, for instance, the ration of a *Jacket width/length* as one of 45-dimensional features can be encoded as 1 = 'wide/short', 2 = 'narrow/short', 3 = 'narrow/long', and 4 = 'wide/long' from the subcategories. All samples (220 s) were encoded into SPSS 24.0 system manually according to each sample's product descriptions on its purchasing website. After eliminating samples with missing features, a number of 200 samples went into the final dataset.

MEANING Dataset. The MEANING data is identified as the second order of meanings, connotation meanings that are generated outside of apparel by people, such as wearers, viewers or designers. It denotes the feelings, thoughts, beliefs and desires of clothes based on mental concepts. General speaking, peak lapel feels formal. Peak lapel with a wider shape and higher position of gorge could formulate a muscular chest look. This study presents connotation meanings by perceptual feelings (style genres, e.g. elegant) and body meanings (to indicate body shapes). These identified meanings will be used to label apparel samples. As mentioned earlier, the second order of meaning is given by people. Therefore, it is necessary to decide which people could contribute to those meanings. The clothes professionals are qualified as skilled people. This group will contribute to data collection in labelling experiments.

The MEANING data was collected through labelling tasks by clothes experts. A 11-class set of style MEANING adjectives was employed as predefined labels including style genres of Casual, Trendy, Professional, Elegant, Classic and Playful, as well as body corrections for Square (broad all over), Circle (chubby all over), Triangle (soft round middle), Invert Triangle (broad shoulders slim waist) and Slim (straight up and down). The annotators were five commercial stylists with at least two-year working experience in a large retail sector. The labelling adjectives of apparel samples were recorded into SPSS 24.0 in accordance with ATTRIBUTE data.

4.2   Data Evaluations

The quality of the proposed two apparel datasets has been evaluated by several statistics and content-based assessments. We employed Internal Consistency and Test-Retest assessments to evaluate the reliability. Exploratory and confirmatory factor analyses measured the validity through statistical correlations of data. A panel of experts inspected the validity through the content structure. Statistics-based assessments confirmed that the element correlations in statistics are consistent with real connections. The feedback of content-based reviews by domain experts supported current structure and provided additional adjustments for further developments.

5   Development of Apparel Learning Models for Recommendation System

5.1   Apparel Learning Model Design

In an apparel communication system, meanings are transmitted through three layers including the apparel sign, denotation meanings and connotation meanings. The last layer, connotation meanings are the final output of communications. Logically, the learning model of connotation meanings could base on the features extracted from raw apparel Images or features represented by ATTRIBUTEs. Therefore, we essentially develop a deep learning model to extract image-based features that are discriminative to ATTRIBUTEs and MEANINGs and make classifications through a multi-task Convolutional Neural Network (CNN). The further training model is using predicted ATTRIBUTEs from the output of CNN to learning the MEANINGs, with two separate classifiers SVM and a newly proposed LKF.

A Deep CNN Model. We proposed a deep CNN structure which serves as a multi-task model considering all of the three levels of predictions, including style genres, denotation attributes, and body shapes. The key configurations of the proposed deep learning model are summarized in Fig. 3. The proposed CNN model is composed of five convolutional layers, five pooling layers and two fully connected layers. The design from the first convolutional layer to the last pooling layer serves for the feature learning. Each convolution level has two layers in the same size, e.g. Conv1 has Conv1_1 and Conv1_2 in the size of 160 × 128. The max-pooling layers between two levels make the size shrink to half of the original size. After Conv5, all of the patches are concatenated and mapped to two layers of fully connection, Fc6 and Fc7. The resultant feature space is in $X(apparel) = [x_n] \in \mathbb{R}^{N \times 4096}$, where 1, ..., n, ...$N$ indexes the training samples. The design from fully connected layers to the final outputs serves for classifications. There are two types of objective functions, Cross Entropy for $Y(style\ genre)$ and $Y(body\ shape)$, and Norm-2 distance for $Y(attribute)$.

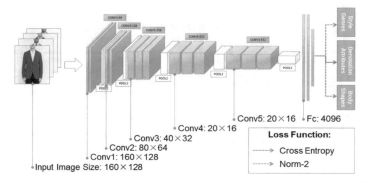

Fig. 3. Key configurations of the proposed deep learning model, Conv: convolution layers; Fc: fully connection layers.

MEANING Prediction Model. Secondly, we train apparel MEANING prediction model using predicted apparel ATTRIBUTEs from the outputs of CNN model. Since CNN feature is obtained by learning the multi-task loss regarding apparel attributes, style genres and body shapes, the predicted attributes may have been adjusted for the loss of style genres and body shapes prediction. Therefore, using the predicted attributes from the output of CNN model could further benefit the style genres and body shapes predictions.

Since the deep learning model is based on a multi-task loss, empirically, the performance of each individual task may have been sacrificed for the sake of balanced overall loss. Therefore, the separate SVM classifiers are introduced for predictions with extracted visual features from CNN model. However, the proposed SVM classifier may encounter the degradation due to the small training size. Because our evaluation is based on a highly random cross-validation scheme, the maximum number of training images is 180, and the samples of each label can be very unbalanced. The weak classes with few samples may be completely sacrificed for the denser ones. In order to mitigate the unbalanced prior, we propose a distribution based classification algorithm that is named as Later Kernel Fusion (LKF). Instead of learning the class boundaries, LKF focuses on maximizing the likelihood function between the test samples.

5.2   Results

Above proposed apparel learning approaches have been implemented through Python scripts. We feed the Image, ATTRIBUTE and MEANING three parts of data as the inputs and outputs to train models. The model testing results are demonstrated in Table 1. The average prediction rate of 11 predefined classes is 74.9% in the deep CNN model. To combine with the conventional SVM classifier is less effective. In comparison, the average predictability with the newly proposed LKF classifier is 9.4% higher than the first introduced CNN model. The MEANING prediction model further improved the performances of eleven classes by introducing the predicted ATTRI-BUTEs in CNN model as feature data. The most significant increases are in the class of 'Circle' from the CNN model (76.4) to the CNN model with LKF (84.4) and finally to the MEANING prediction model with LKF (90.4).

Table 1. Model training and testing results

Predictability of classes (style genres & body shapes) (%)

	Casual	Elegant	Playful	Professional	Trendy	Classic	Circle	Invert triangle	Triangle	Square	Slim	Mean
CNN model	72.8	74.5	69.4	75.8	80.5	72.2	76.4	71.8	67.7	79.6	82.8	74.9
CNN model (with SVM)	72.9	72.6	70.1	76.2	77.5	70.7	75.8	69.5	64.4	79.7	79.2	73.5
CNN model (with LKF)	82.4	83.8	77.9	84.6	91.2	85.7	84.4	82.5	79.6	86.4	89.2	84.3
MEANING prediction model (with SVM)	79.6	85.1	76.9	84.3	91.8	86.9	88.4	84.5	83.8	91.4	93.2	86.0
MEANING prediction model (with LKF)	84.5	85.4	80.1	88.5	92.4	87.1	90.4	86.3	87.8	92.6	93.5	**88.1**

In feature extractions, apparel design attributes are recognisable from images through deep learning method, and the recognised apparel attributes are the most effective feature representations to adapt the tasks of apparel meaning classifications. In classifications, the proposed LKF algorithm is more effective than the baseline SVM in both CNN model and MEANING prediction model. Overall, the proposed MEANING prediction model with LKF classifier reached the best performance with an average prediction accuracy of 88.1%. This is a highly effective and useful model for proto-typing apparel recommendation systems.

The prediction of apparel attributes is made by the regression output of the CNN model. Figure 4 illustrates the Mean Squared Errors (MSE) of overall predicted attri-butes, which estimate the average of the squares of the prediction errors. This is an important parameter to measure the performance of regression model. The value of MSE closer to zero means the higher predictability of apparel attributes. In Fig. 4, the majority apparel attributes are predictable with small errors (MSE < 1.5). The pre-dictions on five attributes are less accurate with the relatively higher MSE, including Jacket outline, jacket pattern, jacket material, shirt hue and shirt pattern. This is caused

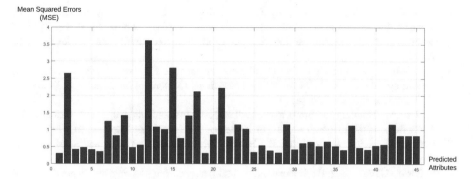

Fig. 4. Mean squared errors of predicted attributes

by the unbalanced distribution of samples in these high-dimensional attributes. Some features due to lack of observing samples are unlikely to be predicted.

6   Conclusion

Knowledge-based apparel recommendation systems could be achieved through the newly proposed apparel datasets and modelling methods. The ATTRIBUTE prediction model can be used for Menswear advanced search system that allows natural and design features to be filtered, and the MEANING prediction model can be used as a Menswear style recommendation engine that distinguishes online clothes by specific style genres or body shapes. These models are superior to other models in recognising in-depth clothes features and connotative meanings like the trained style professionals. Although the proposed apparel learning methods and models are based on small size of training datasets, the evaluation results of predictability on eleven classes are all above 80%, which means the proposed apparel data development and training methods are valid. In the future, it has spaces for improvements. In order to optimise the model preference, the newly proposed methods will be used to adapt a larger size of sample data, a wider category of clothes, more predefined classes regarding apparel meanings, e.g. dandy, body colours, as well as occasional meanings.

References

1. Vuruskan, A., et al.: Intelligent fashion styling using genetic search and neural classification. Int. J. Cloth. Sci. Technol. **27**(2), 283–301 (2015)
2. Liu, S., et al.: Hi, magic closet, tell me what to wear! In: Proceedings of the 20th ACM International Conference on Multimedia. ACM (2012)
3. Vaccaro, K., et al.: The elements of fashion style. In: Proceedings of the 29th Annual Symposium on User Interface Software and Technology. ACM (2016)

4. Kalantidis, Y., Kennedy, L., Li, L.-J.: Getting the look: clothing recognition and segmentation for automatic product suggestions in everyday photos. In: Proceedings of the 3rd ACM Conference on International Conference on Multimedia Retrieval. ACM (2013)
5. Jagadeesh, V., et al.: Large scale visual recommendations from street fashion images. In: Proceedings of the 20th ACM SIGKDD International Conference on Knowledge Discovery and Data Mining. ACM (2014)
6. Landia, N. Building recommender systems for fashion: industry talk abstract. In: Proceedings of the Eleventh ACM Conference on Recommender Systems. ACM (2017)
7. Bracher, C., Heinz, S., Vollgraf, R.: Fashion DNA: merging content and sales data for recommendation and article mapping. arXiv preprint arXiv:1609.02489 (2016)
8. de Barros Costa, E., et al.: Understanding and personalising clothing recommendation for women. In: Rocha, Á., et al. (eds.) Recent Advances in Information Systems and Technologies, vol. 1, pp. 841–850. Springer International Publishing, Cham (2017)
9. Kim, D., Kwon, Y.-B., Park, J.: A scoring model for clothes matching using color harmony and texture analysis. In: Graphics Recognition. New Trends and Challenges, pp. 218–227. Springer (2013)
10. Cheng, C.-I., Liu, D.S.-M.: An intelligent clothes search system based on fashion styles. In: 2008 International Conference on Machine Learning and Cybernetics. IEEE (2008)
11. Iwata, T., Wanatabe, S., Sawada, H.: Fashion coordinates recommender system using photographs from fashion magazines. In: IJCAI Proceedings-International Joint Conference on Artificial Intelligence. Citeseer (2011)
12. Barnard, M.: Fashion Theory: A Reader. Routledge, New York (2007)
13. McNeil, P., Karaminas, V.: The Men's Fashion Reader. Berg Publishers, Oxford (2009)
14. Barnard, M.: Fashion as Communication. Psychology Press, Campbell (2002)
15. Blanc, O.: From battlefield to court: the invention of fashion in the fourteenth century. In: Encountering Medieval Textiles and Dress, pp. 157–172. Springer (2002)
16. Karaminas, V., Übermen: masculinity, costume, and meaning in comic book superheroes. In: McNeil, P., Karaminas, V. (eds.) The Men's Fashion Reader. Berg Publishers (2009)
17. Edwards, T.: Men in the Mirror: Men's Fashion, Masculinity, and Consumer Society. Bloomsbury Publishing, London (2016)
18. Rouse, E.: Understanding Fashion. BSP Professional Books, Oxford (1989)
19. Fiore, A.M.: Understanding Aesthetics for the Merchandising and Design Professional. Fairchild Books, New York (2010)

Non-personalized Fashion Outfit Recommendations

The Problem of Cold Starts

Anna Iliukovich-Strakovskaia[1(✉)] , Victoria Tsvetkova[2] ,
Emeli Dral[1] , and Alexey Dral[1]

[1] Moscow Institute of Physics and Technology,
1 "A" Kerchenskaya street, Moscow 117303, Russian Federation
strakovskaya.am@phystech.edu
[2] Lomonosov Moscow State University,
GSP-1, Leninskie Gory, Moscow 119991, Russian Federation

Abstract. Nowadays the demand for compatible outfit recommendation system is rising up. There are a number of online fashion shops and web applications allowing their customers to observe collages of fashionable items which look good together. Such forms of recommendations help both clients and sellers. On one hand clients can make better choices while shopping and find new items of clothing that will match each other. On the other hand, complex outfit recommendations help sellers to sell more items which will influence their business KPIs (key performance indicators).

We can build fashion outfit recommendation system using various instruments: machine learning (mostly supervised learning), statistics, user modelling, rules construction and many others. The problem is that most of these techniques require a solid amount of data about users' preferences and tastes to create good personal recommendations. But what if there is no such information? In this paper we are proposing the approach of creating fashion outfit recommendations in situations when we lack of that data (the cold start problem). We are going to share (1) the approach of dataset gathering and (2) the results of using external datasets and pre-trained models for fashion outfit recommendation creation.

Keywords: Fashion outfit recommendations · Recommender systems
Machine learning · Neural networks · Transfer learning

1 Introduction

Recommendation system is an obvious choice for effective operating with large amounts of low structured data as it effectively reduces the amount of information shown to users. Personal recommendations, in their turn, may help to interact with each user in a unique way in terms of data: with the help of recommendation system we are building a unique representation of available items for each user. There are numerous examples of recommender system usage in different domain areas these days.

© Springer International Publishing AG, part of Springer Nature 2018
Á. Rocha et al. (Eds.): WorldCIST'18 2018, AISC 747, pp. 41–52, 2018.
https://doi.org/10.1007/978-3-319-77700-9_5

This technique is extremely important in online e-commerce applications (Amazon), streaming media players (Netflix), online radio and music services (Pandora), and many others. Surely, online fashion sphere is not an exception (Polyvore).

Having high-quality personal fashion recommendations (e.g. recommendations about stylish ensembles of clothes) is crucial for online fashion shops. Most of the online apparel stores have a solid amount of items from different categories of the assortment. Their problem is that the more products a store has, the longer time it will take for a client to find a relevant item. Users need more and more time to browse the whole catalog of products that sometimes may go beyond the reasonable time limit.

The trick here is that in fact users do not need to examine the whole range of available products to find what they need. Having at least a bit of information about user's preferences, recommender algorithms help to reduce the amount of products to review significantly.

Fashion outfit recommendations can be considered as a standard recommendation problem; however, there is a difference worth noticing: clothing items cannot be recommended independently. If a user chose, for example, some jeans, the recommendation system should (1) show a top which would be (2) suitable to the jeans s/he chose. Having good recommendations for one clothing category does not automatically lead you to a ready-made stylish outfit. In other words, fashion look generation is a way more complicated problem, as you not only have to recommend items which users will probably like, but recommend them as a set, items of which actually match each other both in terms of fashion and user's preferences.

Also, we should mention another challenge: there are different types of online shops: from big ones with enormous catalogues of clothing to smaller ones which are specialized on selling only one brand. In this paper we propose an approach for building such recommendation system which can be used by smaller shops as well.

2 Related Work

Fashion domain attracts more and more attention in different fields. Existing researches mostly focus on design generation [6, 9], fashion prediction [14] and fashion recommendation [5, 11–13].

Our work falls into the area of fashion recommendation and extends the research [2, 10] where the authors achieved high accuracy on distinguishing fashionable and unfashionable dresses. In this paper we are focusing on differentiating compatible and incompatible outfits. There have been several researches toward that, too. For example, in [5, 11, 12] the authors trained their models on dataset consisting of already created fashion looks from Polyvore.

In [12] the authors trained their model to recognize good and bad outfits containing two objects (top and bottom). In [5] there was suggested an approach to use the pre-trained model to find out whether items from one's wardrobe match each other or not.

Unlike all listed works, in this paper we suggest a new approach which will be based on (1) specializing on triple sets, (2) gather dataset in a different way (using clothes images and crowdsourcing to validate whether the sets are compatible or not). We also would like to test whether it is possible to transfer our model to other shops.

3 Problem Statement

In this paper we aim to build such a fashion outfit recommender system which will generate a set of three matching each other items (a top, a bottom and shoes) as a single recommendation. We also want to determine if the recommendation model trained on our dataset could be transferred to other datasets (belonging to smaller retailers) and, if so, how much external data we should need for it to be transferred successfully.

Obviously, the first question to rise is the following: what outfit can be objectively called "a good one" without any personalization like colour or style preferences? How can we build a system which will be able to decide if these items really look compatible enough to be recommended to any user and how can we validate its performance? This is actually a complicated problem: there is no formal criterion. We can try to do it automatically: for example, generate a number of outfits basing on some rules ("any blouse suits any jeans but not shorts") or use popular items if a retailer has some statistics about the popularity of its products ("combine most popular tops with most popular bottom"). Either way, it doesn't guarantee that the generated outfit would really be good or could be called "a stylish one".

For our recommendation system we compiled several datasets which are described in Sect. 4.

After we have settled the first question, another one is rising up: what should a small shop do to obtain a recommendation system? One of the options is to use a pre-trained model. And how to incorporate external data on clothing sets into a recommender system that will have to make recommendations based on the new products of this smaller shop and its new catalog hierarchy? We have conducted a number of experiments to proof that the idea of using external fashion outfit dataset can be useful and to find the most effective and promising ways of using such data for smaller retailers. We will discuss this in Sect. 6.

The structure of our work is the following:

1. Hypothesis 1 (H1): there is a difference between compatible and incompatible (or "good" and "bad") outfits and we can train a machine learning model to sort them out.
2. Hypothesis 2 (H2): it is possible to apply a model which was trained to differentiate compatible and incompatible outfits using the data from one shop, to another one.
3. Hypothesis 3 (H3): if we could apply the pre-trained model to another shop, there will be an exact moment when we should switch from the pre-trained model to the model trained on the original dataset. There is also an addition question here: how much external data will we need to build a robust model?

Apart from that, we will also describe a design of the datasets we collected and the way people matched it to create positive samples.

4 Data

We can say that there exist two types of fashion retailers: big and small ones. A big fashion retailer (usually an aggregator) can have more resources to build their own fashion models from its catalog and more information about its users' behavior, which can help them build a better personalized model. On the other hand, a small fashion retailer has neither information about its customers nor resources to build a good generated model. In such situation, using external data seems a natural choice.

For our tasks we gathered data from one big retail website and two small ones and created three image collections. Each collected item was defined by an image, a unique name, a level and a category. Raw stats are available in the Tables 1 (big retailer – "B"), 2 (small retailer – "s_0") and 3 (small retailer – "s_1"). Levels (top, middle and bottom) were matched manually. The names for the categories (such as "jeans", "pants", "tunics") were collected from the websites.

Table 1. Big retail Website (B), raw stats.

top (24686)		bottom (17670)	
jumpers	9039	sneakers	3562
shirts	5690	shoes	2783
t-shirts	4343	ankle boots	2591
sweatshirts	2442	boots	2389
tops	2333	high heels	1976
tunics	839	sandals	1394
middle (11366)		slip-ons	888
pants	4976	flats	610
skirts	3348	snowjoggers	385
jeans	2502	uggs	370
shorts	540	espadrilles	351
		clogs	106
		slippers	105
		jackboots	91
		loafers	69

Table 2. Small retail Website (s_0), raw stats.

top (239)		middle (90)	
tops	194	trousers-shorts	51
knitwears	45	skirts	39
		bottom (45)	
		shoes	28
		boots	17

Table 3. Small retail Website (s_1), raw stats.

top (440)		middle (251)	
t-shirts	224	pants	134
shirts	195	shorts & skirts	67
sweatshirts	21	jeans	50
		bottom (76)	
		shoes	76

5 Our Approach

To sort out compatible outfits we used a crowdsourcing approach. For that we developed a simple website where we uploaded our image collections. In order to create a good outfit a user had to follow three steps: first, he or she chose the collection: big retailer "B", small retailers "s_0" or small retailer "s_1" from which he intended to pick up clothing items. Second, s/he chose categories for each level (what exact top he would like to pick up, blouse or t-shorts). Third, the user picked up clothing items and created the look s/he considered compatible.

All collected outfits were saved under the user's login which can be updated at any time. See Fig. 1 for visual example of the interface and Fig. 2 for collected fashionable outfits.

Fig. 1. Visual examples of the interface

Apart from positive samples, to perform out experiments we have to define the concept of what we call a "bad" or incompatible outfit. This task was solved automatically. To generate such outfits we followed a simple rule: combine all items as long as they are matching the three levels (for example, any blouse can go with any jeans and any sneakers, but it is not possible to have an outfit consisting of a blouse, a tunic and a T-shirt). Examples of such "bad" outfits you can find in Fig. 3. In the generated datasets, for one good outfit there is exactly one example of a bad one (50:50 split).

Fig. 2. Examples of "good" fashion outfits generated.

Fig. 3. Examples of "bad" fashion outfits generated.

Of course, there are also other possibilities to generate negative samples. For instance, a user could mark an automatically generated look as a "bad one" or a "good one". We do not follow this approach to reduce "noise" (see the definition in [3]) in implicit feedback as all image grids were reloaded after each outfit generation.

The visual pipeline for a dataset generation is presented in Fig. 4. Overall, we have collected 1812 good fashion looks for the big retailer "B", 255 good outfits for s_0 and 297 for s_1.

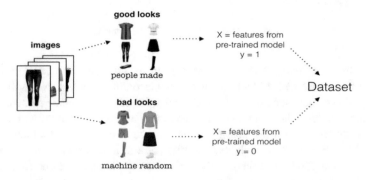

Fig. 4. Diagram explaining the process (pipeline) of dataset generation.

All images were normalized with the help of inception pre-trained deep network (MXNET framework [1, 8]). It has been proven that extracted features from a global pool layer serve as a good image representation [4, 7]. Therefore, each fashion outfit is presented in the form of 1024×3 features (1024 features per each clothing item).

6 Experiments, Evaluation and Conclusions

To validate our first hypothesis (H1) whether we are able to differentiate good looks from bad ones, we have conducted a series of experiments. In such case natural choice is to compare the quality of a random classifier with a machine learning model. After having tried a few other models we chose Random Forest classifier for its better performance. For each collected dataset we ran 5-fold Cross-Validation to estimate the accuracy of the classification. As each dataset contained 50% of good looks and 50% of bad ones, the random classifier gave us 50% accuracy.

After receiving the estimation from the Random Forest model we sorted it out and measured the quality on different TOPs (see Table 4 and Fig. 5). It is obvious to notice that our model cannot recognize negative samples well, but that is not the point. The fact is that for Industrial purpose we only need to be able to generate good recommendations - and not many of them, so we can be sure that TOP 10% will be sufficient enough. In our TOP 10% we have quite good accuracy.

Table 4. Quality metrics across hypothesizes H1 and H2

Top N%	Precision (mean ± std)			
	No transfer learning (experiment Q1)		Pretrained on "B" (experiment Q2)	
	s_0	s_1	s_0	s_1
1	1.0 ± 0.0	1.0 ± 0.0	1.0 ± 0.0	1.0 ± 0.0
2	1.0 ± 0.0	1.0 ± 0.0	1.0 ± 0.0	1.0 ± 0.0
3	0.93 ± 0.13	1.0 ± 0.0	0.93 ± 0.13	1.0 ± 0.0
5	0.92 ± 0.1	0.96 ± 0.08	0.92 ± 0.1	0.88 ± 0.1
10	0.86 ± 0.15	0.96 ± 0.07	0.86 ± 0.05	0.8 ± 0.07
15	0.81 ± 0.11	0.93 ± 0.07	0.77 ± 0.05	0.74 ± 0.08
20	0.78 ± 0.12	0.89 ± 0.05	0.69 ± 0.07	0.69 ± 0.07
30	0.75 ± 0.09	0.82 ± 0.04	0.65 ± 0.06	0.64 ± 0.03
50	0.67 ± 0.04	0.71 ± 0.05	0.57 ± 0.04	0.6 ± 0.03
70	0.6 ± 0.03	0.63 ± 0.03	0.49 ± 0.03	0.54 ± 0.03
100	0.5 ± 0.03	0.5 ± 0.02	0.5 ± 0.03	0.5 ± 0.02

For the next hypothesis (H2) we checked if it is possible to apply a pre-trained fashion outfit classifier from one shop to another one. It is an important question to every new small shop without any data and users' feedback collected. Using the external pre-trained model can speed-up the process of launching a new fashion business without spending any time and effort.

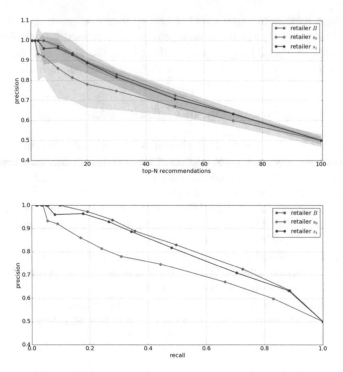

Fig. 5. Quality of recommendations (precision) as a function of recall (bottom image) and a recommendation list size (top-N recommendations task, N is measured in percent %) (top image).

We trained a fashion outfit classifier on dataset "B" and validated its quality on datasets s_0 and s_1. Quality metrics were evaluated on the same folds as in the first experiment using the same 5-fold cross-validation (see Fig. 6). Therefore, we can measure deviation and compare approaches across the experiments (see Table 4 and Fig. 7).

The useful corollary is that if you generate TOP-10% outfit recommendations (50–60 recommendations in our case for s_0 and s_1), then you can get 80+% good outfit recommendation out-of-the-box with the help of external pre-trained fashion outfit model. Hereby, our hypothesis 2 is validated positively. If we have to generate a reasonable number of recommendations, then the pre-trained model is a fine choice. Otherwise, if you need to generate really long listings (100+) of recommendations, then, obviously, the quality of the pre-trained model usage would leave much to be desired.

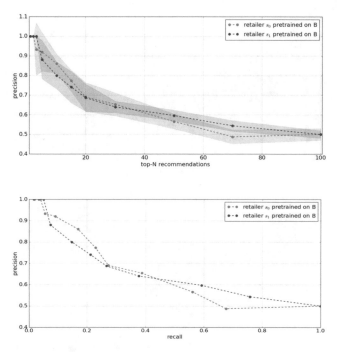

Fig. 6. Quality of recommendations (precision) as a function of recall (bottom image) and a recommendation list size (top-N recommendations task, N is measured in percents %) (top image).

With regard to the previous statement, we conducted an experiment to study how much data has to be gathered by a fashion retailer to switch from a pre-trained model to a specially tuned retail-oriented model (H3). To answer this question, we split each datasets s_0 and s_1 in proportion 80:20 (train:test) five times (the same way we had performed cross-validation in the experiments H1 and H2). Therefore, all the metrics are reported in the same conditions (see Fig. 8).

Overall, we have 500–600 good and bad outfits for each retailer s_0 and s_1 and during the experiments train size was in the range between 50 and 300. As shown in Fig. 9, the quality of recommendation in TOP-10% is the same for pre-trained and retail-oriented models. One fashion designer can collect 100–150 outfits per day. Thus, getting a retail-oriented model which works considerably well for a provided fashion domain is a question of 2–3 days of working.

The experiment also shows that existing amount of data is not enough to prove or reject the hypothesis (H3). We should either collect more data or compare the quality of personal recommendations (and, consequently, attract more volunteers).

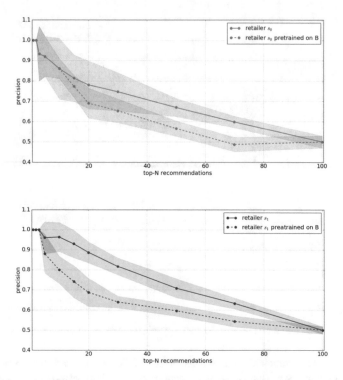

Fig. 7. Quality of models between experiments (with and without usage of pre-trained outfit models).

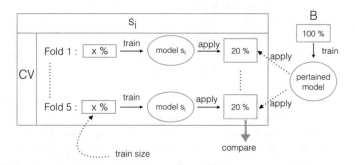

Fig. 8. Diagram explaining the process of comparison pretrained and retail-oriented models. Retail-oriented model is trained on a slice of the retail dataset (train_size).

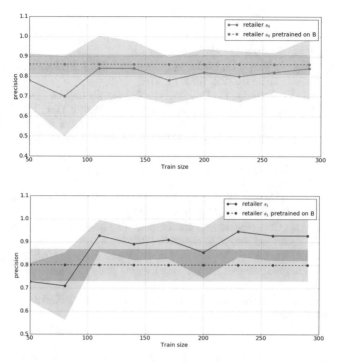

Fig. 9. Precision at Top-10% for s_0 and s_1.

7 Future Work

We consider improving our work in several ways:

(1) by generating a larger dataset using a website with a collection of outfits presented (for example, Polyvore),
(2) by including multi-modal information about each item,
(3) by adding more negative samples to the train set.

Also, we would like to add a virtual trial room to our system, so a user would not only get a recommendation what to buy, but could also try it on and see how it suits him personally.

Acknowledgments. This work has been completed at the chair of Algorithms and Technologies of Programming, Department of Innovation and High Technology, Moscow Institute of Physics and Technology (ATP DIHT MIPT).

References

1. Model. http://data.dmlc.ml/mxnet/models/imagenet/
2. Chen, Q., Wang, G., Tan, C.L.: Modeling fashion. In: 2013 IEEE International Conference on Multimedia and Expo (ICME). IEEE (2013)
3. Hu, Y., Koren, Y., Volinsky, C.: Collaborative filtering for implicit feedback datasets. In: Eighth IEEE International Conference on Data Mining, 2008, ICDM 2008. IEEE (2008)
4. Iliukovich-Strakovskaia, A., Dral, A., Dral, E.: Using pre-trained models for fine-grained image classification in fashion field. In: KDD 2016 Workshop on Machine Learning Meets Fashion (2016)
5. Tangseng, P., Yamaguchi, K., Okatani, T.: Recommending outfits from personal closet. In: The IEEE International Conference on Computer Vision (ICCV) (2017)
6. Zhu, S., Fidler, S., Urtasun, R., Lin, D., Loy, C.C.: Be your own prada: fashion synthesis with structural coherence. arXiv:1710.07346 (2017)
7. Szegedy, C., et al.: Going deeper with convolutions. In: Proceedings of the IEEE Conference on Computer Vision and Pattern Recognition (2015)
8. Chen, T., Li, M., Xiao, T., Li, Y., Xu, B., Lin, M., Wang, N., Wang, M., Zhang, C., Zhang, Z.: MXNet: a flexible and efficient machine learning library for heterogeneous distributed systems (2015)
9. Kang, W.-C., Fang, C., Wang, Z., McAuley, J.: Visually-aware fashion recommendation and design with generative image models. arXiv:1711.02231 (2017)
10. Wang, J., et al.: Towards predicting the likeability of fashion images. arXiv preprint arXiv:1511.05296 (2015)
11. Han, X., Wu, Z., Jiang, Y.-G., Davis, L.S.: Learning fashion compatibility with bidirectional LSTMs. arXiv:1707.05691 (2017)
12. Songy, X., Fengx, F., Liuy, J., Liy, Z., Niey, L.: NeuroStylist: neural compatibility modeling for clothing matching. In: Proceedings of the ACM International Conference on Multimedia, June–May 2017. ACM (2017)
13. Lin, Y., Wang, T.: Dress up like a stylist? Learning from a user-generated fashion network. In: KDD 2017 Workshop on Machine Learning Meets Fashion (2017)
14. Al-Halah, Z., Stiefelhagen, R., Grauman, K.: Fashion forward: forecasting visual style in fashion. arXiv:1705.06394 (2017)
15. Liu, Z., Luo, P., Qiu, S., Wang, X., Tang, X.: DeepFashion: powering robust clothes recognition and retrieval with rich annotations. In: Proceedings of IEEE Conference on Computer Vision and Pattern Recognition (CVPR) (2016)
16. Liu, Z., Yan, S., Luo, P., Wang, X., Tang, X.: Fashion landmark detection in the wild. In: European Conference on Computer Vision (ECCV) (2015)

System for Deduplication of Machine Generated Designs from Fashion Catalog

Rajdeep H. Banerjee$^{(\boxtimes)}$, Anoop K. Rajagopal$^{(\boxtimes)}$, Vikram Garg,
and Sumit Borar

Myntra Designs Pvt. Ltd., Bangalore, India
{rajdeep.banerjee,anoop.kr,vikram.garg,sumit.borar}@myntra.com

Abstract. A crucial step in generating synthetic designs using machine learning algorithms involves filtering out designs based on photographs already present in the catalogue. Fashion photographs on online media are imaged under diverse settings in terms of backgrounds, lighting conditions, ambience, model shoots etc. resulting in varying image distribution across domains. Deduping designs across these distributions require moving image from one domain to another. In this work, we propose an unsupervised domain adaptation method to address the problem of image dedup on an e-commerce platform. We present a deep learning architecture to embed data from two different domains without label information to a common feature space using auto-encoders. Simultaneously an adversarial loss is incorporated to ensure that the learned encoded feature space of these two domains are indistinguishable. We compare our approach with baseline calculated with VGG features and state of art CORAL [19] approach. We show with experiments that features learned with proposed approach generalizes better in terms of retrieval performance and visual similarity.

Keywords: Dedup · Domain adaptation · GAN · Similarity

1 Introduction

Fashion being a visually immersive domain engages consumers on an e-commerce platform through catalog shoots which vary in terms of models, backgrounds, poses, illumination etc. These variations accentuate the visual aesthetics of the product and aid in enriching the consumer experience on the platform. In Fig. 1(a) and (b) we observe that the products are catalogued differently in terms of background, model poses and lighting between the two different e-commerce platforms. As a result of these conditions fashion photographs exhibit different data distributions across different platforms.

Consumer tastes in fashion is ephemeral and changes with trends, season, geographies etc. Brands quickly need to identify changing customer preferences and bring in new designs that customers are looking for. Generating new designs or images, henceforth used interchangeably, through machine learning algorithms

© Springer International Publishing AG, part of Springer Nature 2018
Á. Rocha et al. (Eds.): WorldCIST'18 2018, AISC 747, pp. 53–63, 2018.
https://doi.org/10.1007/978-3-319-77700-9_6

can significantly aid this turnaround cycle. Image generative models like GAN [6] have seen enormous success of late and are being used in the Fashion domain extensively [9]. Synthetic images obtained through generative models like Generative Adversarial Networks (GAN) as shown in Fig. 1(c), in general end up having artifacts like blur, distortions etc.

Fig. 1. Images from different domains

Fig. 2. Our goal is to bring both catalogue images and synthetic images to a common space using domain adaptation to perform dedup.

Dedup of images using generative image models from catalog data is challenging as the images from these two sets do not have similar data distributions.

Thus a machine learning model learned to represent designs of catalog shoots might not work well for designs obtained from machine generated images. The generalization ability of any learning algorithm depends on the fact that the test data follows the same underlying probability distribution as the training data [13]. However in many cases it fails to adhere to this axiom. In this work we develop a novel unsupervised domain adaptation algorithm to bring design representations of catalog (referred to as "source") and generative image models (referred to as "target") to a common ground and learn a single feature representation without any *labelled* information as depicted in Fig. 2. The network architecture consists of a autoencoder unit and a domain classifier unit. The autoencoder unit encodes the input feature space whereas the domain classifier unit ensures that the encoded feature space of the two domains are invariant.

In our experiments we use machine generated images and catalog images as our data domains without loss of generality. We show that features learned through our approach provides better retrieval performance from catalog for query machine generated images. Further analysis show that our domain adaptation approach has low KL divergence between the source and target. We also show our method performs comparatively better than the recent CORAL [19] method. Additionally, we show qualitative results where our domain adapted features maintain good visual coherence between the source and the target sets.

The approach considered is generic and is not dependent on datasets. Our feature representations can also be applied to other applications like visual similarity, grading etc. across domains.

2 Related Work

Dedup of images is a crucial problem in e-commerce. Most algorithms use pretrained features like in [8,18]. In machine generated images like in GAN [15] fully connected layers of the discriminator is used to remove similar images generated by the generator. However, in our application, we attempt to eliminate images generated by a generative model which are similar to the input catalog images but having different data distributions.

In cases where the training and testing test data are drawn from different distributions transfer learning/domain adaptation of data is highly desirable for generalization. An elaborate categorization and review of transfer learning methods is provided in [13]. In [1,10,11,16] domain adaptation is performed by obtaining better feature representations using labelled source and target samples. In [5,7] unsupervised domain adaptation is performed by projecting source and target data on a manifold wherein the distance between their subspaces is minimum. However these approaches require expensive subspace calculations. A comprehensive review of visual domain adaptation is provided in [14].

Recently, there has been works of visual domain adaptation with deep learning in [3,12]. All these works are used in classification settings where they have two units one for source data classification and the other for mitigating domain invariance. In [4] one unit does reconstruction of the target data using autoencoders. Another unit uses the encoded features for source classification. The network is optimized alternately for reconstruction and classification. All these approaches require source labels to be present.

In [19] an unsupervised approach is proposed which minimizes the domain shift between source and target by aligning the second order statistics of source and target distribution. However this may not preserve high order statistics of data. In our work, we propose an architecture combining both reconstruction and adversarial classification for learning domain invariance as in [3].

3   System Architecture

Given machine generated designs our goal is to filter out images that closely resemble the ones in the catalog. However, as shown in Fig. 2 we clearly notice that there is a data distribution difference between the machine generated styles and actual catalog shoot images. We use Domain adaptation to bridge this gap. Our architecture is shown in Fig. 3. We first generate new designs using Generative Adversarial Networks (GAN) [15]. This is followed by Domain Adaptation layer between the GAN and Catalog image features. We use VGG features [18] f_{vgg}, 16 layer network pretrained using Imagenet [2], which are known to be generic high level features, from both source and the target data. These features are passed through our Domain Adaptation framework to obtain domain invariant features which is explained in the next section. We finally perform a dedup using K-nearest neighbour (KNN) approach.

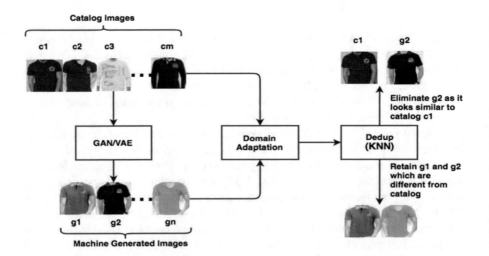

Fig. 3. System architecture for image dedup.

4   Unsupervised Domain Adaptation

In this section, we propose a architecture to perform domain adaptation using autoencoders and a classification unit with adversarial loss. Autoencoder ensures faithful representation of features whereas adversarial unit enables domain invariance. The network architecture is as shown in Fig. 4.

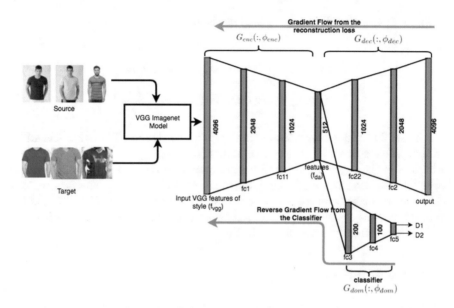

Fig. 4. Our framework for domain adaptation.

Let $\{x_i \in \mathbb{R}^d\}_{i=1}^N$ be the pooled source and target data and $\{y_i\}_{i=1}^N$ where $y_i \in \{S, T\}$ be their respective domain labels representing source or target. Let ϕ_{enc}, ϕ_{dec} be the parameters of the encoder and decoder in the autoencoder shown in Fig. 4. Let ϕ_{dom} be the parameters of the classifier layer. G_{dec}, G_{enc} and G_{dom} are the respective functions in decoder, encoder and classifier layers.

The overall objective of the system is to learn features (f_{da}) which encodes the input data and which is invariant to domain changes. We achieve this by finding the parameters ϕ_{enc} that maximise the loss of the domain classifier, while simultaneously finding ϕ_{dom} that minimise the loss of domain classifier. Also the parameters ϕ_{enc} and ϕ_{dec} try to minimise the reconstruction loss L_{rec}. The reconstruction loss L_{rec} is represented as in Eq. 1. In our work, L_{rec} is a L1 loss function between the input and the output of the autoencoder unit.

$$L^i_{rec}(\phi_{enc}, \phi_{dec}) = L(G_{dec}(G_{enc}(x_i; \phi_{enc}); \phi_{dec}), x_i) \tag{1}$$

The classifier loss L^i_{dom} is represented by Eq. 2. We use a binary cross entropy loss for the domain classification unit L^i_{dom}.

$$L^i_{dom}(\phi_{enc}, \phi_{dom}) = L(G_{dom}(G_{enc}(x_i; \phi_{enc}); \phi_{dom}), y_i) \tag{2}$$

The overall loss is given in Eq. 3 below. Here N_1 and N_2 are the number of images from domain 1 and domain 2 respectively with $N = N_1 + N_2$

$$E(\phi_{enc}, \phi_{dec}, \phi_{dom}) = \frac{1}{N} \sum_i^N L_{rec}^i - \lambda \left(\frac{1}{N_1} \sum_{i=1}^{N1} L_{dom}^i + \frac{1}{N_2} \sum_{i=1}^{N_2} L_{dom}^i \right)$$

As in [3] the overall objective is given below

$$(\hat{\phi}_{enc}, \hat{\phi}_{dec}) = \underset{\phi_{enc}, \phi_{dec}}{\arg\min} \ E(\phi_{enc}, \phi_{dec}, \hat{\phi}_{dom}) \tag{3}$$

$$(\hat{\phi}_{dom}) = \underset{\phi_{dom}}{\arg\max} \ E(\hat{\phi}_{enc}, \hat{\phi}_{dec}, \phi_{dom}) \tag{4}$$

λ in Eq. 3 is a hyper-parameter that controls the trade-off between the adversarial objectives to obtain the saddle points $(\hat{\phi}_{enc}, \hat{\phi}_{dec}, \hat{\phi}_{dom})$. The above adversarial loss is then optimized using backpropagation with the Gradient Reversal Layer [3] between the encoder and classifier. This layer during the forward pass acts as an identity layer and during backward passes it multiplies the gradient from subsequent layer by $-\lambda$ and passes to preceding layer for updates. The gradient flow directions are shown in Fig. 4.

5   Experiments

We use around 44000 images from Myntra catalogue, which are cropped to extract the article (tshirt in this work) using a bounding box algorithm [17] forming our source dataset. We then train a DCGAN model [15] using the source images to generate synthetic images forming our target dataset. For our DCGAN training we use a learning rate of 0.0008, and Adam optimizer (beta1 = 0.5) and generate images of size 256×256 as shown in Fig. 5.

Fig. 5. Sample images generated with DCGAN.

The network parameters of our domain adaptation architecture is shown in Fig. 4. We use only fully connected (fc) layers in our network.

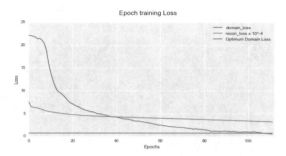

Fig. 6. Our framework for domain adaptation.

For finding the domain adapted features we use a set of 20000 images each from the catalog (source) and GAN generated (target) images. First we calculate 4096 length feature vectors for each image using the VGG-16 pretrained network, which are subsequently passed into the network as shown in Fig. 4. Figure 6 shows both domain and reconstruction loss during training of the network across epochs. We observe that both the losses keep on decreasing with epochs indicating good convergence. We choose those model parameters where the domain loss saturates to an optimum point as shown in red line in Fig. 6 as our final trained model. From a classification perspective since we want the domain classifier to be confused between the target and source domains, the optimum domain classification loss is when the classifier unit outputs 0.5, the binary cross entropy loss then turns out to be $-ln(0.5)$. Note that the scale of domain and reconstruction losses are different. We extract the encoded fully connected layer f_{da} as our domain adapted feature representation which we next assess for dedup. To quantitatively assess dedup, we set up a retrieval experiment. We first generate around 100 images from GAN and manually annotate 100 catalog images for each of the above GAN images as similar. We then use KNN to retrieve similar catalog images for a given GAN image. We compare our domain adapted f_{da} feature representation with VGG f_{vgg} features as our baseline. We also compare our domain adaptation approach with state-of-art CORAL [19] f_{coral} as it is unsupervised and does not necessarily require source labels to be present. We observe that f_{da} shows better precision-recall (PR) curves than f_{coral} and f_{vgg} as shown in Fig. 7.

The t-sne plots of the original (f_{vgg}), CORAL (f_{coral}) and domain adapted features (f_{da}) are shown in Fig. 8a, b and c respectively. In Fig. 8a and b we observe that source and target image data sets have two different principal directions for the original f_{vgg} and f_{coral} features respectively as shown. This indicates that the data distribution of source and target images are very different. However, in Fig. 8c for the domain adapted features f_{da} the principal directions are well aligned for both the source and target image data sets.

We calculate the KL divergence between the domains using f_{vgg}, f_{coral} and f_{da}. To calculate KL divergence we assume that data from both domains come from multivariate normal distribution (of dimension d).

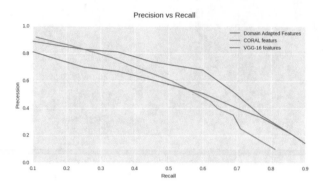

Fig. 7. PR-curves for f_{da} (blue), f_{coral} (green) and f_{vgg} (red) features.

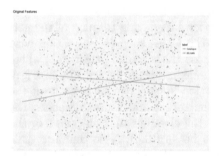

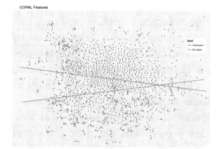

(a) T-SNE plot using original features f_{vgg} of GAN (green) and catalog(red) images. The lines indicate the principal directions.

(b) T-SNE plot for CORAL [19] features f_{coral} of GAN(green) and catalog(red) images. The lines indicate the principal directions.

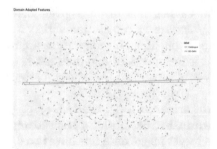

(c) T-SNE plot for domain adapted features f_{da} of GAN(green) and catalog(red) images. The lines indicate the principal directions.

Fig. 8. TSNE plots for features f_{vgg}, f_{coral} and f_{da}

Table 1. KL divergence between catalogue and GAN generated images with f_{vgg} and f_{da} features.

KL	Features		
	f_{vgg}	f_{coral}	f_{da}
$KL\,(source\|target)$	1.36	0.455	0.075

As shown in Table 1 we see that the network has learned features f_{da} that are able to reduce the KL divergence between the source and target image feature spaces significantly. In other words it means that f_{da} have a lot of overlap between either domains compared to the f_{vgg} and f_{coral} features.

Additionally, we qualitatively validate our features for dedup shown in Fig. 9. For *four* query synthetic GAN images in the leftmost column, we show top *four* catalogue images samples retrieved using f_{vgg} in the 4 middle columns and f_{da} in the last 4 columns. We observe that catalog samples retrieved using f_{da} are visually more coherent in terms of color and pattern. For instance f_{da} retrieved samples have similar color in 1st and 2nd row and similar pattern/color in 3rd row.

Fig. 9. Dedup retrieval: Left most column shows query image generated from DCGAN. Central 4 columns show similarly retrieved samples using f_{vgg} features and the last set of 4 columns show similarly retrieved images using f_{da}.

6 Conclusions

In this work we have proposed a framework to dedup machine generated images from catalog images. To efficiently handle the difference in data distributions between the machine generation and catalog image domains we have used a deep unsupervised domain adaptation network. The proposed architecture has an auto encoder unit that ensures efficient feature reconstruction and a classifier unit with adversarial loss ensuring that encoded features are indistinguishable. We have compared our adapted features (f_{da}) with the original VGG features (f_{vgg}) and CORAL [19] features (f_{coral}). We show PR curves wherein f_{da} has better performance than that of f_{vgg} and f_{coral} indicating better dedup with our domain adaptation approach. We further analyzed that principal directions of these two domains come close together using our f_{da} features than the original features f_{vgg} and f_{coral} using the T-SNE plots. Further analysis show that f_{da} achieves lower KL divergence between the two domains over f_{vgg} and f_{coral}. We also show qualitative analysis through visual similarity of the two domains. The analysis shows that the f_{da} of machine generated images are able to retrieve visually coherent images from the catalog dataset in comparison to f_{vgg}. Our approach is generic and can be applied across domains for cross domain feature learning.

References

1. Daumé III, H.: Frustratingly easy domain adaptation. arXiv preprint arXiv:0907.1815 (2009)
2. Deng, J., Dong, W., Socher, R., Li, L.-J., Li, K., Fei-Fei, L.: Imagenet: a large-scale hierarchical image database. In: IEEE Conference on Computer Vision and Pattern Recognition, 2009, CVPR 2009, pp. 248–255. IEEE (2009)
3. Ganin, Y., Lempitsky, V.: Unsupervised domain adaptation by backpropagation. In: International Conference on Machine Learning, pp. 1180–1189 (2015)
4. Ghifary, M., Kleijn, W.B., Zhang, M., Balduzzi, D., Li, W.: Deep reconstruction-classification networks for unsupervised domain adaptation. In: European Conference on Computer Vision, pp. 597–613. Springer (2016)
5. Gong, B., Shi, Y., Sha, F., Grauman, K.: Geodesic flow kernel for unsupervised domain adaptation. In: 2012 IEEE Conference on Computer Vision and Pattern Recognition (CVPR), pp. 2066–2073. IEEE (2012)
6. Goodfellow, I., Pouget-Abadie, J., Mirza, M., Xu, B., Warde-Farley, D., Ozair, S., Courville, A., Bengio, Y.: Generative adversarial nets. In: Advances in Neural Information Processing Systems, pp. 2672–2680 (2014)
7. Gopalan, R., Li, R., Chellappa, R.: Domain adaptation for object recognition: an unsupervised approach. In: 2011 IEEE International Conference on Computer Vision (ICCV), pp. 999–1006. IEEE (2011)
8. He, K., Zhang, X., Ren, S., Sun, J.: Deep residual learning for image recognition. In: Proceedings of the IEEE Conference on Computer Vision and Pattern Recognition, pp. 770–778 (2016)
9. Isola, P., Zhu, J.-Y., Zhou, T., Efros, A.A.: Image-to-image translation with conditional adversarial networks. In: CVPR (2017)
10. Kulis, B., Saenko, K., Darrell, T.: What you saw is not what you get: domain adaptation using asymmetric kernel transforms. In: 2011 IEEE Conference on Computer Vision and Pattern Recognition (CVPR), pp. 1785–1792. IEEE (2011)
11. Lim, J.J., Salakhutdinov, R.R., Torralba, A.: Transfer learning by borrowing examples for multiclass object detection. In: Advances in Neural Information Processing Systems, pp. 118–126 (2011)

12. Liu, M.-Y., Tuzel, O.: Coupled generative adversarial networks. In: Advances in Neural Information Processing Systems, pp. 469–477 (2016)
13. Pan, S.J., Yang, Q.: A survey on transfer learning. IEEE Trans. Knowl. Data Eng. **22**(10), 1345–1359 (2010)
14. Patel, V.M., Gopalan, R., Li, R., Chellappa, R.: Visual domain adaptation: a survey of recent advances. IEEE Signal Process. Mag. **32**(3), 53–69 (2015)
15. Radford, A., Metz, L., Chintala, S.: Unsupervised representation learning with deep convolutional generative adversarial networks. arXiv preprint arXiv:1511.06434 (2015)
16. Rajagopal, A.K., Subramanian, R., Ricci, E., Vieriu, R.L., Lanz, O., Sebe, N., et al.: Exploring transfer learning approaches for head pose classification from multi-view surveillance images. Int. J. Comput. Vision **109**(1–2), 146–167 (2014)
17. Ren, S., He, K., Girshick, R., Sun, J.: Faster R-CNN: towards real-time object detection with region proposal networks. In: Advances in Neural Information Processing Systems, pp. 91–99 (2015)
18. Simonyan, K., Zisserman, A.: Very deep convolutional networks for large-scale image recognition. arXiv preprint arXiv:1409.1556 (2014)
19. Sun, B., Feng, J., Saenko, K.: Correlation alignment for unsupervised domain adaptation. arXiv preprint arXiv:1612.01939 (2016)

Color Naming for Multi-color Fashion Items

Vacit Oguz Yazici[1,2]([⊠]), Joost van de Weijer[1], and Arnau Ramisa[2]

[1] Computer Vision Center, Universitat Autonoma de Barcelona,
Building O Campus UAB, 08193 Bellaterra, Barcelona, Spain
{voyazici,joost}@cvc.uab.cat
[2] Wide Eyes Technologies, Barcelona, Spain
aramisa@wide-eyes.it

Abstract. There exists a significant amount of research on color naming of single colored objects. However in reality many fashion objects consist of multiple colors. Currently, searching in fashion datasets for multi-colored objects can be a laborious task. Therefore, in this paper we focus on color naming for images with multi-color fashion items. We collect a dataset, which consists of images which may have from one up to four colors. We annotate the images with the 11 basic colors of the English language. We experiment with several designs for deep neural networks with different losses. We show that explicitly estimating the number of colors in the fashion item leads to improved results.

Keywords: Deep learning · Color · Multi-label

1 Introduction

Computer vision offers great potential to develop tools to improve interaction between buyers and sellers in the fashion industry [1–3]. Color attributes (in this article referred to as color names) are among the essential properties of fashion items and their understanding is therefore crucial for efficient interaction with users. Therefore, in this article we focus on the automatic estimation of color names of images of fashion items. We will focus on extracting the colors of the fashion items in real-world images with background clutter and without available segmentation masks or bounding boxes which indicate the exact location of the fashion item. The task therefore is twofold, automatic detection of the fashion item, and estimation of its colors.

Color naming is a challenging task due to several reasons, including discrepancies between the physical nature of color and human perception (which is also affected by the cultural context), or external factors like varying illumination and complex backgrounds. Moreover complex background, human skin, or human hair act as clutter that deteriorates the accuracy of models. It is important to minimize the effects of this type of clutter in order to improve accuracy. A further difficulty of color naming in fashion, which is the focus of this paper,

© Springer International Publishing AG, part of Springer Nature 2018
Á. Rocha et al. (Eds.): WorldCIST'18 2018, AISC 747, pp. 64–73, 2018.
https://doi.org/10.1007/978-3-319-77700-9_7

is that many of the objects that we see in the real world have several colors, which complicates the decision making process for algorithms.

Computational color naming has primarily focused on the 11 basic colors of the English language [4,5]. Those 11 basic colors are defined in the seminal work of Berlin and Kay [6] in which they researched the usage of color names in various different languages. Color names have been successfully used in a number of computer vision applications, including action recognition, visual tracking and image classification; see [7] for an overview. In the field of fashion image understanding, Liu et al. [8] do color naming using Markov Random Fields to infer category and color labels for each pixel in fashion images. To the best of our knowledge, all existing work on color naming focuses either on single colored objects or pixel-wise predictions.

Therefore, we address the problem of color name assignment to multi-color fashion items. We design several neural network architectures and experiment with various loss functions. We collect our own multi-label color dataset by crawling data from Internet sources. We show that a network with an additional classification head that explicitly estimates the number of color names improves performance. In addition, we show in a human annotation experiment that multi-color naming is an ambiguous task and human annotation results are only a few percent higher than results obtained by our best network.

The rest of this paper is organized as follows. Related work is discussed in Sect. 2. Section 3 describes details of the dataset that we use for the experiments. Section 4 elaborates the proposed approach. Experiments are presented in Sect. 5. Finally, we conclude this paper in Sect. 6.

2   Related Work

Research papers for fashion firstly focused on the segmentation of fashion products in images. Yamaguchi et al. [1], propose the Fashionista dataset consisting of 158,235 fashion photos with associated text annotations. They use a Conditional Random Field Model (CRF) in order to parse fashion clothes pixel-wise. However, their algorithms require fashion tags during the test time to get good accuracies. Simo-Serra et al. [2] address this issue and also propose a CRF model that exploits different image features such as appearance, figure/ground segmentation, shape and location priors for cloth parsing. They manage to obtain state-of-the-art performance on the Fashionista dataset. Liu et al. [9] propose a novel dataset which consists of 800,000 images with 50 categories, 1,000 descriptive attributes, bounding box and clothing landmarks. Moreover, they also propose a novel neural network architecture which is called FashionNet. The network learns clothing features by jointly predicting clothing attributes and landmarks. They do pooling and gating of feature maps upon estimated landmark locations to alleviate the effect of clothing deformation and occlusion. Recently, Cervantes et al. [3] propose a hierarchical method for the detection of fashion items in images.

For color, Cheng et al. [10] use a modified version of VGG for pixel-wise prediction out of 11 color labels (which are blue, brown, gray, white, red, green, pink,

Fig. 1. Sample images from the dataset with varying content and background clutter. Note that we do not provide segmentation and therefore to estimate the colors, the algorithm needs to implicitly segment the main fashion item.

black, yellow, purple and orange) and a CRF to smooth the prediction. Although their model is robust to background clutter, and can produce pixel-wise prediction, it is not robust to other clutter such as skin and hair color. van de Weijer et al. [4], use probabilistic latent semantic analysis (PLSA) on Lab histograms to learn color names. Benavente et al. [5], present a model for pixel-wise color name prediction by using chromaticity distribution. Wang et al. [11], propose an algorithm which has two stages: in the first stage, which they name self supervised training, they train a shallow network with color histograms of random patches from the dataset. In the second stage, they fine-tune the same network to predict 11 basic colors. Mylonas et al. [12], use a mixture of Gaussian distributions. Schuerte and Fink [13] propose a randomized hue-saturation-lightness (HSL) transformation to get more natural color distributions; secondly, they used probabilistic ranking to remove the outliers. They claim that these steps helps color models accommodate to the variances seen in real-world images. In none of the before mentioned works to task of color naming multi-color objects is addressed.

3   Multi-color Name Dataset

There are several datasets for color name learning. van de Weijer et al. [4] introduced two datasets, constituted of images of objects retrieved from Google and EBAY respectively, and labeled with the 11 basic color names. Liu et al. [8] introduce another dataset which consists of 2682 images with pixel-level color annotations of the 11 basic colors plus a "background" class. However, almost every image in the dataset has a single color. To the best of our knowledge there is no dataset which explicitly considers multi-color objects.

We therefore collect a new dataset for this article, composed of images of fashion objects with one to nine colors (see Table 1). Single colored fashion images are crawled from various online shopping sites, and most of the multicolor labeled images are obtained by querying the Google images search engine with a query term containing a pair of color names and a fashion keyword (e.g. red and blue skirt) and downloading the 100 first images. There are 67 fashion keywords that we use and 55 color pairs that can be obtained with combinations of 11 basic colors. At the end, we remove irrelevant and noisy data and crop the fashion item to prepare the dataset.

Table 1. The number of images for each color category

	One	Two	Three	Four	Five	Six	Seven	Eight	Nine	Total
Train	5556	5431	2178	1203	476	131	19	4	3	15001
Test	50	50	50	50	0	0	0	0	0	200

This process allows us to obtain images with two colors and more colors, as sometimes the search engine also returns images with additional colors not included in the query. Unfortunately, this leads to an imbalance between the number of 2-colored images and multi-colored images. Directly crawling for products with more than two colors using Google Images produces unsatisfactory results.

The dataset includes different types of images of varying complexity: catalog shots with smooth or complex background, images with plain background without any person or images taken by social media users; all labeled with the color names of the main fashion item. Sample images from the dataset can be seen in Fig. 1. It should be noted that we do not use segmentation for the images, and naming the multiple colors of the fashion items includes dealing with clutter from the background, occlusions, and skin and hair of the person. However, if there is more than one fashion item in an image, to avoid any confusion, we provide a bounding box for the correspondent fashion item. In any case, the network has to implicitly segment the fashion item from occlusions and clutters.

4 Networks for Multi-color Name Prediction

Methods on color naming focus on single colored objects. In this work we aim to propose a method for multi-colored fashion items. We evaluate several network architectures and losses for this task.

4.1 Network Design

In principle we believe the mapping from RGB to color names not to be highly complicated and only several layers are required. However, differentiating background from foreground is a highly complex process that requires many layers and should be implicitly done by the network.

First, we propose a shallow network; the truncated version of Alexnet [14]. We keep the first five convolution layers of the architecture and remove the fully connected layers of 4096 dimension. At the end we add a fully connected layer which maps features to the eleven basic color names. As a second network we use the full Alexnet architecture. Both nets are initialized with pretrained weights from ILSVRC 2012 dataset [15]. We think that finetuning from this model can alleviate noise caused by clutter such as complex backgrounds, hair or skin.

4.2   Loss Functions

We consider two loss functions for the purpose of color naming for multi-color fashion items. The first loss we consider to train the network is the softmax cross-entropy loss (SCE). The softmax cross-entropy can be seen in Eq. 1.

$$L_{sce} = -\frac{1}{N} \sum_{i}^{N} P(i) \log Q(i) \tag{1}$$

where Q the predicted color distribution, P is the true color distribution, and N is the number of images. Q is obtained by applying a softmax normalization to the output of the last fully connected layer of the network, and the ground truth P is computed by assigning a uniform probability to all color names annotated for the fashion item (e.g. in case of three annotated color names, P would contain three elements with value 0.33).

While the softmax cross-entropy loss teaches a network to compute color probability distributions for an input fashion item, no decision is made on the actual number of colors. To remedy this, a threshold on the computed probabilities Q, learned from an independent validation set, is used to discard the colors unlikely to be really present.

The second loss we consider is the binary cross-entropy loss (BCE), which inherently supports multi-label classification. This loss is commonly used for attribute detection [9, 16] because it models the presence of multiple labels simultaneously. Therefore it is expected to obtain better results than the softmax-cross entropy. Unlike with the softmax cross-entropy loss, the computed probability for a color name is independent of the others. For example the probability of both 'green' and 'orange' can be one simultaneously, something which is impossible for the softmax cross-entropy loss. Therefore, the loss trains 11 binary classifiers for each color. In Eq. 2, the binary cross-entropy can be seen.

$$L_{bce} = -\frac{1}{N} \sum_{i}^{N} P_i \log Q_i + (1 - P_i) \log(1 - Q_i) \tag{2}$$

Similarly as the softmax-cross entropy loss we determine a threshold on a validation set to decide on the colors which are present in the fashion item. We found this to yield better results than choosing the natural threshold of 0.5.

4.3   Extra Head to Explicitly Estimate Number of Colors

In the previous section we consider two losses to estimate the color names. In principle the binary cross-entropy loss which implements the multi-label softmax cross-entropy loss is more suitable for the estimation of multiple colors. However, the probabilities which are the outcome of these networks both encode information of the number of color names as well as the confidence of the network in its estimation of the color names. Considering a single colored object which the system is not sure to label with either 'orange' or 'red', the algorithm that is based on binary cross-entropy might give both colors a probability of 0.6. Based on this we might conclude that the object is a multi-color object which is both 'orange' and 'red'. However, looking at the object it might be obvious that it only has a single color.

Therefore, we experiment with adding an extra classification head to the network, which explicitly estimates the number of colors in the main object. We model this objective as a classification task, and define four possible classes: one, two, three and four or more colors. A natural choice for this objective is the softmax cross-entropy loss layer, typically used for classification. In the experiments, we add this additional objective both to the networks which use softmax-cross entropy loss and the binary cross-entropy loss. The architecture of the network can be seen in Fig. 2.

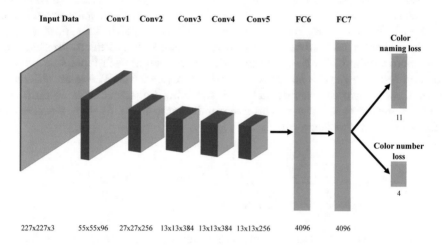

Fig. 2. The architecture of the deep network with the extra head.

4.4 Training Procedure

To train the network, we finetune from an Alexnet model which is trained on ILSVRC 2012 [15] using the Caffe framework [17]. The batch size is 64, the optimization method is SGD with momentum, set to 0.99, and we decrease the learning rate after every 5000 iterations. The initial learning rate is 0.0001 and the maximum iteration number is 20000. We also use data augmentation techniques in order to increase the accuracy of the models. The data augmentation techniques that we use are changing contrast, rescaling image and cropping random parts from images. Rescaling basically consists on changing the resolution of the image before resizing to the required network input size. The probability that any augmentation technique is applied to an image is 50%. We never keep both the original and the augmented image in the same batch, as we have observed that it may negatively impact the accuracy of the learned model. Finally, to avoid aspect ratio distortions caused by the resizing process, we use a padding function in order to make all images square.

Table 2. Results of our models and the human annotators

		Shallow BCE w/ extra head	SCE w/o extra head	SCE w/ extra head	BCE w/o extra head	BCE w/ extra head	Human score
Micro	Precision	77.24%	85.64%	80.27%	82.31%	83.57%	81.39%
	Recall	67.20%	63.20%	70.80%	69.80%	71.20%	81.91%
	F1	71.87%	72.73%	75.24%	75.54%	**76.89%**	81.32%
Macro	Precision	78.40%	84.73%	79.82%	81.41%	83.10%	81.60%
	Recall	65.24%	62.10%	69.78%	67.43%	69.33%	80.89%
	F1	69.32%	69.38%	73.46%	72.51%	**74.19%**	80.18%

5   Experiments

To evaluate the performance we use label based metric methods. We calculate
the micro-precision, micro-recall, micro-F1, macro-precision, macro-recall and
macro-F1. In the micro methods we sum up true positive, false positive and
true negative for each label in order to get micro-recall and micro-precision. In
the macro methods, we calculate precision and recall of each label and average
them in order to get the macro-recall and macro-precision. The main difference
is that the macro metrics do not take the label imbalance into account. To
clarify the difference between the micro and macro methods, here we give the
micro-precision and macro-precision:

$$P_{micro} = \frac{\sum_{j=i}^{L} tp_j}{\sum_{j=0}^{L} tp_j + \sum_{j=0}^{L} fp_j} \qquad P_{macro} = \sum_{j=0}^{L} \frac{tp_j}{tp_j + fp_j} \qquad (3)$$

L is the number of classes, tp_j and fp_j is the true positive and false positive
of class j. All of the results can be seen in Table 2. We focus on the F1-score
which is a fair metric to compare methods. We first evaluate the two network
architectures, namely the shallow and deep network. Both of the models have
the extra head which forces them to learn the number of colors on a fashion item.
The deep model clearly outperforms the shallow model. We attribute this to the
fact that the shallow model is not able to segment the fashion items implicitly,
and therefore fails for the more cluttered cases as can be seen in Fig. 3.

Next we evaluate the different losses, and we verify if the additional head
which explicitly predicts the number of colors contributes to a performance gain.
It can be seen that adding the additional objective improves both the softmax
cross-entropy loss and the binary cross-entropy loss; it forces the network to
learn the number of colors on a fashion item, and also contributes to name them
as can be seen when comparing the columns 4–5 and 6–7 of Table 2. During the
inference, the extra head can predict maximum 4 colors. In case the networks
without extra head predicts more than 4 colors, we get the first 4 with the highest
scores.

In the last column of Table 2, we show the average performance obtained
by humans for the same task. We asked seven annotators of different ages and
backgrounds to provide labels for the images in the test set. The obtained scores

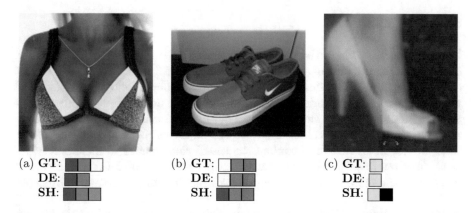

Fig. 3. Qualitative results of the shallow and deep networks. GT, DE, SH denote the ground truth, the predictions of the deep and shallow model respectively. Note that the networks should estimate the colors of the fashion item while ignoring the non-relevant colors present in the background.

for humans show that multi-color labelling is an ambiguous task, and for many objects humans do not agree on the labels. This score can be considered to be an upper bound for computational methods.

The contribution of adding an extra head is shown in Fig. 4. From the ground truth and prediction of the cross entropy models it can be seen that the extra head provides robustness if the color distribution is not uniform in the image. However, it makes the model more conservative and biases it towards predicting a lower number of colors in the image (last two examples on the right).

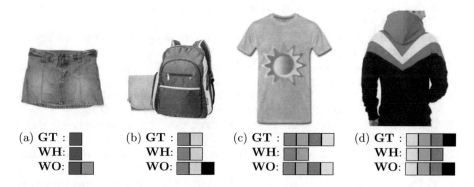

Fig. 4. Qualitative results of the BCE model with (WH) and without (WO) the extra head.

6 Conclusions

In this paper we address the problem of color name estimation in multi-colored objects that, to the best of our knowledge, we are the first to address. We collect a dataset of over 15.000 images with a varying number of colors per object and we evaluate several network architectures for the purpose of multi-color estimation. Preliminary results show that adding an additional objective to explicitly estimate the number of colors in the object improves results. Following recent work we are interested in extending the set of color names to include a wider range of colors [18,19]. We hope that this paper further motivates researchers to investigate the more realistic setting of color naming for multi-colored objects.

Acknowledgements. This work was supported by TIN2016-79717-R of the Spanish Ministry and the CERCA Programme and the Industrial Doctorate Grant 2016 DI 039 of the Ministry of Economy and Knowledge of the Generalitat de Catalunya.

References

1. Yamaguchi, K., Kiapour, M.H., Ortiz, L.E., Berg, T.L.: Parsing clothing in fashion photographs. In: CVPR, pp. 3570–3577. IEEE (2012)
2. Simo-Serra, E., Fidler, S., Moreno-Noguer, F., Urtasun, R.: A high performance CRF model for clothes parsing. In: ACCV (2014)
3. Cervantes, E., Yu, L.L., Bagdanov, A.D., Masana, M., van de Weijer, J.: Hierarchical part detection with deep neural networks. In: ICIP, pp. 1933–1937 (2016)
4. van de Weijer, J., Schmid, C., Verbeek, J., Larlus, D.: Learning color names for real-world applications. IEEE Trans. Image Process. **18**(7), 1512–1523 (2009)
5. Benavente, R., Vanrell, M., Baldrich, R.: Parametric fuzzy sets for automatic color naming. JOSA A **25**(10), 2582–2593 (2008)
6. Berlin, B., Kay, P.: Basic color terms: their universality and evolution. University of California, Berkeley (1969)
7. van de Weijer, J., Khan, F.S.: An overview of color name applications in computer vision. In: CCIW, pp. 16–22 (2015)
8. Liu, S., Feng, J., Domokos, C., Xu, H., Huang, J., Hu, Z., Yan, S.: Fashion parsing with weak color-category labels. IEEE Trans. Multimedia **16**(1), 253–265 (2014)
9. Liu, Z., Luo, P., Qiu, S., Wang, X., Tang, X.: DeepFashion: powering robust clothes recognition and retrieval with rich annotations. In: CVPR (2016)
10. Cheng, Z., Li, X., Loy, C.C.: Pedestrian color naming via convolutional neural network. In: ACCV, pp. 35–51. Springer (2016)
11. Wang, Y., Liu, J., Wang, J., Li, Y., Lu, H.: Color names learning using convolutional neural networks. In: ICIP, pp. 217–221. IEEE (2015)
12. Mylonas, D., MacDonald, L., Wuerger, S.: Towards an online color naming model. In: Color and Imaging Conference, vol. 2010, pp. 140–144. Society for Imaging Science and Technology (2010)
13. Schauerte, B., Fink, G.A.: Web-based learning of naturalized color models for human-machine interaction. In: DICTA, pp. 498–503. IEEE (2010)
14. Krizhevsky, A., Sutskever, I., Hinton, G.E.: ImageNet classification with deep convolutional neural networks. In: NIPS, pp. 1097–1105 (2012)

15. Deng, J., Dong, W., Socher, R., Li, L.J., Li, K., Fei-Fei, L.: ImageNet: a large-scale hierarchical image database. In: CVPR, pp. 248–255. IEEE (2009)
16. Liu, Z., Luo, P., Wang, X., Tang, X.: Deep learning face attributes in the wild. In: ICCV, pp. 3730–3738 (2015)
17. Jia, Y., Shelhamer, E., Donahue, J., Karayev, S., Long, J., Girshick, R., Guadarrama, S., Darrell, T.: Caffe: convolutional architecture for fast feature embedding. In: ACM International Conference on Multimedia, pp. 675–678. ACM (2014)
18. Mylonas, D., MacDonald, L.: Augmenting basic colour terms in English. Color Res. Appl. 41(1), 32–42 (2015)
19. Yu, L., Zhang, L., van de Weijer, J., Khan, F.S., Cheng, Y., Parraga, C.A.: Beyond eleven color names for image understanding. Mach. Vis. Appl. 29(2), 361–373 (2018)

Study on Machine Learning Algorithms to Automatically Identifying Body Type for Clothing Model Recommendation

Evandro Costa[1]([⊠]), Emanuele Silva[1], Hemilis Rocha[2], Artur Maia[1], and Thales Vieira[3]

[1] Institute of Computing, Federal University of Alagoas - UFAL, Maceió, Brazil
{evandro,ets,amp}@ic.ufal.br
[2] Informatics in Campus Viçosa,
Federal Institute of Alagoas - IFAL, Maceió, Brazil
joysesel@gmail.com
[3] Institute of Mathematics, Federal University of Alagoas - UFAL,
Maceió, Brazil
thalesv@gmail.com

Abstract. The task of automatically identify body type with high accuracy is still a relevant problem in clothing fashion settings. This paper addresses such problem, presenting a study on machine learning techniques applied to classify women's body shapes, taking into account a small set of body attributes, in order to further find appropriate clothing models. Thus, we perform a comparative study on such techniques to evaluate the accuracy of four classifiers, aiming at selecting the best of them to be used for clothing model recommendation based on rules. Overall, in the conducted computational experiment, Random Forest and SVM methods had the best performance, but the other two had also very good results, demonstrating their effectiveness to automatically identifying body type, serving as a relevant information to be used in our rule-based system to provide clothing model recommendation.

Keywords: Body type identification · Machine learning
Clothing model recommendation

1 Introduction

In the last years, there has been a rapid growing of fashion industry leading to very positive social and economic impacts worldwide. This fact has particularly demanded academic applied research to address interesting problems whose solutions involves artificial intelligence techniques, mainly including the topics of machine learning, knowledge representation and reasoning, natural language processing, computer vision.

In a broading sense, clothing fashion domain involves cognitive, affective and social aspects, directly affecting women behaviors. In female universe, such behaviors are strongly influenced by intrinsic and extrinsic motivations, leading to most women to search for and use nice clothes to feel good, elegant, confident, primarily thinking in them. Aligned with this women motivation, in a recent work, Adam and Galinsky [1]

© Springer International Publishing AG, part of Springer Nature 2018
Á. Rocha et al. (Eds.): WorldCIST'18 2018, AISC 747, pp. 74–84, 2018.
https://doi.org/10.1007/978-3-319-77700-9_8

introduced the term "Enclothed Cognition" showed how clothes influence wearers' psychological processes, concluding that clothes that we wear directly affect how we think and what we do.

Like in several other domains, fashion industry has followed the search for providing personalization solution, for instance, offering appropriate clothes in accordance with individual characteristics. For this purpose, personal physical characteristics have been considered, mainly because people appear in different shapes and sizes. Thus, the information about body types of a given woman is a very relevant characteristic, which has been required by the fashion domain, particularly to help in supporting the task of clothing personalized recommending.

The task of automatically identify body type with high accuracy is still a relevant problem in clothing fashion settings. In this paper, we address the mentioned challenge by presenting a study on machine learning techniques to classify women's body shapes, taking into account a dataset with a set of body attributes, in order to further find appropriate clothing models. We use expert annotations collected from 61 women, forming a table with a training set and another with a testing set. The used body attributes were defined taking into account the proposal in [2]. Hence, we perform a comparative study on such techniques to evaluate the accuracy of four classifiers [3]: Decision Tree via C5.0 algorithm, Support Vector Machine (SVM) via different kernels, K-Nearest Neighbors (KNN), and Random Forest, aiming at selecting the best of them. With the selected algorithm, we then classify women's body type to be used in a rule-based system for clothing model recommendation, for instance skirt models (pleated skirt, pencil skirt), blouse models (V-neckline, U-neckline), pants models (flare pants, pleated pants).

In summary, the main contribution of this paper is to show the feasibility of machine learning techniques as a suitable resource to provide effective automatic method to determine body type, only using a few body measures as attributes. To the best of our knowledge, there is no other work that uses machine learning technique to that purpose.

The remainder of this paper is organized as follows. In Sect. 2, we provide some background knowledge regarding fashion domain and machine learning techniques. Section 3 discusses the comparative study of classification methods, as well as, presents our approach in order to further find appropriate clothing models. Section 4 discusses the experimental results. We finally conclude the paper in Sect. 5.

2 Background Knowledge

We provided some background knowledge related to body shapes with body measurements, as well as, machine learning techniques. There are five common female body shapes that women will often fall into one of these: Hourglass, inverted triangle, rectangle (or straight), oval (or apple), and triangle (or Pear).

2.1 Body Types

Each woman presents in her body, characteristics, such as: skin color, hair color, height, weight, as well as other body measurements that have influence in distinguish her from other women. However, there is an attempt to group female body, taking into account some metrics, mapping them into body categories. There is no consensus in that categorization, for example, some fashion stylists define four body categories, but other define five, six, seven, but the majority of them just consider five basic body types, as illustrated in Fig. 1: Hourglass, triangle, inverted triangle, rectangle, and oval.

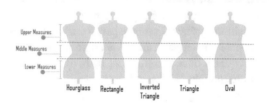

Fig. 1. Body types.

In what follows, there is a discussion about some characteristics of each body type here used, taking into account nomenclature, female body type with its main characteristics, possible links between female body type and clothing, as well as clothing type indicated or not, according which styles best flatter that shape.

i. Hourglass

In this body type, the upper (e.g. shoulders, bust) and lower (e.g. hip) are almost of equal size, but with a narrow the middle part (waist). The main objective here is to highlight the body shape in natural way. In this sense, there is indicated clothing, such as lightweight pants and skirts, straight pants, and non-indicated, such as short sweaters, jackets without marked waist.

ii. Rectangle

In this body type women's waist, hip and shoulders are almost the same, therefore there is no defined waist. The main objective here is to disguise the straight silhouette using curved lines. In this sense, there is non-indicated clothing, such as pants too wide or too narrow, short blouses, and indicated, such as wear styles that have detail on the bust and on the hips. Moreover, it is important to choose designs that have horizontal lines, adding dimension to your shape.

iii. Inverted Triangle

In this body type shoulders are wider than the hip, and hips are small. The main objective here is to disguise the straight silhouette using curved lines. In this sense, there is non-indicated clothing, such as Tops with V-necks, raglan sleeves or diagonal designs, and indicated, such as lightweight pants and skirts

iv. Triangle

In this body type women have broader shoulders and bust, and narrower hips. The main objective here is to provide a balance between shoulders and hips, smoothing the bottom and looking for volume at the top. In this sense, there is non-indicated clothing, such as tops or shirts with hemlines that fall directly on the hips, and indicated, such as to start with monochromatic colors and/or muted shades in complementary colors.

v. Oval

In this body type female body is highlighted by the width of the waist is much larger than the measured shoulders or hips. The main objective here is to disguise the oval shape searching for lines that will lengthen. In this sense, there is non-indicated clothing, such as blocky, bold, contrasting colors and boxy garments, and indicated, such as monochromatic colors and/or muted shades in complementary colors.

2.2   Classification Methods

In what follows, the four used classification methods based on machine-learning techniques are briefly discussed.

2.2.1   Support Vector Machines

It is a machine learning technique, as a kind of supervised learning approach based on kernel notion, developed to solve classification and regression problems, mainly involving binary (two-class) problem, but subsequently multiclass SVMs were also developed [4].

2.2.2   Decision Tree

Induction of Decision Tree is also a supervised classification technique based on a process of building a model of classes from a set of instances in a table structure that contains class labels [5]. It has been used for binary (two-class) and multi-class classification problems. In particular, C5.0 is a decision tree algorithm that implements a predictive modeling approach, going further to the ID3 and C4.5 [6].

2.2.3   Random Forest

This technique is a kind of decision tree algorithm for classification, regression, and other tasks [7], operating by constructing a multitude of decision trees at training time and outputting the class that is the result of the classification or mean prediction of the individual trees [8].

2.2.4   KNN

K-Nearest Neighbors (KNN) is also a kind of supervised learning method for classification and regression. It makes predictions using the training dataset directly, therefore it has no model other than storing the entire dataset, so there is no need for the learning step [7].

3 Our Proposal

The Fig. 2 illustrates our general approach for clothing model recommendation, here focusing on body type detection based on some personal physical attributes. We have worked on dataset involving body attributes as the ones in Fig. 2, proposing a new method for automatically identifying body type from a dataset of body attributes using machine learning algorithms for classifying women into one of the five considered body types. Having detected the woman's body type, this information will be used as input to the clothing model recommender system. In this section, we describe the main steps of the methodology used in our approach, as well as we provide some description the data and the achievements with four classifiers, and also, we discuss the proposed recommendation mechanism.

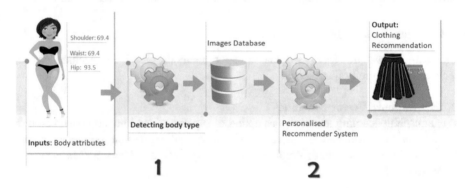

Fig. 2. Approach for clothing model recommendation.

3.1 Data

In our experimental study we use a dataset, which includes body attributes of 61 women. These data were used to perform a comparative analysis between the accuracy of classifier models. In this study we define the following independent variables: (1) body attributes; and (2) classification algorithms. The first variable represents a set of parameters used by classification algorithms, as can be seen in Table 1. The second variable represents the classification algorithms used to create the models. This variable was divided into four levels: C5.0 algorithm; SVM algorithm; KNN algorithm, and Random Forest algorithm. Moreover, we define a dependent variable mean accuracy as a value between 0 and 1.

Table 1. Body attributes

Variable	Description
Hip	2D measure in centimeters
Shoulder	2D measure in centimeters
Waist	2D measure in centimeters

3.2   Body Shape Classifiers

The experimental study result in 4 models, one for each classification algorithms. In order to achieve better results in our work we evaluate some variations in the algorithms SVM, KNN, and Random Forest. All models were created in R language version 3.4.1. To measure the model's effectiveness, the k-fold cross-validation method with k = 10 was used. Moreover, each model was executed 10 times producing 400 random samples.

Finally, to determine if there is a statistically significant difference among the models we use a statistical test called Kruskal and Wallis one-way analysis of variance with a significance level of 5% and the post-hoc pairwise multiple comparison Nemenyi test.

3.2.1   SVM

The R language just implements four kernel types (linear, polynomial, radial and sigmoid) for the SVM algorithm. In our study, we used the radial kernel and several tests were performed to evaluate its mean accuracy for the dataset. We set gamma ranging from 0.05 up to 1 and the default value of the e1017 package from R, which is defined as (1/(data dimension)). The Figs. 3 and 4 present the mean accuracy of the gamma variations tested. Analyzing the Fig. 4, the best performance (accuracy = 0.9394) was achieved for gamma = 0.85.

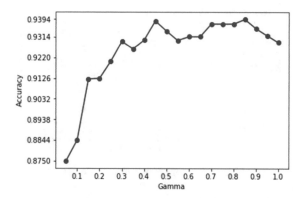

Fig. 3. SVM mean accuracy with radial kernel.

3.2.2   Decision Tree via C5.0

In the tests with the C5.0 algorithm a single decision tree was generated without any type of parameter variation or the use of improvement techniques.

3.2.3   Random Forest

In Random Forest algorithm, several tests were performed to evaluate their mean accuracy for the dataset. These tests present a variation in the number of decision trees generated for classification, ranging from 50 up to 550 trees. The Fig. 5 presents the

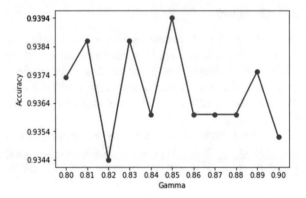

Fig. 4. SVM mean accuracy with gamma range [0.80, 0.90].

mean accuracy for all variations tested in the number of trees. The Fig. 6 presents only the variations from 130 up to 170 where we can see that the best performance (accuracy = 0.9493) was for the trees = 150.

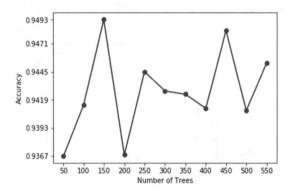

Fig. 5. Random Forest mean accuracy.

3.2.4   KNN

In KNN algorithm we perform variations in the number of k-nearest neighbors to evaluate mean accuracy. Thereby, k was set as: 3, 5, 7, 9, 11, 13, and 15. As can be seen in Fig. 7 the best performance (accuracy = 0.9326) was achieved for k = 3.

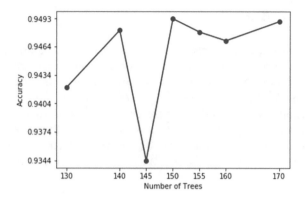

Fig. 6. Random Forest mean accuracy with number of trees range [130, 170].

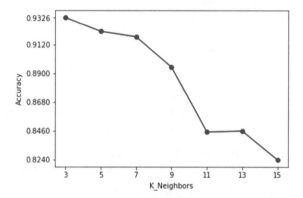

Fig. 7. k-Nearest Neighbors mean accuracy.

3.3 Clothing Model Recommendation

The proposed approach for clothing model recommendation follows the flow in Fig. 2, starting with the task of automatically identifying the body type, by using one of the four adopted classifiers, and then using this information in rule-based recommender system to select appropriate clothing model. It implements a mechanism that is responsible for providing users with clothing model recommendation containing a collection of models expressed in terms of a table containing name, description, and images to illustrate the models. Its knowledge base contains a set of IF-THEN rules, coming from a personal stylist expert, where one rule example is **IF Body_Type = Rectangle THEN Recommend model = V-neckline blouse**, meaning that for one women that has rectangle body type V-neckline blouse is recommended. Its inference engine explores the rules by using a backward chaining strategy in order to obtain its conclusions. Regarding implementation aspects, we developed a prototype using Java for the inference engine, as well as MySQL and the framework Hibernate as a database technologies.

4 Results and Discussion

Table 2 illustrates our results, where we compared the four algorithms in terms of classification performance. From the results, the Random Forest with 150 trees and gaussian kernel SVM presented the best performance, obtained a mean accuracy of approximately 95% and 94% respectively. From now on, we need to determine whether the results observed suggest differences between performance in all classifier models or whether they are just random variations. Based on results from Kolmogorov-smirnov test (p-value = 0.4713), we cannot assume a normal distribution of data. Therefore, we used the Kruskal-Wallis as statistical test with a significance level of 0.05.

Table 2. Performance of classification algorithms using mean.

Algorithm	Accuracy
C5.0	0.9287
SVM	0.9394
KNN	0.9326
Random Forest	0.9493

Given our main question whether machine learning classifiers are adequate to accurately identify body types, one particular question to be answered in this paper is:

- Q: Are there statistically significant differences in the median accuracy in each classifier model (C5.0, SVM, KNN, Random Forest) within dataset?

The hypotheses may be written in the following form:

- H0: The median accuracy is the same in all classifier models. Then, we write MedianC5.0 = MedianSVM = MedianKNN = MedianForest;
- HA: At least one median accuracy is different than others.

If we reject the null hypothesis in favor of the alternative hypothesis, there is a set of hypotheses to identify which classifier has the best accuracy. Thereby, the hypotheses may be written as following:

- H1.1: C5.0 has the best median accuracy;
- H1.2: SVM has the best median accuracy;
- H1.3: KNN has the best median accuracy;
- H1.4: Random Forest has the best median accuracy.

Using the Kruskal-Wallis test a p-value = 0.4713 was obtained that is more than 0.05. Therefore, we cannot reject the null hypothesis (H0). Thus, there is no significant difference between the medians of all classifier models. Moreover, using the Nemenyi multiple comparisons test the results indicate there is no significant difference between the medians of SVM and C5.0 (p-value = 0.96), SVM and Random Forest (p-value = 0.94), SVM and KNN (p-value = 0.87), KNN and Random Forest (p-value = 0.54), KNN and C5.0 (p-value = 0.99), and Random Forest and C5.0

(p-value = 0.71). Finally, we cannot determine which algorithm presented the best performance, this way, we reject the hypotheses H1.1, H1.2, H1.3 and H1.4.

Concerning our approach to the rule-based clothing model recommendation, we performed a preliminary evaluation to measure the quality of the recommendation by comparing the system output and the opinion of the fashion expert. To this end, the participants were 30 women, where almost all of them are between 22 and 30 years of age. We also have a support of a Personal Stylist. We explained to each participant the procedure of the experiment and what kind of data would be collected, as well as, the aim of the experiment. The experiment consisted in comparing the input and output of the system and of the stylist opinion, concerning the agreements in terms of recommendation. There was a total agreement with regard to the clothing model recommendation. Of course, there was little data in the experiment, but it represents a first positive result.

5 Conclusion

We have presented a result of a study on machine learning techniques in a computational approach to automatically classifying women's body shapes, taking into account a set of body attributes, aiming at providing personalized clothing model recommendation. We explored and trained four machine learning methods for this body type classification task, thus obtaining good accuracy performance. Random Forest and SVM classifiers demonstrated the best classification performance, with respectively 95% and 94%, outperforming in a very little value the other two methods. Therefore, we show that machine learning classifiers are adequate to accurately identify body shapes, using the proposed body measures as attributes, using one of the here adopted classifiers. Moreover, we show first results on some quality aspects of the provided rule-based clothing model recommendation, only considering the information about body type.

As immediate future work, we planned to carry out more detailed evaluation, as well as searching for more improvement in the accuracy for body shape determination. It will involve other studies with other data sources from different places, allowing us to better test and explore our approach.

References

1. Adam, H., Galinksy, A.: Enclothed cognition. J. Exp. Soc. Psychol. **48**(4), 918–925 (2012). https://doi.org/10.1016/j.jesp.2012.02.008
2. Aguiar, T.: Personal Stylist: guia para consultores de imagem. Editora Senac São Paulo, São Paulo (2015)
3. Clarke, B., Fokoue, E., Zhang, H.: Principles and Theory for Data Mining and Machine Learning. Springer (2009). https://doi.org/10.1007/978-0-387-98135-2
4. Cortes, C., Vapnik, V.: Support vector networks. Mach. Learn. **20**(3), 273–297 (1995). https://doi.org/10.1023/A:1022627411411
5. Quinlan, J.: Induction of decision trees. Mach. Learn. **1**(1), 81–106 (1986). https://doi.org/10.1007/BF00116251

6. Patil, N., Lathi, R., Chitre, V.: Comparison of C5.0 & CART Classification algorithms using pruning technique. Int. J. Eng. Res. Technol. (IJERT) **1**(4), June 2012, ISSN 2278-0181
7. Witten, I., Frank, E., Hall, M.: Data Mining: Practical Machine Learning Tools and Techniques, 3rd edn. Morgan Kaufmann Publishers Inc., San Francisco (2011)
8. Breiman, L.: Random forests. Mach. Learn. **45**(1), 5–32 (2001). https://doi.org/10.1023/A: 1010933404324

Emerging Trends and Challenges in Business Process Management

How Do BPM Maturity and Innovation Relate in Large Companies?

Veronika Widmann, Benny M. E. de Waal[(⊠)],
and Pascal Ravesteyn[(⊠)]

Research Group Process Innovation and Information Systems,
HU University of Applied Sciences Utrecht, PO Box 85029
3508 AA Utrecht, The Netherlands
Veronika.widmann@iscparis.com,
{benny.dewaal,pascal.ravesteijn}@hu.nl

Abstract. Due to the increasing level of globalization competition between companies is growing. As a consequence, large companies need to enhance both their productivity and innovation simultaneously. Where business process management methods, such as BPM maturity models, are typically seen as a means to improve performance and productivity, their impact on innovation is unclear. Therefor the objective of this study is to determine what the relation is between business process management maturity and Innovation in large companies. A research model is developed based on existing theory on innovation adoption, innovation value chain and BPM maturity models. Subsequently a questionnaire is constructed to gather data at four large European organizations. Based on both a correlation and regression analysis of the data provided by 143 respondents a moderate relation between the overall Innovation construct and BPM maturity is shown. The proportion of variance in innovation that can be explained by BPM maturity amounts to 22,4%. This means that investing in BPM capabilities is not enough to increase the innovation capability of an organization.

Keywords: Process management · BPM maturity · Innovation adoption
Innovation value chain

1 A Need for Innovation

The process of globalization has given rise to an increased competition between companies on a global level. As a consequence, large companies are under pressure to enhance productivity and innovation simultaneously (Sanders Jones and Linderman 2014). In particular companies based in developed countries are increasingly challenged by competitors from developing countries, which are quickly picking up in the quality of their offerings while exporting globally at lower prices. To be able to compete, existing processes need to be continuously managed and improved in line with strategic aims, to enhance efficiency and time-to-market (Hung 2006). Moreover, the increased competition calls for higher levels of innovativeness (Tidd 2001). Business process management (BPM) is a management discipline including the recognition, definition, analysis, repeated improvement, automation, execution,

© Springer International Publishing AG, part of Springer Nature 2018
Á. Rocha et al. (Eds.): WorldCIST'18 2018, AISC 747, pp. 87–96, 2018.
https://doi.org/10.1007/978-3-319-77700-9_9

measurement and tracking of business processes (Scott 2007). By means of maturity models companies can assess their process architecture and distinguish capabilities that require enhancement (Forstner et al. 2014). Subsequently innovation is the process through which enterprises convert ideas into new or enhanced products, services or processes (Baregheh et al. 2009). It has been much argued that successful innovation requires a degree of flexibility that contrasts the efficiency orientation of BPM (Sanders Jones and Linderman 2014; Dijkman et al. 2016). In light of the need for successful performance in both disciplines, this study aims to expand existing studies by examining the relation between BPM maturity and innovation in large companies. The focus of this research is on large, multi-national companies. The motivation for this is that, though increasingly preoccupied and devoted to innovation, large companies are found to struggle with successful innovation (Christensen 2013). Based on the above this study is committed to answering the following research question: *What is the relation between business process management maturity and innovation in large companies?*

The remainder of this paper is organized as follows, first the literature that is the foundation to the conceptual model is described. Section 3 provides insight in the data collection for this research and in Sect. 4 the analysis of the data is discussed. Finally, conclusions and implications are given in Sect. 5.

2   Literature

In this section a brief overview is provided on the theory and methods that are the foundation to the conceptual model constructed for this research.

2.1   Innovation

In a most basic definition of innovation the notion of novelty must be included (Gupta et al. 2007). Furthermore, in the context of organizations there is also the need of commercialization and/or successful implementation of innovations (Popadiuk and Choo 2006). It is therefore generally conceptualized that innovation is a process starting from idea creation, through to implementation/ commercialization. Innovation can be seen as a multi-level process through which organizations convert ideas into new or enhanced products, services or processes to increase their competitive advantage (Baregheh et al. 2009). Besides innovation being a driver of change in products or services, it can also be a driver of change for enterprises (Smit 2015). Organizations can adopt external innovations to change the way they operate or are organized. Based on this we divide innovation in two concepts (1) Innovation Value Chain (IVC) and (2) Innovation Adoption.

The Innovation Value Chain is described by Hansen and Birkinshaw (2007). According to them the IVC consists of three phases: Idea Generation, Idea Conversion and Diffusion. Innovations typically start with a new idea. The generation of an idea can occur in teams within the organization, across teams, or externally to the organization (Hansen and Birkinshaw 2007). During the conversion stage ideas are transformed into new products, service, processes etc. In the final phase, diffusion, the innovation is put into exploitation. From this it is clear that the IVC provides an interlinked and linear, three phased description of how innovation presents itself in an organization.

Innovations that have been created but not adopted and used have no value. The adoption of innovation studies the factors and sentiment that may be related to the adoption or non-adoption of new products and services. Several studies show that such factors can be both internal and external to an organization (Tan and Teo 2000; Zheng et al. 2008; Udo et al. 2016). External factors are for example social norms, government policies, or rules that will make the innovation illegal or hard to get (Tan and Teo 2000; Udo et al. 2016), while internal factors relate to aspects such as technical ability, resources and intention to adopt (Tan and Teo 2000; Zheng et al. 2008).

Based on the above the innovation concept in our conceptual model is based on both the IVC theory of Hansen and Birkinshaw (2007) and the innovation adoption model by Tan and Teo (2000).

2.2 Business Process Management Maturity

Maturity models allow deductions on which capabilities an organization should improve (Forstner et al. 2014). Thus, a BPM maturity model helps organizations enhance their business process architecture to reach company goals. The number of studies on BPM maturity models has increased alongside the growing interest for BPM, which itself has its roots in both total quality management and business process re-engineering (Ravesteyn and Versendaal 2007; Plattfaut et al. 2011). Currently there are several theoretical models for studying BPM maturity and developing BPM capabilities (De Bruin et al. 2005; Bucher and Winter 2010). However, it is still not clear how organizations best achieve maturity (Plattfaut et al. 2011). Also, there is a discussion on the optimal level of maturity for BPM since a maximization of BPM maturity need not be necessary for realizing the organization's objectives according to the business strategy (Rosemann et al. 2004). Furthermore, the majority of available BPM Maturity models is descriptive (Tarhan et al. 2016) and cannot be used in a prescriptive manner. In their study Tarhan et al. (2016) found that most models only measure the BPM maturity and just three of the models also measured the (organizational) performance. One model that is prescriptive in its nature is the OMG BPMM model. This model was developed by Curtis and Weber and is based on the architecture of the Capability Maturity Model (Heller and Varney 2013). It is made up of five maturity levels and 30 capability areas that are called process areas. In comparison to most models it gives clear instructions on which process areas have to be improved in order to reach the next maturity level (Roeglinger et al. 2012). In this study the aim is to measure BPM maturity and investigate the relationship with innovation. Therefore, a model that is descriptive and only measures maturity will suffice. For this research the BPM maturity model by Ravesteyn et al. (2012) is used (see Fig. 1). In this model the maturity levels are not linear, i.e. it is not necessary for an organization to "complete" a maturity level in order to reach another. Instead, the model recognizes that organizations will perform over all levels simultaneously, merely at different quality. This better reflects to dynamics of organizational change and addresses criticism on other BPM models (Pöppelbuss et al. 2011).

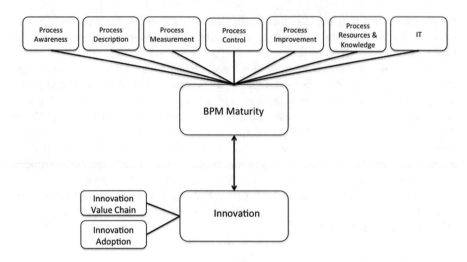

Fig. 1. Conceptual model.

2.3   The Conceptual Model and Hypothesis

To study the relation between BPM maturity and innovation and in accordance to above literature review the research framework for this study builds on a total of three conceptual models of the core elements of BPM maturity and innovation (see Fig. 1). The various constructs from the conceptual model are derived from the BPM maturity model suggested by Ravesteyn et al. (2012), the Innovation Value Chain by Hansen and Birkinshaw (2007) and Innovation Adoption as conceptualized by Tan and Teo (2000).

The BPM maturity construct in the conceptual model is divided in to the seven dimensions of business process management that resemble the BPM-lifecycle (process awareness, process description, process measurements, process control, and process improvement) and the supporting dimensions 'process resources and knowledge' and 'IT usage', as described in De Waal et al. (2017). For each dimension several BPM capabilities have been defined (in total 37).

The Innovation construct is conceptualized in seven items that measure the Innovation Value Chain, which according to Hansen and Birkinshaw (2007) consists of the phases Idea Generation, Idea Conversion and Idea Diffusion. Furthermore, six items measure Innovation Adoption, this includes items relating to Attitude to Innovation (Relative Advantage and Risk), Subjective Norms (Customers and Competitors) and Perceived Behavioral Controls (Self-Efficacy and Facilitating Conditions) (Tan and Teo 2000).

3   Research Methodology

This section describes the procedure to collect data and the outcomes of the validation of the BPM Maturity and Innovation scales. For the analysis the collected data was transferred to SPSS 23 for factor analysis, descriptive statistics, correlation analysis and regression analysis.

3.1 Data Collection

As described above the conceptual model for this research is based on a comprehensive literature study on the concepts of BPM maturity and Innovation both individually and in combination. Subsequently, quantitative data was gathered in four European multinational organizations and analysed statistically. For this a questionnaire was developed that consisted of 55 questions related to the core elements of the conceptual model namely BPM maturity (37 items) and innovation (13 items) as well as five general questions to capture supporting variables such as size, sector, and knowledge and experience in BPM. The questions on BPM maturity follow the BPM dimensions described in Ravesteyn et al. (2012). The respondents selected the degree to which they agreed or disagreed with the given statements according a five-point Likert scale (1 = fully disagree, 5 = fully agree).

The data was collected as part of internships by master students of Innovation in European Business in the academic years of 2015 and 2016. All data was collected by means of a questionnaire, which was shared with employees at the companies of the internships. In most organisations data collection was conducted via an online survey with mandatory questions. The link to the questionnaire was sent via email. The survey could only be completed when all questions were answered. In one large organization the data was collected on paper instead of online. In this sample a total of 33 questionnaires were handed in, however six questionnaires were aborted midway. These questionnaires have proven to be unusable as too many questions were left unanswered, resulting in 27 valid responses from that company.

The total data set amounts to 143 respondents, obtained from the four large, multinational companies. The respondents have different business functions within their organizations, ranging from IT, marketing and sales through to procurement and quality management, amongst others. The variable size was classified by the question: approximately how many employees are there in your company? In accordance with often cited research, large is arbitrarily defined as a minimum of 5,000 employees (Porter 1963). An overview of the complete data set of the large companies can be found in Table 1.

Table 1. Overview of data set large companies (N = 143).

Company	Headquarter	Sector	Sample size
1	Ireland	IT	55
2	Ireland	Utilities	27
3	Austria	Jewellery manufacturing	33
4	Belgium	Fast moving consumer goods	28

3.2 Measurement and Validation

The validity of the scales was tested by means of a factor analysis in SPSS 23 with varimax rotation to maximize the dispersion of loadings within factors. Therefore, it tries to load smaller number of variables highly on each factor, resulting in more interpretable clusters of factors (Field 2013). This approach simplifies the interpretation of factors and is thus chosen for the scope of this research. Principal Component Analysis (PCA) with varimax rotation of the 37 items of BPM maturity resulted in a seven-factor solution, accounting for 64.6% of the overall variance. Although this analysis shows that the seven dimensions of BPM maturity are represented in seven rotated factors, some items have a low factor loading. After removing seven of these items, the PCA with varimax rotation resulted again in a seven-factor solution, accounting for 69.0% of the overall variance. This supports the seven dimensions of the conceptual model of this study.

Similarly, for the innovation part of the conceptual model, a PCA with varimax rotation on the 13 items, resulted in a three-factor, accounting for 56.0% of the overall variance. The results demonstrated that the seven items loading moderate to first factor (Innovation Value Chain), four items loading moderately to highly with the second factor (Innovation Adoption without Perceived Behavioural Control) and two items correlating with the concept of Perceived Behavioural Control. Because the factor loadings were relatively high and the concepts of Innovation Value Chain and Innovation Adoption have been tested and verified in previous studies (Smit 2015; Tan and Teo 2000), no adjustments were made to the scales for the purpose of this study.

To further test the reliability of the constructs in the conceptual model a reliability test was conducted for each dimension. The results are presented in Table 2. As can be seen, the factor loadings were between 0.851 and 0.499, which can be considered as being significant (Hair et al. 1998). The reliability of the scales was confirmed by

Table 2. Factor analysis and reliability of BPM maturity and innovation scales (N = 143).

Dimension	Number of items	Own value	Explained variance	Factor loading (Max.)	Factor loading (Min.)	Cronbach's alpha
Process awareness	3	1.23	4.1	.807	.499	.765
Process description	5	3.26	10.9	.782	.520	.871
Process measurement	4	1.39	4.6	.813	.565	.825
Process control	3	1.05	3.5	.766	.183	.730
Process improvement	5	1.84	6.1	.736	.605	.806
Process resources	3	1.06	3.5	.756	.686	.786
Process IT tools	7	10.88	36.3	.851	.653	.907
Innovation value chain	7	3.37	48.2	.793	.577	.817
Innovation adoption	6	2.63	43.8	.764	.561	.738

Cronbach's alpha value of 0.907 to 0.730. In accordance with the study of Kline, a Cronbach's alpha coefficient above 0.7 is interpreted to imply a reliability of the scales (Kline 2000; Urdan 2011). The seven dimensions of BPM maturity and the two constructs that make up Innovation have an alpha coefficient of 0.863 resp. 0.623. This suggests the dimensions used to measure each construct have a relatively high and moderate internal consistency.

4 Findings and Discussion

To answer the research question on the relation between BPM maturity and innovation in large companies, a Pearson correlation analysis was performed on the 143 respondents from the four large organizations. The results of the test are depicted in Table 3. The findings show a moderate significant relation between respectively Innovation Value Chain (0.40) and BPM maturity and Innovation Adoption and BPM maturity (0.41). The relation between the overall Innovation construct and BPM maturity is slightly stronger than with the individual concepts but can still be interpreted as mere moderate (0.48), according to statistical research theory (Field 2013).

Table 3. Correlations between BPM maturity (dimensions) and innovation (dimensions) (**p < .01; *p < .05; N = 143).

	BPM Maturity	Process awareness	Process description	Process measurement	Process control	Process improvement	Process resources	Process tools
Innovation	,479**	,324**	,248**	,229**	,320**	,605**	,557**	,243**
Innovation value chain	,403**	,304**	,181**	,231**	,219**	,589**	,475**	0,138
Innovation adoption	,414**	,249**	,240**	0,160	,325**	,445**	,475**	,273**

To further analyse the relation between the concepts a (multiple) regression analysis is conducted with Innovation (dimensions) being the dependent variable and BPM maturity (dimensions) as the independent variable. The results are shown in Table 4. As is shown all the results of the multiple regression analysis were significant. The proportion of variance in Innovation that can be explained by BPM maturity amounts to 22,4%. Therefore, it can be concluded that in large companies BPM maturity has a positive effect on innovation. This finding is in line with previous studies (Benner and Tushman 2002; Dijkman et al. 2016). The analysis of the dimensions of BPM maturity does show that two dimensions are stronger predictors for innovation than the overall BPM maturity level. Process Improvement is the main predictor for Innovation Value Chain and Process Resources and Knowledge is the main predictor for Innovation Adoption.

If we compare these findings with the research of Ravesteyn et al. (2016) there are some interesting conclusions. Though the relation in this data set is significant and moderate, it does not show the strength of the relation found for the large enterprise of the research by Ravesteyn et al. which was 0.64. However, the relation found in this

Table 4. Multiple regression analysis between BPM maturity (dimensions) and innovation (dimensions) (N = 143).

Dependent variable	Predictor	Beta	p	Adjusted R^2	F	df	p
Innovation	BPM maturity	.48	.000	22,4	42,014	142	.000
Innovation	Process improvement	.42	.000	41,0	50,348	142	.000
	Process resources	.29	.001				
Innovation value chain	Process improvement	.48	.000	35,7	40,337	142	.000
	Process resources	.17	.047				
Innovation adoption	Process resources	.32	.001	25,1	24,788	142	.000
	Process improvement	.24	.010				

study is significantly stronger than the relation found for the SME and start-ups from the research of Ravesteyn et al. in 2016, which was negative and weak, respectively. Therefore, this seems to support previous findings that company size is related to the relation between BPM maturity and innovation (Ravesteyn et al. 2016; Tang et al. 2013).

5   Conclusion and Limitations

The objective of this study is to determine whether there is a relation between BPM maturity and innovation in large organizations. For this the following research question was formulated: *What is the relation between business process management maturity and Innovation in large companies?*

Based on analysis of the collected data of large organisations, it is possible to conclude that there is a moderate correlation between respectively the concept of the Innovation Value Chain (0,40) and BPM maturity, and Innovation Adoption (0,41) and BPM maturity. The correlation between the overall Innovation construct and BPM maturity is slightly stronger than with the individual concepts but can still be interpreted a mere moderate (0,48). The regression analysis that is conducted shows that the proportion of variance in innovation that can be explained by BPM maturity amounts to 22,4%. Together these analyses confirm that in large companies BPM maturity has a positive effect on innovation. However, this does not mean that investing in BPM capabilities to increase BPM maturity is also enough to increase the innovation capability of an organization.

To further understand the relationship between business process management and innovation more research is needed. One way to validate the findings is to estimate the conceptual model using SEM (Structural Equation Modelling). Second, we suggest to keep focusing on large organizations, as analysis of data collected at smaller organizations seems to suggest that the relationship between BPM maturity and Innovation is very weak. As this research only looked at profit organizations, future studies should also include large not for profit organizations such as governmental organizations. Furthermore, the impact of culture on the relation between BPM and innovation could also be considered as a topic of research.

References

Baregheh, A., Rowley, J., Sambrook, S.: Towards a multidisciplinary definition of innovation. Manage. Decis. **47**(8), 1323–1339 (2009)

Benner, M.J., Tushman, M.: Process management and technological innovation: a longitudinal study of the photography and paint industries. Adm. Sci. Q. **47**(4), 676–707 (2002)

Bucher, T., Winter, R.: Taxonomy of business process management approaches. In: vom Brocke, J., Rosemann, M. (eds.) Handbook on Business Process Management 2, pp. 93–114, Springer, Heidelberg (2010)

Christensen, C.: The Innovator's Dilemma: When New Technologies Cause Great Firms to Fail. Harvard Business Review Press, Boston (2013)

De Bruin, T., Rosemann, M., Freeze, R., Kulkarni, U.: Understanding the main phases of developing a maturity assessment model, In: Campbell, B., Underwood, J., Bunker, D. (eds.) Australasian Conference on Information Systems (ACIS), Sydney (2005)

De Waal, B.M.E., Valladares, R., Ravesteyn, P.: BPM maturity and process performance: the case of the Peruvian Air Force. In: Proceedings of the 23rd Americas Conference on Information Systems AMCIS 2017, Boston, USA, 10–12 August, pp. 1–10 (2017)

Dijkman, R., Lammers, S.V., de Jong, A.: Properties that influence business process management maturity and its effect on organizational performance. Inf. Syst. Front., **18**(4), 717–734 (2016)

Field, A.: Discovering Statistics using IBM SPSS Statistics. Sage, Thousand Oaks (2013)

Forstner, E., Kamprath, N., Röglinger, M.: Capability development with process maturity models – decision framework and economic analysis. J. Decis. Syst. **23**(2), 127–150 (2014)

Gupta, A.K., Tesluk, P.E., Taylor, M.S.: Innovation at and across multiple levels of analysis. Organ. Sci. **18**(6), 885–897 (2007)

Hair, J., Anderson, R., Tatham, R., Black, W.: Multivariate Data Analysis, 5th edn. Prentice Hall, Englewood Cliffs (1998)

Hansen, M.T., Birkinshaw, J.: The innovation value chain. Harvard Bus. Rev. **6**(85), 121–130 (2007)

Heller, A., Varney, J.: Using process management maturity models. APQC (2013). https://www.apqc.org/knowledge-base/documents/using-process-management-maturity-models. Accessed 13 Nov 2015

Hung, R.Y.Y.: Business process management as competitive advantage: a review and empirical study. Total Quality Manage. Bus. Excellence **17**(1), 21–40 (2006)

Kline, P.: Handbook of Psychological Testing, 2nd edn. Routledge Kline, London (2000)

Plattfaut, R., Niehaves, B., Pöppelbuß, J., Becker, J.: Development of BPM capabilities – is maturity the right path? In: European Conference on Information Systems Paper, vol. 27, pp. 1–13 (2011)

Popadiuk, S., Choo, C.W.: Innovation and knowledge creation: how are these concepts related? Int. J. Inf. Manage. **26**(4), 302–312 (2006)

Pöppelbuss, J., Niehaves, B., Simons, A., Becker, J.: Maturity models in information systems research: literature search and analysis. Commun. Assoc. Inf. Syst. **29**(27), 505–532 (2011)

Porter, L.W.: Job attitudes in management: IV. Perceived deficiencies in need fulfilment as a function of size of company. J. Appl. Psychol. **47**(6), 386–397 (1963)

Ravesteyn, P., Versendaal, J.: Success factors of business process management systems implementation. In: Proceedings of the 18th Australasian Conference on Information Systems, Toowoomba, Australia (2007)

Ravesteyn, P., Zoet, M., Spekschoor, J., Loggen, R.: Is there dependence between process maturity and process performance? Commun. IIMA **12**(2), 65–79 (2012)

Ravesteyn, P., Smit, J., McGuinness, B.: A study on the relation between business process management maturity and innovation. In: Proceedings of the AMCIS 2016 Conference, Paper 340 (2016). http://aisel.aisnet.org/amcis2016/Adoption/Presentations/8/

Roeglinger, M., Poeppelbuss, J., Becker, J.: Maturity models in business process management. Bus. Process Manage. J. **18**(2), 328–346 (2012)

Rosemann, M., Bruin, T., Hueffner, T.: A model for business process management maturity. In: ACIS Proceedings, Paper 6 (2004). http://aisel.aisnet.org/acis2004/6

Sander Jones, J.L., Linderman, K.: Process management, innovation and efficiency performance: the moderating effect of competitive intensity. Bus. Process Manage. J. **20**(2), 335–358 (2014)

Scott, J.E.: Mobility, business process management, software sourcing, and maturity model trends: Propositions for the IS organization of the future. Inf. Syst. Manag. **24**(2), 139–145 (2007)

Smit, J.: The innovation value chain and adaptability of organizations. J. Int. Technol. Inf. Manage. **24**(3), 57–73 (2015)

Tang, J., Pee, L.G., Iijima, J.: Investigating the effects of business process orientation on organizational innovation performance. Inf. Manage. **50**(8), 650–660 (2013)

Tan, M., Teo, T.S.H.: Factors influencing the adoption of internet banking. J. Assoc. Inf. Syst. **1** (5), 1–44 (2000)

Tarhan, A., Turetken, O., Reijers, H.A.: Business process maturity models: a systematic literature review. Inf. Soft. Technol. **75**, 122–134 (2016)

Tidd, J.: Innovation management in context: environment, organization and performance. Int. J. Manage. Rev. **3**(3), 169–183 (2001)

Udo, G., Bagchi, K., Maity, M.: Exploring factors affecting digital piracy using the norm activation and UTAUT models: the role of national culture. J. Bus. Ethics, **135**(3), 517–541 (2016)

Urdan, A.T., Huertas, M.K.Z.: Success determinants of new products launched by foreign companies in Brazil. Innovative Mark. **7**(4), 49–59 (2011)

Zheng, N., Guo, X., Chen, G.: IDT-TAM integrated model for IT adoption. Tsinghua Sci. Technol. **3**(13), 306–311 (2008)

Multilingual Low-Resourced Prototype System for Voice-Controlled Intelligent Building Applications

Alexandru Caranica[1(✉)], Lucian Georgescu[1], Alexandru Vulpe[2], and Horia Cucu[1]

[1] Speech and Dialogue Research Laboratory, University Politehnica of Bucharest,
Bucharest, Romania
{alexandru.caranica,lucian.georgescu}@speed.pub.ro,
horia.cucu@upb.ro
[2] Telecommunications Department, University Politehnica of Bucharest, Bucharest, Romania
alex.vulpe@radio.pub.ro

Abstract. With speech recognition databases spanning most of the widely used languages around the globe, there is a lot of incentive to build linguistically diverse, voice-driven applications, in different languages and in diverse acoustic conditions. Although state of the art speech processing has achieved great performance for most widely used languages, little efforts were made for under-resourced languages, such as Romanian. Moreover, most of these systems are not focused in supporting specific voice recognition scenarios, such as assistive applications for elder or disabled people, or consider a triggered close talking voice interaction. This paper focuses in building a prototype system for Romanian language, to be used in distant speech recognition scenarios, for voice driven speech applications in intelligent homes or buildings. Previously acquired speech databases in Romanian language are used, recorded in real life conditions, by our research group. For a baseline comparison, an English recognition engine is also implemented.

Keywords: Voice drive applications · Automatic speech recognition
Distant speech recognition · Home automation · Multilingual recognition

1 Introduction

One of the issues of a market driven economy, over the recent years, was the almost exclusive use of English as the means of communication between man and machine. But speech is the most natural way of communication for all human beings, no matter their mother tongue, or where in the world they live.

A great deal has been written recently, on the growing importance of consumer voice-driven computing devices, such as Amazon's Echo, Google Home and others like them. Here, the effort made by Google and Amazon, to rapidly push into the consumer market their vision of home personal assistants, can be mentioned [2]. Both popular systems rely on the "cloud" model to push relevant information to the user, and will not work without setting up a personal account, that aggregates data from multiple proprietary

© Springer International Publishing AG, part of Springer Nature 2018
Á. Rocha et al. (Eds.): WorldCIST'18 2018, AISC 747, pp. 97–107, 2018.
https://doi.org/10.1007/978-3-319-77700-9_10

online services. Nevertheless, they brought "always on" hands-free convenience assistance with voice-control across the room, with distant speech recognition (DSR), and are able to answer spoken questions, read audiobooks and the news, report traffic information and weather, give info on local businesses or provides sports scores and schedules. Turning on the lights with a remote control or smartphone is pretty interesting, but one can imagine doing it simply by saying it out loud, in spoken language: "Turn the lights on in the living room". The impact of this type of interaction is big, and might take some time to accommodate.

However, they lack customization ability, and interoperability with existing appliances or applications is not guaranteed. There are also custom scenarios where a more general, open system approach is desired, or required, in order to interface with existing appliances inside a house, or systems inside a building. As emphasized in [2], standards ensure compatibility between devices and ease the maintenance as well as orient the smart home design toward cheaper solutions. Another custom scenario, targeting specific user groups, which may greatly benefit from having voice controlled system at their own disposal, is in the home-care domain, also known as ambient assisted living (AAL) [3], where voice control can be a great asset for people in need, and can significantly improve their quality of life.

Moreover, although state of the art speech processing has achieved great performance for most widely used languages (English, French, German or Chinese), little to no efforts were made for under-resourced languages, such as Romanian, still considered a low resourced language [4]. One of the main reason for this was the lack of an annotated speech database from real life conditions [5].

In this context, this paper also describes a set of experiments in evaluating and building reliable noise-robust acoustic and grammar models, to be used in a voice controlled prototype, for intelligent homes or buildings. Section 3 describes the system integration and architecture, with a brief description of the modules and peripherals used. In Sect. 4, we describe the evaluation methodology and present an experimental validation of the prototype. Results are also presented in the same section. Finally, Sect. 5 is dedicated to the closing remarks and conclusions.

2   Related Work

Home automation is one of the most common Internet of Things (IoT) applications. As classical input methods, via touchscreen panels and remote controlled systems reached a certain level of maturity, voice control will play a central role in the development of IoT, which seeks to connect everyday devices with their users. From open source standards, to proprietary protocols and applications, there are various implementations that support the smart home concept.

Most smart homes or buildings will not be able to accomplish much without appliances or devices to control, nor will they be able to communicate with these devices in the absence of a control server and a network. Thus, many previous studies focused on protocols and network types [6, 7], later on putting an accent on user interfaces [8, 9]. The Open Interconnect Consortium (OIC) tried to create an open source universal project called IoTivity [10], where other technologies can be easily integrated in this project using plugins, according to the consortium.

Lately, there is also much interest in securing these networks, in the light of firmware breaches in a lot of embedded sensors and routers, used in these networks [11]. In [12], authors dissect the behavior of household appliances in connected homes, highlighting the ease on how security and privacy can be compromised, thus enforcing the concern of security and privacy of smart home devices.

Regarding speech interfaces for home or building appliances, very few smart home research projects integrated this type of input into their design, mainly because of the complexity of setting up this technology to work in real life environments, as many challenges need to be solved, such as distant speech recognition (DSR), noise robustness, etc. All previous research projects mainly focus on one language, as the difficulty in moving to another language is posed by the expensive process of acquiring new acoustic, phonetic and language resources, at least for low resourced languages. Hence, the proposed solutions might not be scalable to other languages. Development of such systems was boosted as soon as real life speech databases were acquired and used in experiments [13, 17, 18, 20, 21].

For Romanian language, there has been some efforts in acquiring noisy data sets, but these efforts where limited, until recently, to just TV recording, which are not suitable for the smart home distant speech recognition task [21]. Current paper addresses this issue by using previously acquired corpora, by our research group, specifically tailored for this task.

3 System Integration and Architecture

This section describes the steps followed to integrate our system and presents the final prototype architecture, with a brief description of the modules and peripherals used.

3.1 Proposed System Architecture

From an architectural point of view, the system can be viewed as an assembly of multiple components and interconnected modules. Considering the functionality of each component, they can be divided into 3 major categories, as can be seen in Fig. 1.

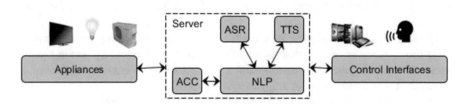

Fig. 1. Conceptual system architecture

The first category is represented by user control interfaces, such as mobile devices (mobile phone, tablet), or fixed devices incorporating microphones. They have the role of taking over the voice command of the users. The second category is the server component that processes and interprets the incoming voice commands. It can be connected to the Internet to allow outside control of the house. The third category is the smart devices to be controlled (lights, air conditioning, doors, etc.). The server consists of several modules, which will be described in the following sections.

3.2 Speech Recognition Module (ASR)

The main purpose of this module is to detect a vocal command forwarded by user interaction modules and to convert it into a text message.

Once the module is started, it begins a keyword detection process, which identifies the occurrence of the "Casandra" keyword in the vocal signal. This indicates the beginning of a valid command, and the process switches to an automatic speech recognition process that uses a finite state grammar, obtaining the spoken command in text format. Furthermore, this is sent as a HTTP POST request to the Natural Language Processing (NLP) module. The response returned by this request contains another text message that informs about the status resulting from the actual execution of the command. This text confirmation is given as a parameter to a HTTP GET request to the speech synthesis module, which returns an audio file with the synthesized message. The audio file content is played on the speaker device.

The implementation was accomplished using a Raspberry PI 3 development kit which runs a mixed system, performing both keyword detection and automatic speech recognition. It was developed using the C programming language, on the basis of the continuous speech decoder PocketSphinx API. The development kit connects both a microphone to retrieve the vocal signal and a speaker device to confirm to the user that his command has been made.

3.3 Voice Synthesis Module (TTS)

This module translates text into speech in order to give the user an acoustic acknowledgment in response to the spoken command.

From the point of view of integration with the speech recognition module, the speech synthesis module receives a HTTP GET request with the following parameters: the text message, the language in which the synthesis is desired (English or Romanian) and the type of spectral parameters used. The resulting audio file is sent back as an HTTP request response.

It performs a statistical parameter synthesis of speech, supporting both MLSA and STRAIGHT analysis. The system is a corpus based one, performing text analysis to extract relevant linguistic information and a synthesis system based on the linguistic information obtained by the text analysis system.

The extraction of linguistic information is a complex process that aims to convert a sentence/paragraph into a discrete attribute sequence that serves the process of selecting/generating acoustic units necessary in the voice generation process. The attributes extracted from the text are: the phonetic context, the current syllable, the following punctuation/sentence type, morphosyntactic information.

Determining how attributes affect the prosody is done automatically by using decision trees. Three decision trees are used, aiming at spectral modeling, the fundamental frequency of the voice as well as the speaking time.

The analysis is done as follows: it is used a phoneme level aligner to get the correspondence between the text and the recorded voice. Each phoneme is modeled with Markov Models with 5 states. Spectral parameters (MLSA or STRAIGHT), fundamental duration and frequency are extracted from the acoustic files. Based on alignments, decision trees are used [22].

From the point of view of integration with the speech recognition module, the speech synthesis module receives a HTTP GET request with the following parameters: the text message, the language in which the synthesis is desired (English or Romanian) and the type of spectral parameters used (MLSA or STRAIGHT). The resulting audio file is sent back as the HTTP request response.

The module was developed using the Java programming language and it runs on a Linux virtual machine running on an Intel NUC mini PC. The HTS toolkit was used to generate the decision trees.

3.4 Natural Language Processing Module (NLP)

This is the server's core module. It receives and sends messages from/to all other modules. The main task is to understand the incoming messages (text mode) so they can "translate" them to ACC (Appliance Configuration and Control) mode.

The module is based on scenarios, each scenario being defined by a name and three components. The first component is represented by a list of sentences consisting of fixed parts and dynamic parts. For example, "Set the temperature to * degrees", where "*" is a variable parameter, the rest of the words being the fixed part. The second component is a list of actions performed by the system after it identifies the previously defined command. Actions are defined in JSON format and include smart device ID as well as command parameters (e.g.: light on = 1). The third component is the feedback sent back to the user.

Scenarios identification and parameter extraction are performed according to a well-established methodology. The scenario is identified using a decision tree. Once the scenarios are defined, the sentences for each scenario are extracted. The decision tree is created by extracting in a bag-of-words format each word from each command, having associated the actual scenario [23]. For each scenario there is at least one command (usually minimum 3–4). The training algorithm analyzes each command in part, generating a tree. The tree size is reduced to guarantee a certain level of universality. Extracting parameters from the command is also performed using a decision tree. First, the position of the parameters must be identified, this fact being done in two steps. The first step identifies the start position, the word index that begins the parameter. A parameter can have more than one word, for example "the temperature at twenty-seven degrees" has a single parameter of 4 words. Once a parameter start has been identified, another decision tree, trained to predict stop (e.g.: "seven"). The command parameter is represented by the words between start-stop. Once a stop word is marked, it starts the search with the tree that predicts start because a command may have several parameters. The context used for trees consists of 4 words before the analyzed word for the start tree, respectively 4 words after the analyzed word for the stop tree. Once the parameters are identified, it should be found the type of each one. Another decision tree is used for this task, determining whether the parameter is numeric, string, predefined, etc.

Regarding to the interaction with the other modules, the NLP module receives from the speech recognition module a HTTP POST that contains the user spoken command in text format. The type of the order, the device to which it is intended, and the parameters are identified. This information is forwarded to Appliance Configuration and Control Module. The NLP module receives a feedback from it and it returns a confirmation message in its HTTP response.

Similar to the speech synthesis module, the NLP module was written in the Java programming language and runs on an Intel NUC mini PC.

3.5   Appliance Configuration and Control Module (ACC)

This module performs initial configuration of existing smart devices. The communication protocol used is KNX, the interconnection being made via TCP/IP (Ethernet or Wireless where the device allows configuration).

The KNX bus provides communication with the smart devices, independent of manufacturer and application domain. In the KNX standard there is no controller, each device being directly connected to the bus, and their configuration is software based. In this project some of the equipment was connected directly to KNX, while others with Comfortclick [14] via HTTP. The smart devices used in this project are a series of lights whose color and intensity can me adjusted, automatic climate control system, security system with multiple individual activation partitions.

4   Experimental Validation

This section covers the experimental validation of our group's prototype ASR system in Romanian, with self-acquired resources and the English system, with external resources. NLP module is treated independently and can be consulted in paper [23].

4.1   Training and Evaluation Resources

For Romanian language, Table 1 presents the acquired corpuses by our group, used to train our decoders and build acoustic models. More in-depth information of the self-acquired database can be consulted in our previous paper [5], and on the project page [13].

Table 1. Romanian corpus resources

Set	Subset (model name)	Duration	Type
Training	RSC (AM001)	94 h 46 min 14 s	Read speech
	SSC (AM002)	27 h 27 min 21 s	Spontaneous speech (noisy)
	SSC-clean (AM003)	103 h 17 min 56 s	Clean spontaneous speech
Evaluation	RSC (AM004)	5 h 29 min 6 s	Read speech (noisy)
	SSC (AM005)	3 h 29 min 3 s	Spontaneous speech (noisy)

For our English engine we used the corpus available here: http://www-lium.univ-lemans.fr/en/content/ted-lium-corpus (English conversational speech, from Ted conferences all over the world). Characteristics can be consulter in Table 2.

Table 2. TED-LIUM English corpus

	Training	Evaluation
Nr. of conversations	774	11
Total duration	118 h 4 min 48 s	3 h 4 min 5 s
• Male	81 h 53 min 7 s	2 h 21 min 35 s
• Female	36 h 11 min 41 s	42 min 29 s
Nr. of unique speakers	666	11

4.2 Speech Features, Acoustic and Grammar Models

For our baseline models, we used the traditional MFCC speech features plus temporal derivatives (13 MFCC + Δ + $\Delta\Delta$). For noisy channels, we experimented with Power Normalized Cepstral Coefficients (PNCCs), which tend to bring the most important gains in accuracy.

The CMU-Sphinx speech recognition toolkit [15] was chosen as our main ASR system, as it also offers an embedded decoder that fit our purpose of building a small "appliance-like" prototype. Our local room end-nodes run on PochetSphinx, a lightweight variant of the Sphinx speech recognition engine. It is specifically tuned for handheld and low powered devices, such as our ARM boards used for keyword spotting of the word "Casandra", before uttering the main command. For English language, a Kaldi [16] based DNN system was implemented, that offers better adaptation to non-native English speakers.

The acoustic models used in this study are 5-state HMMs with output probabilities modeled with GMMs, with the speech features detailed in previous section. In all cases the 36 phonemes in Romanian were modeled as context dependent phonemes, with 4000 HMM senones. The number of Gaussian mixtures per senone state varied (64/128), depending on the acoustic model setup, in order to adapt to the size and variability of the training speech corpus.

A Finite State Grammar (FSG) was used for the uttered audio commands database. This offers better results for a hardware limited device, like a Raspberry Pi. Moreover, there are studies that show most users still prefer precise short sentences over more natural long sentences, when controlling their environment, believing this mode of interaction is still the quickest and most efficient. These commands have been split into five categories, depending on which part of the building they wish to interact with (exterior, security, multimedia, hvac and electric).

4.3 Experimental Results

Figure 2 presents the current hardware architecture of the prototype, as described in previous sections. The server part resides on a low-powered Intel NUC platform, with end-points spread through the building rooms, attached to speakers and microphones. KNX routers offers compatibility with existing infrastructure inside a building, with the possibility to control from the server also new home devices that offer an API, such as Philips HUE.

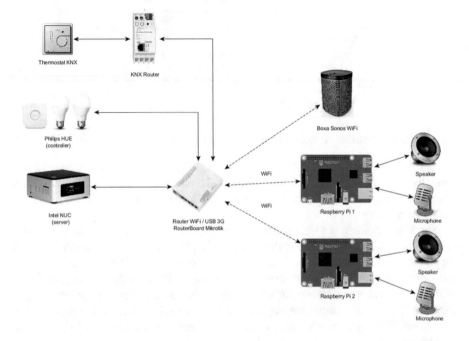

Fig. 2. Prototype architecture for voice-controlled intelligent building applications.

In Table 3, results are presented for the baseline ASR English system. As stated in previous sections, Kaldi toolkit offers the best results in this scenario (line 4 in Table 3), as deep learning algorithms present in this toolkit have the ability to capture utterance variation better, in a non-native speaker scenario. For both systems (English and Romanian), the standard Word Error Rate (WER) metric was used.

Table 3. System evaluation on English dataset

Line	Toolkit	Features configuration	WER [%]
1	Kaldi	Tri 1	55
2	Kaldi	Tri 2 (LDA + MLLT)	54
3	Kaldi	Tri 3 (LDA + MLLT + SAT)	18
4	Kaldi	Tri3 (MMI)	17
5	Kaldi	NNET2	16
6	Sphinx	GMM-LM	49
7	Sphinx	GMM-FSG	49

Table 4 presents results for the ASR in Romanian language, obtained with a set of five acoustic models (AM001 to AM005) in CMU Sphinx. We varied the training corpus (subset) from Table 1. As results show, the best performing model (AM003) was trained on clean spontaneous speech, on the biggest dataset of Romanian speech assigned for this task (see Table 1, more than 100 h of speech). Second is AM001, trained with read speech, on about 94 h of speech. The third model contains spontaneous speech, but with noisier datasets, which explains the worst results. Finally, AM005 is the worst performing model, using a small dataset and noisier speech.

Table 4. System evaluation on Romanian dataset

Line	Toolkit	Subset	WER [%]
1	Sphinx	AM001	16.9
2	Sphinx	AM002	18.9
3	Sphinx	AM003	7.5
4	Sphinx	AM004	20.6
5	Sphinx	AM005	29.8

5   Conclusion

This paper overviewed the steps and preliminary evaluations our group took to build a fully functional, multilingual, low-resourced voice controlled system for intelligent buildings and applications. Using previously acquired speech corpora in Romanian language, in real world conditions, allowed us to better predict the performance of our models in this low-resourced scenario.

Additionally, this paper presented a set of experiments in building a series of acoustic and grammar models, together with the integration of several hardware and software components, to build the functional prototype.

Our best performing model, for Romanian language, with self-acquired corpora, offered very good performance, with a WER of just 7.5% in a controlled environment. For English, a reasonable results was obtained, with a WER of 16%, using a DNN toolkit, with non-native speaker training and decoding.

The final hardware prototype can be demoed in our lab.

Acknowledgements. This work was supported by the Ministry of Innovation and Research, UEFISCDI, partly through project number 5 Sol/2017 within PNCDI III and partly through the programme "Partnerships in priority areas", "Collaborative Applied Research Projects", project ID: PN-II-PT-PCCA-2013-4-0789, contract number 32/2014.

References

1. Swapnil, B.: Amazon Echo vs. Google Home: the choice is obvious. Online Review, CIO - IDG Communications, Inc. (2016)
2. Mäyrä, F., Soronen, A., Vanhala, J., Mikkonen, J., Zakrzewski, M., Koskinen, I., Kuusela, K.: Probing a proactive home: challenges in researching and designing everyday smart environments. Hum. Technol. **2**, 158–186 (2006)
3. Vacher, M., Lecouteux, B., S-Romero, J., Ajili, M., Portet, F.: Speech and speaker recognition for home automation: preliminary results. In: Proceedings of the 8th International Conference on Speech Technology and Human-Computer Dialogue, "SpeD 2015", Bucharest, Romania, pp. 181–190. IEEE (2015)
4. Cucu, H., Buzo, A., Petrică, L., Burileanu, D., Burileanu, C.: Recent improvements of the SpeeD Romanian LVCSR system. In: The Proceedings of the 10th International Conference on Communications (COMM), Bucharest, pp. 111–114 (2014)
5. Dogariu, M., Cucu, H., Buzo, A., Burileanu, D., Fratu, O.: Speech database acquisition for assisted living environment applications. In: The Proceedings of the 8th Conference on Speech Technology and Human-Computer Dialogue (SpeD), Bucharest (2015)
6. Badica, C., Brezovan, M., Badica, A.: An overview of smart home environments: architectures, technologies and applications. In: Proceedings of the Sixth Balkan Conference in Informatics, Greece (2013)
7. Chang, C.-Y., Kuo, C.-H., Chen, J.-C., Wang, T.C.: Design and implementation of an IoT access point for smart home. J. Appl. Sci. **5**, 1882–1902 (2015)
8. Bhuiyan, M., Picking, R.: Gesture-controlles user interfaces, what have we done and what's next? In: Proceedings of the Fifth Collaborative Research Symposium on Security, E-Learning, Internet and Networking (2009)
9. Lee, H., Jeong, H., Lee, J., Yeom, K.-W., Shin, H., Park, J.-H.: Select-and-point: a novel interface for multi-device connection and control based on simple hand gestures. In: CHI 2008, Florence, Italy, 5–10–April 2008
10. IoTivity. http://www.iotivity.org/. Accessed Nov 2017
11. Fernandes, E., Jung, J., Prakash, A.: Security analysis of emerging smart home applications. In: Proceedings of 37th IEEE Symposium on Security and Privacy, San Jose, California (2016)
12. Notra, S., Siddiqi, M., Gharakheili, H.H., Sivaraman, V., Boreli, R.: An experimental study of security and privacy risks with emerging household appliances. In: Proceedings of International Workshop on Security and Privacy in Machine-to-Machine Communications (2014)
13. SpeeD Group, University Politehnica of Bucharest: Natural-language, Voice-controlled Assistive System for Intelligent Buildings (ANVSIB). http://speed.pub.ro/anvsib
14. ComfortClick: KNX Building Automation Toolkit. https://www.comfortclick.com/. Accessed Nov 2017
15. CMU Sphinx Toolkit. http://cmusphinx.sourceforge.net
16. Kaldi ASR Toolkit. http://kaldi-asr.org/
17. Mostefa, D., et al.: The CHIL audiovisual corpus for lecture and meeting analysis inside smart rooms. Lang. Resourc. Eval. **41**, 389–407 (2007)
18. Carletta, J., et al.: The AMI meeting corpus: a pre-announcement. In: Proceedings of the Second International Conference on Machine Learning for Multimodal Interaction (2006)
19. Cooke, M., Barker, J., Cunningham, S., Shao, X.: An audio-visual corpus for speech perception and automatic speech recognition. J. Acoust. Soc. Am. **120**(5), 2421–2424 (2006)

20. Vacher, M., Caffiau, S., Portet, F., Meillon, B., Roux, C., Elias, E., Lecouteux, B., Chahuara, P.: Evaluation of a context-aware voice interface for ambient assisted living: qualitative user study vs. quantitative system evaluation. ACM Trans. Access. Comput. (TACCESS) (2015)
21. Cucu, H., Buzo, A., Burileanu, C.: Unsupervised acoustic model training using multiple seed ASR systems. In: Proceedings of SLTU, St. Petersburg, Russia (2014)
22. Boros, T., Dumitrescu, S.D.: Cassandra smart-home system description. In: Proceedings of SpeD 2017 (2017)
23. Dumitrescu, S.D., Boros, T., Tufis, D.: RACAI's natural language processing pipeline for universal dependencies. In: Proceedings of CONLL Workshop (2017)

SCL-Based Analysis for 4G-Compliant System in Indoor/Pedestrian Environments

Ioana M. Marcu$^{(\boxtimes)}$ and Ionuț Pirnog

Telecommunication Department, ETTI, UPB, Bucharest, Romania
ioana.marcu@upb.ro

Abstract. The future of mobile communication systems challenges researchers to constantly improve efficiency and data rate by performing numerous analyses in lab conditions. The novelty of this analysis is represented by the evaluation of the performances of the systems using two imposed statistical confidence levels (SCLs), parameter which can further provide enough information for reliable and quantitative comparisons. Based on SCLs and BERs, the number of bits (ENB) necessary to achieve certain BER thresholds can be computed increasing thus the efficiency of transmission. The 4G-compliant system will be tested in scenarios according to ITU recommendation for mobile communication environment including indoor building channel. The effects over the necessary ENBs of two techniques (convolutional and LDPC) will be evaluated and several conclusions will be highlighted in the final chapter.

Keywords: Estimated number of bits · Statistical confidence level
BER · Performances

1 Introduction

Current researches in the area of developing the next 5G communication system rely on efficient joining of multiple access techniques (such as Orthogonal frequency-division multiple access (OFDMA), Filter Bank Multi-Carrier (FBMC), Generalized Frequency Division Multiplexing (GFDM), etc.), communication techniques (from multiuser systems to massive multiuser systems), on security methods (including coding/decoding techniques, encryption methods), etc. In [1] the research is focused on maximization of energy efficiency in next-generation multiuser MIMO-OFDM networks and there can be achieved up to 600% gain in energy efficiency over uniform power allocation policies. To control the high Peak-to-Average Power Ratio (PAPR) at the output of MIMO-OFDM systems, iterative clipping and filtering, and partial Transmit sequence are united or combined in [2]. The error correction methods in 4G systems are analyzed in [3–6]. In [3] four error correction codes (convolutional code (CC), Reed-Solomon (RSC)+CC, low density parity check (LDPC)+CC, Turbo+CC) are studied under three channel models (additive white Gaussian noise (AWGN), Rayleigh, Rician) and in this analysis Turbo+CC code in 4×4 configuration exhibits better performance. A turbo equalization model for LDPC codes able to apply MIMO systems combined with Space Time Trellis (STTC) codes is considered in [4] and simulations show that the performance is improved with 2.5 dB than that of turbo codes and with 10.5 dB compared to

© Springer International Publishing AG, part of Springer Nature 2018
Á. Rocha et al. (Eds.): WorldCIST'18 2018, AISC 747, pp. 108–117, 2018.
https://doi.org/10.1007/978-3-319-77700-9_11

conventional scheme. To overcome fading effects on communication channel in [5, 6] LDPC technique is considered in a MIMO-CDMA system and performances are evaluated as bit error rate (BER) vs Signal-to-Noise Ratio (SNRs). For large domain of SNRs (0–30 dB) significant coding gains can be achieved.

Though BER is considered to be the most reliable parameter when evaluating the performances of a communication system, statistical confidence level (SCL) can be used as an efficient tool for establishing the correct number of bits necessary to ensure efficient transmission. Practical analysis using SCL can be found in [7]. The required number of bits determined based on SCL and the error probabilities are estimated for a system using pattern generator and errors detector. SCLs for MIMO systems in [8] increase by counting more errors and/or transmitting more symbols.

The present paper illustrates the performances of a 4G-compliant OFDM-based MIMO system, performances evaluated in different communication environments (ITU-R M.1225 models). Considering each achieved BER as threshold BER values, ENBs will be computed for each case and conclusions related to transmission efficiency will be outlined. The paper is organized as follows: Sect. 2 comprises the description of the communication systems, implementation parameters and main challenges as well as performances reached testing ITU-R M.1225 models; simulation results for every specified case are illustrated in Sect. 3 and Sect. 4 contains the conclusions highlighted following the achieved results.

2   4G-Compliant Communication System

The model of the implemented 4G-compliant system is shown in Fig. 1.

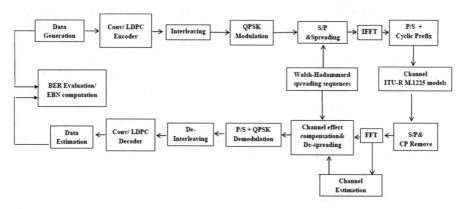

Fig. 1. 4G-compliant OFDM-based MIMO system: corresponding operations on the communication chain [6]

2.1   Implementation and Parameters

The system is designed for 32 users which transmit random data of 32400 bits. To ensure data protection the coding of the source is performed using convolutional or LDPC technique. The rate of conventional codes is ½ and the convolutional LDPC codes are irregular with rate ½. In addition to classic OFDM signal processing (Quadrature Phase-Shift Keying (QPSK) modulation, Inverse Fast Fourier Transform (IFFT) computation, parallel to serial and cyclic prefix adding), the signals are spread using 32 bits length Walsh-Hadamard sequences to differentiate the users (CDMA-based system).

At the receiver the reverse operations are performed for data recovery: Fast Fourier Transform (FFT), De-spreading, De-Interleaving, Demodulation. Additionally, the receiver estimates the communication channel based on the symbols transmitted on the pilot sub-carriers [9]. The convolutional decoding algorithm is Viterbi and LDPC decoder uses Min-Sum decoding for efficiency and low implementation complexity.

The radio channel is modelled using the ITU-R M.1225 standard which defines three types of environments that can be used for testing purposes: indoor office—inside a building, outdoor to indoor pedestrian for low speed users and vehicular high antenna for high speed users [8]. The channel parameters are given in [10].

2.2 Challenges and Performances in ITU Environment (Indoor/Pedestrian Models)

The complexity of the system is given by the large amount of data that need to be processed. Because 32 users transmit $3.24 * 10^4$ bits/user, the LDPC technique requires a large amount of time to perform. The implementation of ITU channel models implies definition of delays on every path of the channel, the power of the signals and Doppler frequencies for every specified case (indoor/pedestrian models). The use of these recommendations is widely spread among researchers in this area [11–13]. ITU Pedestrian-A and Vehicular-A channel models are taken into consideration in [11] when evaluating the performances of a 2-users MIMO system in the presence of CP-OFDM and FBMC. The same channel models from ITU M.1225 are used in [12] to validate the applicability of the implemented block frequency spreading approach for OFDM and FBMC MIMO system. It has been concluded that if the channel delay spread is not too high, block frequency spreading becomes an efficient method to restore complex orthogonality in FBMC. In [8] a similar analysis of OFDM-based MIMO system is given but the simulation includes only the performances of 2 users with the possibility of extension to 30 users (due to the increase complexity of the system). ITU vehicular test A and ITU pedestrian test B models are used to test performances of MIMO system as a combination of nonlinear decision feedback receiver (DFE) in [13] and for this analysis the non-linear receiver is best suitable.

3 Simulation Results

The scenarios for simulations are divided into three parts including analysis for Indoor/Pedestrian recommended channel models in ITU M.1225:

A. no source data coding/decoding
B. convolutional technique
C. LDPC technique.

The statistical confidence level of a communications system can be computed as [14, 15]:

$$SCL = 1 - e^{-\frac{N_{bits}}{BER_S}} \cdot \sum_{k=0}^{E} \frac{(N_{bits} \cdot BER_S)^k}{k!} \tag{1}$$

where N_{bits} = estimated number of bits; SCL = statistical confidence level; E = number of error bits and k = number of simulations. Following Eq. (1) SCL $\times$ 100 is the percent confidence that the system's true BER (i.e. if N = infinity) is less than the specified BER (e.g. BER_S) [15].

By imposing two values for SCL we can compute the estimated number of bits (EBN) necessary to achieve certain BER thresholds.

Following the simulation there will be noticed that all 32 users behave almost similar in the given conditions, therefore the graphical representation of BER vs SNR contains only one situation which represents a mean BER for all 32 values achieved. The percentage column in all tables represents a quantitative estimation of how much of the transmitted bits are actually necessary to achieve certain BER thresholds in the given conditions. Monte Carlo simulation has been performed but still this technique proved to be not efficient for low BERs since it led to increased runtimes.

A. No Source Data Coding/Decoding
Figure 2 illustrates the performances in terms of BER vs SNR for the implemented system when no error protection is used during the transmission.

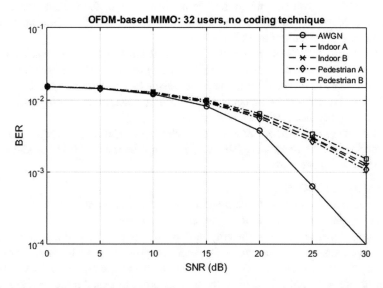

Fig. 2. BER vs SNR for OFDM-based MIMO system: 32 users for ITU recommendation (no source data coding/decoding)

Considering each BER value in Fig. 2 as a potential threshold for the corresponding SNR value and imposing two testing SCLs (95% and 99%) there can be computed the ENB necessary to reach those thresholds. The results are given in Table 1.

Table 1. ENBs for no coding case in AWGN channel.

SNRs (dB)	BER values	SCL (%)	ENB	Percentage (%)
0	$15 * 10^{-3}$	95	200	0.6
		99	307	0.94
5	$14 * 10^{-3}$	95	214	0.66
		99	329	1
10	$11 * 10^{-3}$	95	272	0.8
		99	418	1.3
15	$8 * 10^{-3}$	95	375	1.15
		99	575	1.77
20	$3 * 10^{-3}$	95	998	3
		99	1533	4.7
25	$6.3 * 10^{-4}$	95	4754	14.67
		99	7320	22.59
30	10^{-4}	95	29950	**92.43**
		99	46000	**141.97**

From Table 1 it can be noticed that the performances of the 4G-compliant system are poor regardless the channel modelling in ITU recommendation since no coding is performed at the transmitter and/or receiver.

By computing the ENBs necessary to reach BER thresholds there can be said that in order to achieve a good performance (BER $\approx 10^{-4}$) for a confidence level of 95% the length of the data transmitted by the users is satisfactory but imposing a threshold for SCL of 99% the users must transmit data with larger length increasing thus the processing complexity (for AWGN environment).

For all other cases (Indoor & Pedestrian channel modelling) regardless the transmitted number of bits reasonable BER values cannot be reached (in 95% and 99% of the time of the studied cases).

B. Convolutional Technique

Considering that convolutional technique was involved in data protection and recovery, Fig. 3 shows the performances of the system for this case.

The performances are significantly improved when convolutional technique is involved in data recovery. The curves for BER decrease fast as SNRs increase even for Pedestrian channel where communication parameters change rapidly. As expected the best results are achieved for AWGN channel. In order to observe the behavior of the system in the area where BER $< 10^{-6}$ data with much higher rate should be transmitted. Considering as thresholds the values achieved for BER at every SNR, there can be computed the ENB necessary to reach those thresholds (Table 2).

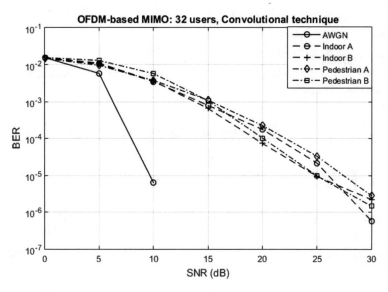

Fig. 3. BER vs SNR for OFDM-based MIMO system: 32 users for ITU recommendation (convolutional technique)

Table 2. ENBs for convolutional case in Indoor/Pedestrian channel.

SNRs (dB)	BER values	SCL (%)	ENB	Percentage (%)
0	$15 * 10^{-3}$	95	200	0.6
		99	307	0.94
5	$12.9 * 10^{-3}$	95	232	0.71
		99	356	1.09
10	$4 * 10^{-3}$	95	749	2.31
		99	1150	3.54
15	10^{-3}	95	2995	9.24
		99	4600	14.2
20	$1.7 * 10^{-4}$	95	17617	54.37
		99	27058	8.51
25	$2 * 10^{-5}$	95	149750	–
		99	230000	–
30	$2 * 10^{-6}$	95	1497500	–
		99	2300000	–

From Table 2 it can be noticed that the data length of 32400 bits is enough to achieve good BER values (approx. 10^{-4}) even in drastic conditions (Indoor/Pedestrian environment). Therefore, the system can be designed in a more efficient manner when convolutional technique is used by reducing the number of transmitted bits (only 10% are actually used to reach such BERs) or by performing analysis for other 5G proposed coding/decoding techniques. Very low BER thresholds can be achieved in 95% and

99% of the cases using a very high data rate, condition which cannot always be fulfilled for indoor or pedestrian environment.

C. *LDPC Technique*
Figure 4 contains the variations for BER versus SNR when LDPC technique is used on the communication path.

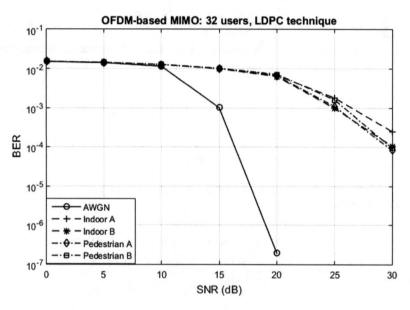

Fig. 4. BER vs SNR for OFDM-based MIMO system: 32 users for ITU recommendation (LDPC technique)

Using LDPC technique the overall performances of the system are improved compared to performances in Fig. 2 when data (de)coding is not available. There can be achieved coding gains up to 10 dB for critical cases (Indoor or Pedestrian models) but the increase of the number of users lead to limitation of performances (a disadvantage in CDMA systems).

In the ideal case of AWGN channels BERs decrease very fast as SNRs increase (especially in case where the power of the signal is similar to the power of the noise).

In AWGN channel LDPC technique leads to efficient BERs up to 10^{-7} for SNRs = 20 dB. Compared to convolutional case the performance achieved is a good one since convolutional technique implies transmission of data with much higher rates in similar case of AWGN channel.

Considering the obtained values for BERs and for the same SCLs as in previous cases there can be computed the ENBs necessary to achieve certain BER thresholds. The values are given in Tables 3 and 4.

In Table 3 there can be observed that good BERs ($1.92 * 10^{-7}$) can be achieved for high SNRs when LDPC is used in AWGN channels but ENBs computation will lead to a very large value which is not a viable solution for other communication environment

Table 3. ENBs for LDPC technique in AWGN channel.

SNRs (dB)	BER values	SCL (%)	ENB	Percentage (%)
0	$15 * 10^{-3}$	95	200	0.6
		99	307	0.94
5	$14 * 10^{-3}$	95	214	0.66
		99	329	1
10	$12 * 10^{-3}$	95	250	0.7
		99	383	1.1
15	10^{-3}	95	2995	9.2
		99	4600	14.2
20	$1.92 * 10^{-7}$	95	–	–
		99	–	–

Table 4. ENBs for LDPC technique in Indoor/Pedestrian channels.

SNRs (dB)	BER values	SCL (%)	ENB	Percentage (%)
0	$15 * 10^{-3}$	95	200	0.6
		99	307	0.94
5	$14 * 10^{-3}$	95	214	0.66
		99	329	1
10	10^{-2}	95	230	0.7
		99	460	1.4
15	$7 * 10^{-3}$	95	428	1.3
		99	657	2
20	$1,5 * 10^{-3}$	95	1997	6.1
		99	3067	9.4
25	10^{-3}	95	2995	9.2
		99	4600	14.1
30	$9.54 * 10^{-5}$	95	31394	**96.9**
		99	48218	**148.8**

(Indoor or Pedestrian). For SNRs < 20 dB a percentage of less than 10% of the transmitted bits is required to obtain lower BERs (10^{-2} or 10^{-3}).

From Table 4 it can be noticed that BERs $\approx 10^{-5}$ can be achieved in Indoor/Pedestrian channels but the number of necessary bits exceed the number of bits used in these simulations (with over 48%).

4   Conclusions and Future Work

This paper aims to present a method to estimate the necessary number of bits that should be transmitted in a multiuser OFDM-based MIMO system when different propagation channels according to ITU-R M.1225 models are taken into consideration for two imposed SCL values. Another goal of the analysis is to achieve low BER values for the simulated transmission.

Analyzing three cases (when source data are not coded or are convolutional/LDPC coded) there has been concluded that low BERs can be achieved in AWGN channels in which data rate can be decreased under 32400 bits to obtain BER around 10^{-4}. Still, to get good BER values when LDPC technique is used for data protection, the necessary ENBs reach a very high value (more than 10^6 bits). This cannot be easily achieved in severe communication channels like indoor buildings in which the probability of interference is high.

Therefore, future work implies the analysis of other data source coding/decoding techniques (for ex. polar codes or other type of LDPC codes proposed for 5G communication systems), data transmission with higher rates according to the results achieved in this analysis, etc.

Acknowledgments. This work has been funded by University Politehnica of Bucharest, through the "Excellence Research Grants" Program, UPB – GEX 2017. Identifier: UPB- GEX2017, Ctr. No. 43/2017".

References

1. Mertikopoulos, P., Belmega, E.V.: Learning to be green: robust energy efficiency maximization in dynamic MIMO–OFDM systems. IEEE J. Sel. Areas Commun. **34**(4), 743–757 (2016). ISSN 0733-8716
2. Kakkar, A., Nitesh-Garsha, S., Jain, O., Kritika: Improvisation in BER and PAPR by using hybrid reduction techniques in MIMO-OFDM employing channel estimation techniques. In: IEEE 7th International Advance Computing Conference (IACC), India (2017). ISSN 2473-3571
3. Agarwal, A., Mehta, S.N.: Performance analysis and design of MIMO-OFDM system using concatenated forward error correction codes. J. Cent. South Univ. **24**(6), 1322–1343 (2017). Print ISSN 2095-2899
4. Baek, C.-U., Jung, J.-W.: LDPC coded turbo equalization for MIMO system. J. Commun. **12**(1), 49–54 (2017)
5. Marcu, I., Voicu, C., Halunga, S., Preda, R.: LDPC encoding performances for fading suppression in MIMO-CDMA wireless networks. In: 2nd EAI International Conference on Future Access Enablers of Ubiquitous and Intelligent Infrastructures (FABULOUS 2016), Belgrade, Serbia (2016)
6. Marcu, I., Voicu, C., Halunga, S.: LDPC performances in multi-carrier systems. In: IEEE 11th International Conference on Communications (COMM 2016), Bucharest, Romania, pp. 209–212 (2016). ISBN 978-1-4673-8197-0
7. Mitić, D., Lebl, A., Markov, Z.: Calculating the required number of bits in the function of confidence level and error probability estimation. Serb. J. Electr. Eng. **9**(3), 361–375 (2012)
8. Zhao, Y., Qi, L., Dou, Z., Zhou, R.: MIMO waveform design using Spectrally Modulated Spectrally Encoded (SMSE) framework. In: 5th International Conference on Computer Science and Network Technology (ICCSNT), China (2016). e-ISBN 978-1-5090-2129-1
9. Craciunescu, R., Manea, O., Halunga, S., Fratu, O., Vizireanu, D.N.: Considerations on CDMA–OFDM system performances in different channel environments for different modulation and coding scenarios. Wirel. Pers. Commun. **2014**(78), 1667–1682 (2014)

10. IEEE Computer Society, IEEE Microwave Theory and Techniques Society, IEEE Standard for Air Interface for Broadband Wireless Access Systems—IEEE Std 802.16™-2012. Revision of IEEE Std 802.16 (2009)
11. Cheng, Y., Li, P., Haardt, M.: FBMC/OQAM for the asynchronous multi-user MIMO uplink. In: 18th International ITG Workshop on Smart Antennas (WSA), Germany (2014). ISBN 978-3-8007-3584-6
12. Nissel, R., Blumenstein, J., Rupp, M.: Block frequency spreading: a method for low-complexity MIMO in FBMC-OQAM. In: IEEE SPAWC 2017, Japan (2017). ISBN 978-1-5090-3009-5
13. Sur, S.N., Bera, R., Maji, B.: Feedback equalizer for vehicular channel. Int. J. Smart Sens. Intell. Syst. **10**(1), 50–68 (2017)
14. Hamkins, J.: Confidence intervals for error rates observed in coded communications systems. In: Communications Architectures and Research Section, IPN Progress Report, vol. 42-201 (2015)
15. https://www.jitterlabs.com/support/calculators/ber-confidence-level-calculator

Tenable Smart Building Security Flow Architecture Using Open Source Tools

Alexandru Caranica[1(✉)], Alexandru Vulpe[2], and Octavian Fratu[2]

[1] Speech and Dialogue Research Laboratory, University Politehnica of Bucharest,
Bucharest, Romania
alexandru.caranica@speed.pub.ro
[2] Telecommunications Department, University Politehnica of Bucharest,
Bucharest, Romania
alex.vulpe@radio.pub.ro, ofratu@elcom.pub.ro

Abstract. Nowadays it's rare to find newly opened buildings or under construction sites that don't aspire to be "smart" or "intelligent". There are wired systems through the building that offer the basic infrastructure for VoIP systems, TCP/IP communication, IP or CCTV cameras, systems that require low latency and great amounts of bandwidth. Complementary to these systems, we also find a variety of wireless devices, from data collection sensors to access points, environmental panels or light control systems. Buildings today are certainly "smarter" and more power efficient than they were five or ten years ago, and all these new devices, that comprise the Internet of Things (IoT) category, pose new issues to all building managers: securing and enforcing digital policies to all these IoT nodes and devices. This paper focuses on building a "vendor neutral" security architecture, based on open source tools, suitable for a wide range of scenarios and building types: school campuses, small or home offices, small building shops, etc. The proposed system architecture is described, together with a preliminary evaluation of the prototype system.

Keywords: Building security · Unified threat systems
Proxy servers · Intelligent buildings

1 Introduction

The term "intelligent" refers to a building that integrates technology and processes to create a housing facility that is safer, comfortable, more productive and more operationally efficient for its owners, occupants and workers, in terms of power usage, climate control or waste disposal. Recent studies see smart buildings spending increasing every year. One, from IDC Energy Insights, predicted in March 2015 that spending on smart building technology, following a recent period of surprisingly slow growth, could advance at a compounded annual rate of 23% through 2019. The research firm put 2014 spending at $6.3 billion and theorized it could hit $17.4 billion in five years [1].

© Springer International Publishing AG, part of Springer Nature 2018
Á. Rocha et al. (Eds.): WorldCIST'18 2018, AISC 747, pp. 118–127, 2018.
https://doi.org/10.1007/978-3-319-77700-9_12

Part of these new technologies are the so-called IoT ("Internet of Things") devices, that consist of uniquely identifiable endpoints (called "things"), that contain embedded technology to sense, collect, communicate and exchange data locally or with external environments, without human interaction, directly affecting the daily life of its occupants.

There are many applications that reside inside an intelligent building, such as (a) occupancy sensors; (b) intelligent LED lighting controls; (c) climate sensors; (d) access control systems; (e) security cameras; (f) wireless access points; (g) IP Phones; (h) data outlets; etc. Those systems, coupled with employees' work computers, IP phones, storage devices, printers and scanners generate a lot of infrastructure challenges for a building manager or owner. One can mention the following issues: (a) IoT and PoE (Power over Ethernet) requirements. How do we efficiently power all these devices?; (b) increased density demands for data-center space to host on-premise systems; (c) availability, latency and bandwidth requirements for all these IP based systems; and lastly, a more and more demanding requirement: (d) security and IAM (Identity Management) of all these "Things". Moreover, we also have to take into "account" the BYOD (Bring Your Own Device) policy, where more and more workers have increased mobility demands and start using their personal devices every day, for work or leisure.

Nowadays, most of these devices rely on "cloud" availability, or a TCP/IP Internet network to communicate with on-premise systems. But recent data breaches [2] and security compromises of internal networks and IoT devices not only makes it necessary, but mandatory, for building owners and IT managers to implement a security flow architecture and IAM procedures, to better secure their networks and intellectual properties.

In this context, this paper describes the steps taken by our group to build a tenable Unified Threat Management (UTM) system architecture, using open source tools, in order to offer baseline security to a cost-sensitive party, such as a university campus. This system also offers flexibility and great customization options and avoids vendor lock-in, perfect for academic institutions.

The paper is organized as follows. Section 2 presents previous work related to securing buildings, IoT/M2M devices and building UTMs. Section 3 describes the proposed architecture and security flow, with a brief description of the modules integrated in this system. In Sect. 4, a brief validation of the concept and prototype is presented. Preliminary results are also presented in this section. Finally, Sect. 5 is dedicated to the closing remarks and conclusions.

2   Related Work

There are many companies and research groups working towards securing and developing standard protocols, for the end-users to obtain private communication and secure access, to devices and their data, on-premises or in the cloud. In the 90s, after the introduction of almost instantaneous communication over the Internet, there was a real revolution in many engineering domains, with a wide variety of popular networked applications resulted as such.

The concept of the Internet of Things (IoT) was introduced in 1999 [3], after the explosion of the wireless device market and the introduction of the Radio Frequency Identification (RFID) technology, with a multitude of Wireless Sensor Networks (WSN) [4]. The IoT concept aims to connect anything with anyone, anytime, and anywhere. This might pose a lot of security issues, such as data breaches, man-in-the-middle attacks, as stated in the introduction.

As such, many previous studies tackled the protocol and framework area, such as [5], where the authors proposed to use a TCP/IP based encryption for a web-based M2M (machine-to-machine) application, over IPSec VPN tunnels. In [6], the target was to address two key issues. First, how to abstract the technological heterogeneity that derives from the vast amount of heterogeneous devices. Second, to study the views of different end-users, in order to ensure proper application provisioning, business integrity and therefore, maximize the usage of the IoT devices. In [7], the authors proposed a secure storage and communication framework, based on IPv6/6LoWPAN protocols. IPv6/6LoWPAN defines IPsec/ESP (Encapsulating Security Payload), that provides encryption and authentication of transmitted data packets. The author uses cryptographic methods and data formats defined by the ESP protocol, for data processing before saving information to storage.

Related to network protection systems, there are both commercial and open-source solutions, that are worth mentioning. In the area of network protection, highly established UTM solutions from vendors such as Cisco [8], Fortinet [9], CheckPoint [10] and Cyberoam [11] can be mentioned. Their systems score highly in every major antivirus test and provide proven support for big corporate networks. Open-source integrated solutions are also available, and we can mention a few projects here, like pfSense [12] and SME Server [13].

3   System Integration and Architecture

This section presents the proposed architecture to secure a small "smart" building, that accommodates a multitude of devices, such as guest mobile phones and laptops, infrastructure computers and servers, with additional Heating, Ventilation, and Air Conditioning (HVAC) control panels and sensors required by the building complementary systems. Later subsections present a brief description of the central part of the filtering system ("router and filtering server") together with the main principal integrated modules, to obtain a central UTM system.

3.1   Proposed Security Architecture

From an architectural point of view, the proposed security flow comprises of logically separating network segments, based on device types, amount of traffic generated in/out by these devices, and the security risk associated with a possible data breach. A quick overview of the architecture can be consulted in Fig. 1.

For a basic small university campus building, where students and staff reside, we proposed and implemented the architecture illustrated in Fig. 1. Each IT

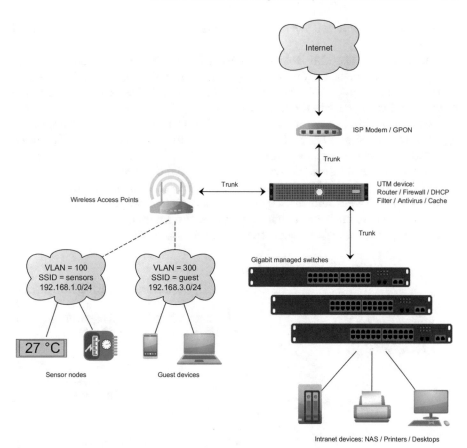

Fig. 1. Proposed building security architecture

department should analyze, before implementing an architecture, the types of IoT devices that will be also accommodated in the near future, the existing devices and current risks associated with these systems.

The central part consists of a Linux based server (UTM device), that will serve as the main router, with the following roles:

– IPTables based routing and firewall [14]
– Squid Proxy Server [15]
– SquidGuard domain filtering [16]
– IPS/IDS intrusion detection using SNORT [17]
– C-Icap proxy extension for HTTPs scanning [18]
– ClamAV Antivirus module [19]

Each network segment is isolated via VLAN (Virtual LAN) functionality. Multiple VLANs can we spawned, in the near future, if necessary.

As an example, the sensors can be isolated in VLAN 100, with no communication to devices in others VLANs or network segments. Policies can be applied to communicate only with necessary on-premise servers, or cloud services, if required.

Guest devices are also isolated, in VLAN 300, with no communication to other network segments. Traffic shaping is applied, to limit bandwidth consumed by these guest devices. Antivirus and domain filtering to suspecting domains are enabled and enforced, via the UTM server.

Intranet devices reside on a separate VLAN also, connected to Gigabit managed switches, using 802.1X, EAP & RADIUS authentication [20]. This way, unauthorized devices cannot just plug a network cable into an RJ45 socket and gain access to the internal protected network of the building.

For logging, a separate log-collector can be installed, that aggregates traffic graphs via RRD tools and saves syslogs on persistent storage.

Figure 2 shows the final traffic flow inside the Linux server (UTM device), and how a packet travels through the system for inspection.

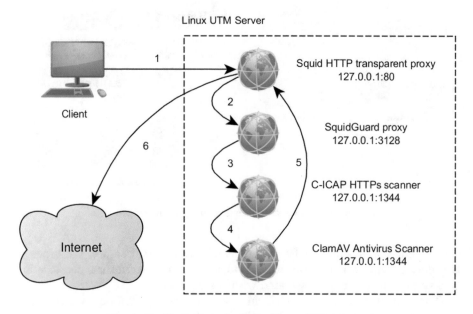

Fig. 2. Traffic flow through the Linux UTM Server

Sections to follow offer a quick description of each open-source module integrated into the final UTM Server, the central filtering part of our architecture.

3.2   Squid Proxy Module

Squid is a HTTP proxy server and web cache daemon [15]. It has a wide variety of uses, from speeding up a web server by caching repeated requests to caching web, DNS and other computer network lookups for a group of devices sharing a gateway or network resource. It also aids security by enabling filtering of common web traffic. Although primarily used for HTTP and FTP, Squid has limited support for encrypted traffic, such as TLS, SSL, HTTPS. As a result, C-ICAP daemon [18] is integrated into the architecture, to aid HTTPs traffic inspection via self-signed Trusted Root Certificates, that need to be installed on devices residing on the internal network.

3.3 SquidGuard Module

SquidGuard [16] is a URL redirector software, which can be used for content control of websites that users can access in an intranet. It is written as a plug-in for Squid and uses lists to define sites for which access is redirected. Lists are offered as blacklists for malware, or with general categories to filter by site categories, so one can block a category such as "general time wasters", which contains sites such as Facebook, YouTube, games, etc. There are free non-commercial lists available, listed on the project page [16].

3.4 ClamAV Antivirus

ClamAV [19] is an open source antivirus engine designed for detecting Trojans, viruses, malware and other malicious threats. It is the de-facto standard for mail gateway scanning on all major open-source mail systems. It provides a high performance multi-threaded scanning daemon, command line utilities for on demand file scanning, and an intelligent tool for automatic signature updates. The core ClamAV library provides numerous file format detection mechanisms, file unpacking support, archive support, and multiple signature languages for detecting threats. Open-source baseline signatures can be used, but one can enhance them using third-party definitions, such as Sanesecurity signatures [21].

3.5 IPS/IDS Detection Using Snort

This module is separate from the traffic flow architecture, as it operates in parallel and does not provide feedback to the user, but to the network admin. Snort [17] is an open source network intrusion prevention system, capable of performing real-time traffic analysis and packet logging on IP networks. It can perform protocol analysis, content searching/matching, and can be used to detect a variety of attacks and probes, such as buffer overflows, stealth port scans, CGI attacks, SMB probes, OS fingerprinting attempts, and much more. Lately, it provides an OpenAppID initiative, that allows for granular control over application access and usage. A library of over 1,000 OpenAppID detectors are already available, at no charge, contributed by Sourcefire and Cisco [22].

4 Evaluation

This section covers the initial experimental validation of our prototype integrated filtering system, using open-source tools. We tested four aspects, related to (a) the systems ability to block malware; (b) filter blocked domains; (c) cache http traffic; (d) inspect encrypted web traffic.

4.1 Preliminary Results

For (a), we used the standard EICAR malware test [23]. EICAR test is an application developed by the European Institute for Computer Antivirus Research (EICAR) and Computer Antivirus Research Organization (CARO), to test the response of antivirus systems, instead of using real malware, which could cause real damage. We visited all eight malware infected URLs provided by the test group. Default behavior should be a blocked page for every hit of the malware sample test. The system passed this test, as it can be seen in Fig. 3.

For the second test, and to test all four aspects, we ran the CheckPoint UTM filter test, available at [24]. This tests an UTM capability to block ransomware payloads, zero day vulnerability malware, use of proxy software to bypass the filters and sensitive data leakage, such as bank account information.

Table 1 summarizes the CheckPoint UTM tests, available at [24]. *Ransomware* is a type of malware that encrypts users' files and require ransom for their decryption. *Identity Theft/Phishing* attack captures personal information by fake websites that appears to be legitimate. *Zero Day attack* uses the element of surprise and exploits a hole in the software that is unknown to the vendor. *Bots* perform malicious attacks that let attackers take complete control over an infected computer. *Browser attack* injects malicious script into web sites to steal cookies from victims for the purpose of impersonating the victims. *Anonymous surfing* allows users to hide their online activity. It can open backdoors into an organization's network. *Data leakage* is the transfer of classified or sensitive information outside an organization's network by theft or accidental exposure.

Table 1. Checkpoint detailed tests results

Test type	Passed
Ransomware	Yes
Identity theft/phishing	Yes
Zero day vulnerability	Yes
Bot infection	No
Browser attack	No
Anonymizer usage	Yes
Sensitive data leakage	No

Table 2 summarizes all validation tests. This paper focused on integrating an open-source anti-virus into the UTM server, then testing the correct triggering of it by our devices. Therefore, in Table 2 we also summarized independent tests run by Sannescurity [21] on a set of in-the-wild malware, compared to the baseline

Table 2. Integrated system initial validation tests

Test	Score
EICAR detection test	100%
CheckPoint automated test	57%
ClamAV default virus signature test	13.82%
Sannesecurity virus signature test	97%
Categories block test	90%

ClamAV default signatures [21]. Figures 3 and 4 shows the correct triggering of the antivirus, the scanning of encrypted traffic and blocking of unwanted domains (ex: warez domains).

This section offered an initial test of our systems ability to filter unwanted domains list, block virus payloads or sample malware, to validate the successful integration of all modules.

SquidClamav 6.16 : Virus detected!

The requested URL https://secure.eicar.org/eicar.com contains a virus
Virus name: Eicar-Test-Signature

This file cannot be downloaded.

Origin: 192.168.1.100 / -

Powered by SquidClamav 6.16

Fig. 3. Server's response to a virus detection over the visited encrypted web-page

Request denied by proxy: 403 Forbidden

Reason:

Client address: 192.168.1.100
Client name: 192.168.1.100
Client group: default
Target group: blk_BL_warez
URL: http://thepiratebay.org/

Fig. 4. Server blocking an unwanted website

4.2 Discussions

Although there are plenty of commercial systems available from big vendors, in the UTM market, one of the advantages of this system is that it uses completely open-source components, ran on "off-the-shelf" hardware. This offers no vendor lock-in scenarios, lower hardware upfront cost with no reoccurring licenses or upgrades.

Another advantage is the ability to customize the system to every level, offering a higher grade of integration with existing systems or upgrades, if needed. For a University Campus Building, that means integration with on-site LDAP (Lightweight Directory Access Protocol) servers for user authentication, detection of unwanted P2P (peer-to-peer) traffic, in order to send students warning notifications and so on.

As disadvantages, such a system requires highly skilled IT staff to implement and administer the system. This might not be such an issue in a university campus, for example, but might pose an issue for other scenarios. Moreover, detection rate of the antivirus component is lower than other established commercial antivirus systems.

5 Conclusion

This paper overviewed the steps and preliminary evaluations taken to build a fully functional, UTM-like, security flow architecture, that can be used to enhance security and compliance in an intelligent building.

Future work will focus on testing the final implementation of the system on standardized malware sets and focus more on OpenAppId detection. This will allow us to identify device types, for example, and apply a certain custom policy, such as QoS (Quality of Service) shaping to these devices only.

Acknowledgements. This work was supported by a grant of the Ministry of Innovation and Research, UEFISCDI, project number 5 Sol/2017 within PNCDI III.

References

1. Firstpost: Smart building tech spending forecast to grow to \$17.4 bn in 2019: IDC. Insight Report (2015). http://www.firstpost.com/business/smart-building-tech-spending-forecast-grow-17-4-bn-2019-idc-2180865.html
2. ZDNet: 2017's biggest hacks, leaks, and data breaches. Online article (2017). http://www.zdnet.com/pictures/biggest-hacks-leaks-and-data-breaches-2017/
3. Gubbi, J., Buyya, R., Marusic, S., Palaniswami, M.: Internet of Things (IoT): a vision, architectural elements, and future directions. Future Gener. Comput. Syst. **29**(7), 1645–1660 (2013)
4. Sfar, A.R., Natalizio, E., Challal, Y., Chtourou, Z.: A roadmap for security challenges in the Internet of Things. In: Digital Communications and Networks (2017)
5. Weber, R.: Accountability in the Internet of things. Comput. Law Secur. Rev. **27**, 133–138 (2011)

6. Raza, S. Duquennoy, S., Chung, T., Yazar, D., Voigt, T., Roedig, U.: Securing Communication in 6LoWPAN with compressed IPsec. In: Proceedings of 7th IEEE International Conference on Distributed Computing in Sensor Systems (DCOSS 2011), Barcelona, Spain, 7–29 June 2011
7. Bagci, I.E., Raza, S., Chung, T., Roedig, U., Voigt, T.: Combined secure storage and communication for the Internet of Things. In: IEEE International Conference on Sensing, Communications and Networking (SECON), New Orleans, LA, pp. 523–531 (2013)
8. Cisco Corporate website (2017). https://www.cisco.com/
9. Fortinet Corporate website (2017). https://www.fortinet.com/
10. CheckPoint Corporate website (2017). https://www.checkpoint.com/
11. Cyberoam Corporate website (2017). https://www.cyberoam.com/
12. pfSense open-source firewall (2017). https://www.pfsense.org/
13. SME Server Linux distribution (2017). https://wiki.contribs.org/Main_Page
14. Netfilter, home of IPTables. http://www.netfilter.org/. Accessed 2017
15. Squid Open Source Proxy Server. http://www.squid-cache.org/. Accessed 2017
16. Squidguard Open Source Filter. http://www.squidguard.org/. Accessed 2017
17. Snord Open Source IPS/IDS. https://www.snort.org/. Accessed 2017
18. C-ICAP server. http://c-icap.sourceforge.net/. Accessed 2017
19. ClamAV Open Source Antivirus. https://www.clamav.net/. Accessed 2017
20. IEEE 802.1X standard. https://en.wikipedia.org/wiki/IEEE_802.1X. Accessed 2017
21. Sanesecurity ClamAV addon signatures. http://sanesecurity.com/. Accessed 2017
22. Cisco Blog: Cisco announces OpenAppID the next open source game changer in cybersecurity. https://blogs.cisco.com/security/cisco-announces-openappid-the-next-open-source-game-changer-in-cybersecurity. Accessed 2017
23. European Expert Group for IT-Security: EICAR malware test download. http://www.eicar.org/86-0-Intended-use.html. Accessed 2017
24. Check Point Software CheckMe test: Accessed 2017

Face Detection in Internet of Things Using Blackfin Microcomputers Family

Sorin Zoican, Marius Vochin[✉], Roxana Zoican, and Dan Galațchi

University Politehnica of Bucharest, Bucharest, Romania
{sorin,marius.vochin,roxana,dg}@elcom.pub.ro

Abstract. This paper describes a face detection system based on the Blackfin microcomputer architecture that may be used in an Internet of Things (IoT) context. The face detection algorithm is based on skin detection and scanning binary images to determine the face area. Further image processing may determine the eyes and mouth in order to extract main face characteristics. The face detection algorithm may be used in context of IoT to determine the searching area for eyes and mouth (e.g. for face recognition and emotion detection). The face detection algorithm is implemented using the Visual DSP++ integrated development environment and face detection is achieved in real time.

Keywords: Face detection · Internet of Things · Real time · DSP processor

1 Introduction

In Internet of Things, smart devices can be connected one to another using high technology that could detect the human faces in order to permit person authentication or person displaying in other security use case scenarios [1, 2].

One typical face detection scenario in respect to the context of IoT is illustrated in Fig. 1.

This paper describes a face detection algorithm that can be implemented in real time (with processing delays of seconds or even less). Taking into consideration the limited resources in digital signal processors (especially memory), new approaches should be investigated to detect faces in a picture or a video frame. Many methods to detect a face in an image exists in current literature, the Viola – Jones algorithm being one of the most popular of them [3, 4]. One major disadvantage of it is its highly time-consuming characteristic, therefore being implemented usually on desktop computers. Other simpler methods, but more efficient, are based on skin detection followed by image segmentation to find the face region [5]. The collected data is preprocessed in the device (e.g. prefiltered, edge detection, segmentation) then further processing may be done on a server with a higher computing capacity. Main problems to tackle are: the expression of the person, and the expression correlations (e.g. eyes and mouth correlations).

Generally, mobile devices used in IoT may have limited memory amount and use fixed point processors, therefore it is necessary to use specific hardware techniques

© Springer International Publishing AG, part of Springer Nature 2018
Á. Rocha et al. (Eds.): WorldCIST'18 2018, AISC 747, pp. 128–135, 2018.
https://doi.org/10.1007/978-3-319-77700-9_13

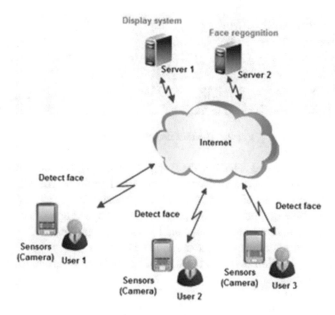

Fig. 1. Face detection scenario.

(hardware loops, multifunction instructions, multi-core processors) and simpler but more effective face detection algorithms.

This paper focus is on the face detection algorithms possible to be implemented on a wearable device without sensors to be attached to the body. The only input devices to be used in face detection will be the camera existing on the portable device so the face recognition will use face changes (specifically eyes and mouth). Based on the best algorithms described in literature, we derive a new and simplified algorithm to detect the face in real time, without significant performance decrease.

The proposed algorithm has been implemented using Blackfin processor BF533 from Analog Devices and Visual DSP++ Integrated Development Environment.

The next section of this paper presents the face detection algorithm. Section 3 presents main results of this work together with some implementation details, while complexity and performance evaluation details are shown in Sect. 4. Section 5 concludes this work.

2 Face Detection Algorithm

The face detection algorithm takes several steps as it can be observed in Fig. 2.

The main tasks in the above algorithm are [6]: red (R), green (G) and blue (B) image components separation, verifying a set of rules to determine if a pixel is skin pixel or a non-skin pixel, and determining the face area.

Image acquisition
Image acquisition
Image component separation (R, G, B)
Verify the rules for skin pixels
Create binary image
Scan binary image to crop the face area
Feature extraction (eyes, mouth)

Fig. 2. The face detection algorithm

The components image separation is straightforward [7]: the video codec acquires the image and stores it in three successive memory locations. The image separation processing will access the locations $3k$ as red component, $(3k + 1)$ as green components and $(3k + 2)$ as blue component. This process may be speed up by using the circular addressing facility in almost all digital signal processors families.

The next step is to compute image normalized components.

The color equalization involves the normalization in RGB space (R, G, B are the values of the pixel) as:

$$r = \frac{R}{R+G+B}, \ g = \frac{G}{R+G+B}, \ b = \frac{B}{R+G+B}. \tag{1}$$

This equalization ignores the ambient and is invariant to changes of face orientation relatively to the light source. The third step calculates two factors f_{upper} and f_{lower} that will be involved in rules which determine skin pixel or non-skin pixel.

$$\begin{aligned} f_{upper}(r) &= -1,3767r^2 + 1,0743r + 0,1452 \\ f_{lower}(r) &= -0,776r^2 + 0,560r + 0.1452 \end{aligned} \tag{2}$$

The rules to determine skin pixels are:

$$R1: g > f_{lower}(r) \text{ and } g < f_{upper}(r) \tag{3}$$

$$R2: W = (r - 0,33)^2 + (g - 0,33)^2 \geq 0.0004. \tag{4}$$

$$R3: R > G > B \tag{5}$$

$$R4: R - G >= 45 \tag{6}$$

The skin or non-skin pixel is determined accordingly with the following rule:

$$\begin{aligned} &if (R1 \ and \ R2 \ and \ R3 \ and \ R4) \\ &\quad then \ pixel \ is \ skin \ pixel \ (S = 1) \\ &else \\ &\quad pixel \ is \ non - skin \ pixel \ (S = 0) \end{aligned} \tag{7}$$

The face detection algorithm is illustrated in Fig. 3:

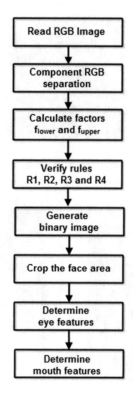

Fig. 3. The face detection algorithm.

The last two blocks are not in the scope of this work, but they could be implemented separately on a remote server (due to the complexity of such algorithms). It should be mentioned that the face detection algorithm will reduce the computational effort of eyes and mouth features determination. These algorithms are based on finding of eyes circles and ellipses and mouth ellipse axis which involve Hough transformation for circle and ellipses [8]. Future work will investigate how the computational effort to implement such algorithms can be reduced.

3 Main Results

The above presented emotion detection algorithm was validated in MATLAB environment and then implemented using the EZ-Kit Lite BF 533 [9, 10] evaluation board. The code was written in C programming language using Visual DSP++ 5.1 as Integrated Development Environment [10]. The execution time was measured using the macros defined in Blackfin Visual DSP++ library: CYCLES_INIT - initializes the cycles counter, CYCLES_START – begins measuring cycles, and CYCLES_STOP - stops measuring cycles. The number of cycles was converted in time units accordingly with microcontroller frequency. Figures 4, 5 and 6 illustrate the main results.

Fig. 4. The image processing: original image (left), processed image after rules R1, R2 (middle) and processed image after R3, R4 (right).

Fig. 5. The binary image and face cropping.

In the above figures, one can observe how relations (2)–(6) are applied to detect skin, how the binary image is obtained by applying relation (7), and how the face cropping is done. The algorithm for cropping the face involves the horizontal and vertical scanning of the binary image in order to determine large areas with white pixels. The larger area is the face area and its coordinates will be retained.

The results show that the algorithm detects face in a simple manner.

4   Complexity and Performance Evaluation

The performance criterion chosen was the execution time (that is the time necessary to detect the face) measured as it was indicated in previous section. The algorithm implementation can be optimized using [11]:

- compiler optimization (enable hardware loops control, loop unrolling, inter procedural analysis)
- using the fractional arithmetic `fract16` and the fractional functions `mult_fr1x16` and `divq` instead the default C arithmetic functions.

A comparison between these optimization methods are shown in Fig. 6.

Fig. 6. The differences between float and fractional mode

One can observe slight differences between the two implementation modes, but it should be mentioned that the execution time for the implementation that is not using the fractional approach is very high (tens of seconds) and the implementation is not in real time. However, these differences may affect the face recognition process.

The face detection has been tested in presence of salt-and-piper noise (generated using [11]). The main results are illustrated in Figs. 7 and 8 and one can observe that the algorithm works for relatively small noise levels, but some improvements are necessarily (e.g. a initial de-noising of the image).

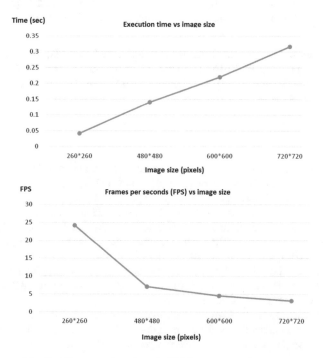

Fig. 7. The execution time and FPS rate versus image size

Fig. 8. The influence of noise present in image (left to right: noise 2%, 5% and 10%).

5   Conclusion

The paper presents an algorithm for detection of faces that can be used in context of IoT in systems such as remote displaying system and person recognition. This algorithm uses the frontal image of a person and involves an image pre-processing followed by skin detection and face cropping. Further processing, on a server, can be done in order to extract eyes and mouth features to be used in person recognition. The obtained results indicate that such algorithm can detect faces in real time (e.g. in seconds or less). Future works will investigate the communication between devices to implement scenarios described in Fig. 1 and will investigate more complex processing techniques to eliminate constrains imposed to frontal images (e.g. moustache, beard, glasses).

Acknowledgments. This work has been partially funded by UEFISCDI Romania under Bridge Grant project grant no. 60BG/2016 "Intelligent communications system based on integrated infrastructure, with dynamic display and alerting - SICIAD. The authors would like to thank to Florin Rosulescu for his support to this work.

References

1. Lin, S.H., Kung, S.Y., Lin, L.J.: Face recognition/detection by probabilistic decision-based neural network. IEEE Trans. Neural Netw. **8**(1), 114–132 (1997)
2. Chiang, C.-C., Tai, W.-K., Yang, M.-T., Huang, Y.-T., Huang, C.-J.: A novel method for detecting lips, eyes and faces in real time. Real-Time Imaging **9**, 277–287 (2003)
3. Viola, P., Jones, M.J.: Real-time face detection. Int. J. Comput. Vis. **57**(2), 137–154 (2004)
4. Wang, Y.-Q.: An analysis of the viola-jones face detection algorithm. Image Process. Line **4**, 129–148 (2014). https://doi.org/10.5201/ipol.2014.104
5. Kim, M.H., Joo, Y.H., Park, J.B.: Emotion detection algorithm using frontal face image. In: 2015 International Conference on Control Automation and Systems (ICCAS 2005), 2–5 June 2005, Kintex, Gyeong Gi, Korea, pp. 2373–2378 (2005)
6. Soriano, M., Huovinen, S., Martinkauppi, B., Laaksonen, M.: Using the skin locus to cope with changing illumination conditions in color-based face tracking. In: IEEE Nordic Signal Processing Symposium, Kolmarden, Suedia, pp. 383–386 (2000)

7. Gan, W.-S., Kuo, S.M.: Embedded Signal Processing with Micro Signal Architecture. Wiley-IEEE Press, Hoboken (2007)
8. Generalized Hough Transform. http://www.cs.cmu.edu/~16385/spring15/lectures/Lecture6.pdf
9. Analog Devices, Blackfin BF533 EZ-Kit Lite evaluation board. http://www.analog.com/en/design-center/evaluation-hardware-and-software/evaluation-boards-kits/BF533-EZLITE.html#eb-overview
10. ADSP-BF537 Blackfin® Processor Hardware Reference. http://www.analog.com/media/en/dsp-documentation/processor-manuals/ADSP-BF537_hwr_rev3.4.pdf
11. http://www.inoisify.com
12. VisualDSP++ 5.0 User's Guide, Revision 3.0, August 2007

Intelligent Low-Power Displaying and Alerting Infrastructure for Smart Buildings

Laurentiu Boicescu[1(✉)], Marius Vochin[1(✉)], Alexandru Vulpe[1], and George Suciu[2]

[1] University Politehnica of Bucharest - UPB Bucharest, Bucharest, Romania
{laurentiu.boicescu,marius.vochin}@elcom.pub.ro,
alex.vulpe@radio.pub.ro
[2] BEIA Consult International, Bucharest, Romania
george@beia.ro

Abstract. The paper presents the architecture of an intelligent displaying and alerting system, based on a scalable integrated communication infrastructure. The system is envisioned to offer dynamic display capabilities using the wireless ePaper technology, as well as to enable indoor location-based services such as visitor guidance and alerting through iBeacon-compatible mobile smart devices. The system will include a secure web management console, along with the software interfaces and procedures necessary for collecting and automatically displaying any relevant information and notifications. As the system is designed primarily for educational and research institutions, remote authentication will be allowed through eduroam technology. Tests regarding the performance of the used wireless ePaper Displays are performed, specifically regarding their response times, and recommendations are made based on the results.

Keywords: ePaper · iBeacon · Alerting system · Indoor positioning
Low power display

1 Introduction

The emergence and continuous development of new communication and display technologies with low power requirements has opened new opportunities in creating a sustainable infrastructure for digital room signage for conferences, meeting rooms, or even modern office buildings. The new focus on electronic, digital technologies, rather than the use of paper, simplifies the management of displayed information. For example, the effort of manually changing thousands, or tens of thousands of paper-based labels can be transferred to a scalable digital display system, with tens of thousands of screens connected to it; as a result, countless man-hours of work can be saved, while also reducing the time required to update all paper labels.

The new low-power display solutions rely on the electronic paper (ePaper) technology, while the low-power wireless communication is based on the Bluetooth Low Energy (BLE) technology, along others like it. Recent advances in the field of monitoring networks of sensors have been marked by the emergence and integration of new technologies, like the above, in the field of Internet of Things (IoT) interactions.

© Springer International Publishing AG, part of Springer Nature 2018
Á. Rocha et al. (Eds.): WorldCIST'18 2018, AISC 747, pp. 136–145, 2018.
https://doi.org/10.1007/978-3-319-77700-9_14

The paper proposes an intelligent displaying and alerting system (SICIAD) that relies on the wireless ePaper and iBeacon technologies [1] in order to provide a framework for displaying both static and dynamic information, as well as to ease the indoor orientation of guests. At the same time, the system aims to provide real-time alerting facilities for emergency situations (including the case of having to direct the evacuation of entire office buildings).

The system's primary objective is to build on the existing technology and simplify its use, in order to provide the custom automatic update of all displayed information, in a secure fashion. SICIAD is primarily designed as a display and notification framework for public institutions like universities and government buildings. However, its applications may include public transport companies, airports, expositions and commercial centres, museums or even indoor and outdoor amusement parks. Any organization that relies on conventional paper-based (or even digital) signage can, in fact, benefit from the use of a centrally-managed low power display system.

The paper's content is organised as follows: the Sect. 2 provides some background information regarding the wireless ePaper technology, while Sect. 3 presents the architecture, design considerations, and several implementation details of SICIAD. Section 4 presents the preliminary evaluation results for the system, while the final section draws the conclusions.

2   Wireless ePaper Technology

The electronic paper (mostly referred to as ePaper or sometimes e-Ink) can be seen as a display medium with memory, one that can be re-written multiple times in order to display various content. Therefore, the concept of ePaper can be defined as a display technology that simulates the appearance of text written on a traditional physical paper [2].

The technology relies on ambient light reflection instead of a backlight, as well as a screen that requires a significant amount of energy only during the update phase, along with a near-zero power consumption [3] in the idle phase. Such digital displays offer good visibility of information in all light conditions, with the benefit of a low power consumption.

Today's applications of ePaper displays include electronic shelf labels, digital signage, time schedules for public transportation, billboards, portable signs, although some of their first wide-scale commercial applications were electronic newspapers and e-book readers.

However, because these technologies are still relatively new, their use requires extensive computer programming skills to access and manage the displayed information. As we stand, the current level of technology relies on the user either micro-managing individual displays, or writing complex scripts for the dissemination of multiple information flows and dynamic update of these displays.

Currently, there are several companies that offer tools for digital room signage. All these solutions are generally comprised of three components:

- the display panels, based of ePaper technology, or other technologies like LCD (Liquid Crystal Display) or LED (Light-Emitting Diode) and their variants;
- the wireless communication infrastructure, which can be based on BLE, ZigBee, GSM (Global System for Mobile communications), WiFi, or even the newer addition to the field, LoRa;
- the content management system, for the signage infrastructure.

SICIAD is envisioned to take the place of the third layer in the above system, as a manager for the ePaper labels and the information being displayed on them. The first two layers are based on the wireless ePaper technology provided by Lancom [4].

The wireless ePaper solution functions on 2.5 GHz radio channels, in the same frequency spectrum as WiFi or Bluetooth, but on much smaller channels. The displays are controlled through special WiFi Access Points, designed specifically to integrate the wireless ePaper communication technology, and all control and data information flows pass through a software ePaper Server [5]. Moreover, since the platform is designed for low-power applications, covering large office spaces will require the use of several access points, configured on different wireless ePaper channels in order to avoid interference.

Lancom's ePaper Server software offers complete access to all management and signage functions through an XML (eXtensible Markup Language) API (Application Programming Interface), including ePaper monitoring and operation, alongside the signage API. The interface permits directly uploading binary image information (encoded in the Base64 format), as well as generating it through the integrated image render and a set of image templates.

At the same time, the manufacturer offers a small number of tools, capable of extracting singular information from external data sources, such as Google Calendar or Office365. Complex interactions, such as the ones described in the current paper, require a more complex management software, described in the following sections.

Complete information regarding the performance of ePaper displays, most notably the time required to display an image, is still unavailable from most manufacturers. Even from the used platform, the manufacturer only provides relevant information related to the lifetime of the display's batteries: around five years, as long as the displays are updated no more than twice every day [4], for the three-colour, red, black and white displays. Other manufacturers indicate display times between 3 and 8 s for black and white displays, and 15 or more seconds for three colour displays of similar sizes [6]. Older research also shows response times in the order of seconds, for ePaper display technology [7]. More information, including results of performance tests on the wireless ePaper devices, will be found in Sect. 4.

One key functionality consists of the fact that the users can remotely update the displayed content in real time. To allow a highly flexible use, the wireless ePaper displays eliminate the need for an external power supply or a physical network connection due to the fact that the devices are battery powered and radio controlled. Furthermore, the data transmission process can be protected by a 128-bit key, allowing secure encryption and authentication standards for eduroam [8, 9].

3   System Architecture

SICIAD is meant to capitalize on recent developments in the field of digital room signage, using Bluetooth Low Energy (BLE) for indoor-positioning services, as well as wireless ePaper devices for remotely displaying information. The project's main objective is to provide an integrated system for the dynamic display of information in educational and research institutes. At the same time, the system aims to provide the automatic display of notifications and alerts, in order to aid in the emergency evacuation of covered spaces, should the necessity ever arise.

The system aims to integrate the Lancom wireless ePaper solution, described in the previous section, in order to dynamically display information and provide notification on certain events. For this, it proposes the development of an integrated management application for the infrastructure and wireless ePaper displays, along with the software interfaces to integrate information from multiple external data sources. The general concept of the target system is described in Fig. 1.

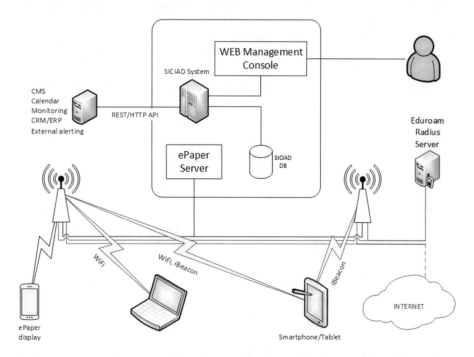

Fig. 1. Concept for the low-power displaying and alerting infrastructure

The framework provided by Lancom includes the wireless ePaper-enabled access points, capable of providing internet access to the communication systems covered by them, as well as of controlling the configured ePaper displays. All ePaper displays are controlled by Lancom's ePaper Server software, which enables user interaction through its XML API.

The use of wireless technology simplifies the deployment of the system, since additional wiring or power sources will not be necessary. However, several facts should be mentioned.

First off, the ePaper displays (including the Lancom ePaper displays proposed for use in the system's architecture) are designed to work at a low level of power consumption. This means a slow refresh rate of the displayed information, when changes occur. For static information, this is not an issue: all the displays can be updated in the off hours, when no one is using them. However, an issue may arise when trying to display urgent dynamic notifications, like an emergency evacuation alert. For such cases, further study is needed regarding the wireless ePaper response time.

The specifications of Lancom ePaper displays indicate a battery life time between 5 and 7 years, if the displayed information is changed four times a day [4]. For more interactive applications (like the first use case), or improving response times for emergency situations, additional power sources may be needed, which may come in the form of solar panels in outdoor deployments, (when battery replacement may become an issue).

Secondly, the Lancom Wi-Fi Access Points integrate iBeacon technology. This offers a simple means of determining whether a mobile smart device is in close range of the access point, but a larger iBeacon network is necessary for determining exact indoor position. As such, further work may relate to the integration of stand-alone iBeacon devices in the SICIAD architecture.

The Server's interfaces, from the project's point of view, are described in Fig. 2.

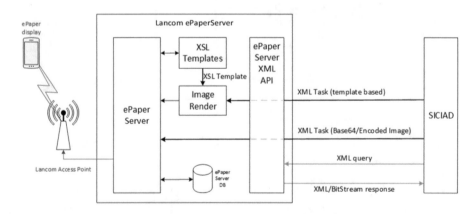

Fig. 2. Lancom ePaper Server XML API

The ePaper Server offers the ability to remotely configure, monitor and control the wireless ePaper devices, offering complete status information regarding all connected access points and displays, alongside information regarding the status of all previously executed tasks. All commands to display information on the ePaper displays are issued in an XML format, either with Base64 encoded images pre-rendered by the issuer (user) to exactly fit on the desired devices, or using an internal rendering engine. The server's own rendering functionality allows images to be generated, based on a set of XSL (eXtensible Stylesheet Language) templates. All information, including generated images, task status changes, monitoring data, is stored in the server's internal database and available through the XML API.

Based of the above-stated objectives for the system, as well as the wireless ePaper system's features, the following three general use cases for the system have been determined:

1. **Dynamic display system** – which updates the ePaper devices based on a set predefined rules, either automatically or at the user's request,
2. **Visitor guidance system** – relying on BLE iBeacon emitters to pin-point the users' locations and provide indoor navigation assistance and location based services, and
3. **Dynamic alerting system** – relying on components from the previous use cases and superseding their displayed/broadcasted information. In case of an alert, the system will store the previous normal running state and, once the alert has passed, restore it.

The BLE-based indoor visitor guidance facility will not be discussed in the current paper, as it is not within its scope.

As previously stated, the system aims at the development of an intelligent system that can dynamically display information and provide notification on certain events. For this, it proposes the development of an integrated management application for the infrastructure and wireless ePaper displays, along with a software interface for connecting to Internet calendar or other cloud services, several access levels and e-mail message programming. The proposed architecture for the SICIAD system, based on the above observations, is described in Fig. 3.

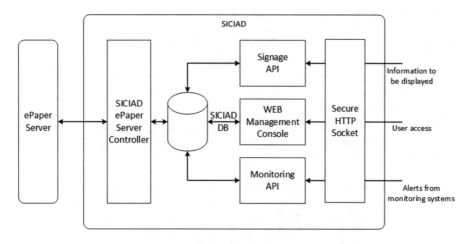

Fig. 3. SICIAD architecture

The task of communicating with the ePaper Server, retrieving all pertinent information through the API, as well as issuing display commands for the wireless ePaper devices will fall to the *ePaper Server Controller*. The software module will function as a background task, continuously checking for any information changes. A specific use-case for the module will be to control the display of alerts in case of emergency situations. For that to correctly function, it must store complete status information regarding the labels. Should an emergency situation appear, all other displayed information will be overwritten (since the displays have a memory of up to 12 pages, one or two can be reserved for alerts, separating the two operation modes). Once the emergency situation ends, the system will revert to its normal operating mode.

The *Signage API* will function as an endpoint for any external data flows of information, while the *Monitoring API* will receive data from monitoring systems or IoT sensors (temperature, CO2, smoke, gas) to identify threats, send alerts through its available means, or even automatically initiate emergency evacuation and facilitate the avoidance of problem areas and to provide guidance towards safe exits (including aid for the hearing impaired). All status information will be stored in SICIAD's internal database.

Finally, the *Web Management Console* will enable the dynamic display of information, either on ePaper devices connected to the infrastructure and without wired power supplies, or on the users' cell phones, using beacons based on the iBeacon technology. All modules, including the management console and the API will be secured in order to prevent unwanted intrusions in the wireless signage system.

Currently, of the above components, most of the ePaper Server Controller, along with a portion of the management console, have been implemented, allowing several performance tests on the ePaper devices to be performed.

4   Preliminary System Evaluation

Several tests have been performed in order to determine the delays that occur when updating the information on the ePaper displays from the wireless ePaper server. To that end, two L-151E access points were used together with one WDG-1 7.4″ ePaper display. The Access Points were registered to a remote ePaper Server installed on a openSUSE Linux machine. Current version of LANconfig was installed on a virtual Windows Server 2013 machine, in order to be able to configure the Access Points, register and update ePaper displays.

In a first experimental test scenario, the delay of changing ePaper content was determined for simple operations, such as image delete, change, rotate and show ID. The results are summarized in Table 1:

Table 1.   Common operations and corresponding delays

Operation	Details	Delay [s]
Delete	Delete image	330–473
Change	Change image 480 × 800 (7.4″)	27
Rotate	Image rotation	2
Show ID	Show label display ID	14

The primary issue with ePaper display testing has been battery longevity. As previously mentioned, the devices consume significant amounts of power when performing operations (displaying new images). In order to limit battery use, each test was performed with only a small batch of operations. The tests have been conducted with

Lancom's Wireless ePaper management software, as well as a first-stage version of SICIAD's management console, capable of sending XML tasks to the ePaper Server.

Because ePaper displays communicate on a wireless channel, interference with nearby radio networks has been noted, and several test results (with extremely large values) have been excluded. During testing, it has also been noted that if two Lancom access point operate on the same wireless ePaper channel and are within reach of the same ePaper display, they will interfere with each other, increasing task execution times, or even preventing it.

However, the first test only offers a limited view of Lancom's Wireless ePaper platform. To test its viability against the above use cases, images varying in size (from less than **1 KB** to over **200 KB**) have been uploaded, and the execution times for each *image display*, *preload* and *switch* operation has been recorded. The main concern in this scenario is determining whether it is possible to display emergency alerts on wireless ePaper displays, within the use cases' time constraints.

The results are shown in Table 2, and include the wireless average transmission time, as well total task execution delays. Excluding extreme values, average Image display time of **14** to **20** s has been obtained, with usual execution times not exceeding one minute.

Table 2. Display and preload operation delays

Operation	Avg. transmission time (s)	Delay [s]		
		Min	Avg	Max
Image	1.6	1.6	14.47	46.21
Preload	1.8	1.8	16.42	23.50
Switch	0.059	0.059	20.84	49.50

Analysis of the dependency between the image transmission time (on the wireless channel) and image size, suggests that small images (less than 100 kB), optimized for B/W ePaper devices, could be transmitted in less than 2 s. In the absence of radio interference, such images could load in less than 30 s.

Based on these tests, it can be said that an average **15 s** delay may be deemed acceptable for displaying alerts on single devices. However, when working with significant numbers of wireless displays, the response times may rise to unacceptable levels (several minutes if only the average case is considered, tens of minutes in the worst-case scenario).

In order to respect the constraints of displaying emergency alerts, the messages can be preloaded on the displays and then the system can use a more rapid switch operation (which only takes a fixed **59 ms**, a short enough time to avoid errors due to radio interference).

Further testing on the wireless ePaper display side will focus on the relationship between transmission power and transmission time, along a more accurate estimation of the relationship between image size and transmission time/total duration. At the same time, future tests will attempt to determine whether it is possible reduce transmission time even further, through use of the ePaper image rendering system.

5   Conclusion and Future Work

The paper presents a displaying and alerting system, based on an integrated communication infrastructure. The system offers dynamic display capabilities using the ePaper technology, as well as enables indoor location-based services such as visitor guidance and alerting using iBeacon-compatible mobile devices.

Starting from the system's architecture (described in the third chapter), most of the ePaper Server Controller has been implemented, enabling several performance tests of the used display devices.

Measurements taken in the evaluation phase show that the ePaper displays' response times are relatively short, being suitable to proposed use cases. Tests show that the largest delays are obtained in the case of deleting images. However, further study will be needed to guarantee rapid response times in case of emergencies.

A production installation of the system would consider avoiding tunnelling between ePaper server and Access points, by collocating all the components of the systems in the same LAN, in order to exclude extra delay and other potential issues introduced by the tunnelling technique. In such a case, availability and performance would improve.

The batteries provided by the manufacturer for the ePaper displays are sufficient for most of the use cases, whilst being easy to replace for indoor applications. For outdoor applications, ePaper systems can be recharged via solar cells and, due to their low power consumption, may function entire seasons without sunlight, offering a long-term solution for displaying information in remote areas.

Future work with the project will include the development of the system's web management console and signage and monitoring APIs, along with the further investigation of ePaper response times and iBeacon functional range, as well as the proposed architecture's scalability, performance and security.

Acknowledgments. This work has been funded by UEFISCDI Romania under grant no. 60BG/2016 "Intelligent communications system based on integrated infrastructure, with dynamic display and alerting - SICIAD.

References

1. Suciu, G., Vochin, M., Diaconu, C., Suciu, V., Butca, C.: Convergence of software defined radio: WiFi, ibeacon and epaper. In: IEEE 15th RoEduNet Conference: Networking in Education and Research, pp. 1–5 (2016)
2. The technology behind LANCOM Wireless ePaper Displays. https://www.lancom-systems.com/fileadmin/download/reference_story/PDF/Wireless_ePaper_Solution_at_a_private_college,_Germany__EN.pdf. Accessed 20 Nov 2017
3. Heikenfeld, J., Drzaic, P., Yeo, J.-S., Koch, T.: Review paper: a critical review of the present and future prospects for electronic paper. J. Soc. Inform. Display **19**, 129–156 (2011). https://doi.org/10.1889/JSID19.2.129
4. Lancom Wireless ePaper Displays Specification sheet. https://www.lancom-systems.com//fileadmin//download/Wireless-ePaper/LANCOM-Wireless-ePaper-Displays-WDG2-EN.pdf. Accessed 10 Dec 2017

5. Lancom Wireless ePaper Server Manual. https://www.lancom-systems.com//fileadmin/download/LANCOM-Wireless-ePaper-Server/LANCOM-Wireless-ePaper-Server-Manual-EN.pdf. Accessed 10 Dec 2017
6. Technical specifications of various Waveshare ePaper display panels. https://www.waveshare.com/w/upload/7/7f/4.2inch-e-paper-b-specification.pdf, https://www.waveshare.com/w/upload/6/6a/4.2inch-e-paper-specification.pdf, https://www.waveshare.com/w/upload/b/b6/7.5inch-e-paper-specification.pdf. Accessed 10 Dec 2017
7. Rogers, J., et al.: Paper-like electronic displays: large-area rubber-stamped plastic sheets of electronics and microencapsulated electrophoretic inks. PNAS **98**(9), 4835–4840 (2001). https://doi.org/10.1073/pnas.091588098
8. University of Belgrade implements eduroam. https://www.lancom.de/fileadmin/download/reference_story/PDF/University_of_Belgrade_EN.pdf. Accessed 20 Nov 2017
9. Winter, S., Wolniewicz, T., Thomson, I.: Deliverable DJ3. 1.1: RadSec Standardisation and Definition of eduroam Extensions. GN3 JRA3, GEANT3 (2009)

Sensors Fusion Approach Using UAVs and Body Sensors

George Suciu[1], Andrei Scheianu[1(✉)], Cristina Mihaela Bălăceanu[1], Ioana Petre[1], Mihaela Dragu[1], Marius Vochin[2], and Alexandru Vulpe[2(✉)]

[1] R&D Department, BEIA Consult International, Bucharest, Romania
{george,andrei.scheianu,cristina.balaceanu,ioana.petre,
mihaela.dragu}@beia.ro
[2] Telecommunications Department, University Politehnica of Bucharest, Bucharest, Romania
{marius.vochin,alexandru.vulpe}@upb.ro

Abstract. Current technological progress allows the implementation of an innovative solution that would aid the work of emergency first responder personnel. However, managing crisis situations is highly dependable on the situational awareness factors. Therefore, this paper proposes a conceptual model of a crisis management platform which integrates several technologies such as UAVs, heat cameras, toxicity sensors, and also body wearable sensors, which are capable of aggregating all the information from the above-mentioned devices. By using the proposed innovative platform, we analyze how a decrease of tragical events within the emergency sites can be achieved.

Keywords: Body sensors · UAV · Heat cameras · Data Fusion

1 Introduction

Unmanned Aerial Systems (UASs) and Unmanned Aerial Vehicles (UAVs) technologies have spread widely and become popular in various military and civilian fields [1]. Some of the popular applications are radar localization [2], wildfire management, observation support [3], agricultural monitoring, border surveillance and monitoring, environmental and meteorological monitoring, and aerial photography [4], as well as for search and rescue missions [5]. The most noticeable benefits compared with conventional manned vehicles are low cost, improved safety for humans, and easy deployment.

One of the biggest advantages of using general purpose drones is the independence between the drone itself and the hardware devices which can be fixed on them. For example, cameras can be attached to drones so that they can be used for certain purposes. At present, drones are constantly developing, various models existing on the market [6]. It is possible to obtain high resolution at low altitudes, SLR (Single-Lens Reflex) cameras offering a superior degree of precision and stability in order to extract data. Other devices that can be attached to drones include imaging systems which cover visible to thermal spectrum, RADARs, LiDARs, or passive microwave radiometers [7].

Civil Drones have many uses, from making good quality video recordings or aerial shots to border guards and event capturing offenders [8]. Borders can be monitored with

© Springer International Publishing AG, part of Springer Nature 2018
Á. Rocha et al. (Eds.): WorldCIST'18 2018, AISC 747, pp. 146–153, 2018.
https://doi.org/10.1007/978-3-319-77700-9_15

the help of drones to avoid unpleasant situations. In case of events such as floods, earth-quakes or landslides, drones prove to be useful, by even providing the capability of food distribution in hard-to-reach places or looking for survivors [9]. Recently, drones have begun to be used to monitor the use of agricultural subsidies, as the Switzerland law requires annual inspection of at least 5% of the areas cultivated with these funds. Initially, these inspections were carried out by authorized personnel, but satellite surveillance has begun to be used to verify that farmers comply with the conventions and meet the requirements for granting subsidies [10]. Drones have become useful in this field, espe-cially for agricultural inspections, as they can take high quality shots from different angles. Drones can be equipped with particle sensors that allow the detection of pollu-tants, namely the measurement of sulfur oxides, nitrogen and ammonia. Other sensors can measure light or sound [11]. With the help of these equipped drones, improvements can be made to the quality of the environment. Sensor-equipped drones are currently used in agriculture to monitor crop growth, biomass estimation, plant disease testing, and air quality assessment. In meteorology, drones can take data such as humidity, wind speed, radiation, etc. [12]. Also, drones are useful in case of tornado or hurricane alerts, in order to evacuate people in a timely manner and save people lives. Drones are also used in the scientific field to collect certain types of data used for research purposes [13].

The purpose of this system is to improve first responder situational awareness by using multi-modal heat cameras, depth, toxicity, acoustic and video sensors mounted on emergency first responders such as firefighters, police, emergency medical units. Responders' location will be tracked via GPS (Global Positioning System) and will be in permanent connection to a central unit. Responders can also wirelessly share the information provided by their equipment/sensors, like digital images and text reports including all the data they gathered, both with other responders and with the incident commander. This can occur in a situation when the responder is inside a building and can also work with the building's sensors (i.e. smart building). We propose a platform which will integrate all the data with the purpose of giving an overview for the chief commander to take decisions regarding real-life dangers and critical situations.

Also, in this article, we will study the solutions available on the market that can be integrated into a smart warning platform to implement a set of technological innovations such as:

- Heat, toxicity, acoustics and video sensors mounted on unmanned aerial vehicles;
- Integration of sensor data, which will result in 3D maps of various perimeters under monitoring;
- Layer vision of the sensors and the possibility of analyzing the data on the control center displays;
- The processing and transmission of data will be based on a dynamic storage network.

The paper is organized in the next three main sections, as follows: Sect. 2 provides a general description of other activities undertaken in order to approach the purpose of this paper. The platform developed in order to retrieve environmental data packages using the Waspmote sensor device, is shown in Sect. 3. Finally, Sect. 4 concludes the article.

2   Related Work

The problem of optimal waypoint planning for UAVs under different constraints has been studied for over a decade. Depending on the applications, the objective functions and optimization methods are different.

Drones contain detection equipment to provide a variety of functions. Geological research, agriculture, archeology and many other industries can greatly benefit from countless sensors that can be placed on the body of the drones. For example, heat sensors can be used to detect water temperature, different animals, etc. One efficient possibility of acquiring different such parameters is to use the integrated sensors in Waspmote [14].

These sensors could be used to monitor the multiple environmental parameters ranges, with applications starting from development analysis to weather observation. This monitoring system can also be used in the reduction of intensities of resources and requirement of labors on site during the growth stage of food. At crucial stages of food production phase, the crop cultivation systems are automatically based upon reading of sensors and parameters for each crop, that can be defined in advance. For collective management of crops, it is possible to work together to maximize and protect or sustain crop cultivation in urban settings [15].

Zarco-Tejada et al. [16] proposed a helicopter based on a UAV which is equipped with commercial-off-the-shelf (COTS) thermal and multispectral imaging sensors in order to generate remote sensing products. The platform has different applications, including precision farming and irrigation scheduling, where time management is a priority.

In 2012, a new approach is presented by Watts et al. [17]. LASE (Low altitude, short endurance) and MALE (Medium altitude, long endurance) UASs are used in the missions in order to detect fire. LASE and MALE UAS provide cost-effective solutions in resource mapping and monitoring applications. The data requirements and the sensors used should be considered. The size of the UAS depends on the size, weight and power of the sensor, but also on the flight duration.

A low-cost thermal scanning UAV platform was designed in 2014 by Aden et al. [18]. The platform can georeference thermal variations with less than one Celsius degree. Also, proper calibration of the Infrared (IR) camera can be an effective air reconnaissance tool which can be utilized in photogrammetric purposes. Based on the temperature, a certain region of interest can be recognized. The temperature measurements were related to a GPS value and by extrapolating the data, a heat map of the scanned area could be created. Google Earth and Matlab are used to process the collected data.

In [19], the authors present a real-time multimodal sensor system which is used for object recognition and tracking. The application strengthens the correlation between data collection and data processing tasks. Furthermore, the proposed solution is able to handle unwanted situations encountered in object-tracking, such as losing targets.

Accurate positioning systems, which are important in military or emergency situations, are referred in [20]. In order to work with accurate data, additional sensors need to be attached, such as barometers and ultrasonic sensors. In this paper, different scenarios are discussed to avoid actions based on partial positioning information.

The current state of UAV used in geomatics applications providing an overview of different platforms, applications and developments in information processing is discussed in [21]. UAVs are used in the geomatics engineering applications such as micro-aircraft vehicles, unmanned combat vehicles, small UAVs and long-lasting low altitude UAVs that depend on the propulsion system, altitude/strength and level of automation. In civil applications, private enterprises invest to provide photogrammetric products such as DSMs which have better resolution and shorter flight times. Even though the authors do not explain what geomatics means, geomatics represents the field of science that deals with gathering, processing of geographic information.

Also, a scalable and adaptable Emergency Management Data Fusion (EMDF) network concept which integrates hardware and software sensor data to enhance real time efficiency of assets in Emergency Management (EM) situations is shown in [22]. Over the years, Data Fusion (DF) has witnessed an evolution from identification and target tracking for military purposes to several non-military applications. By using DF to supply a framework to EM, a loosely coupled EMDF wireless network consisting of fusion centers could be utilized to gather automated collaboration.

In [23], the authors present a comparison between an offline and online neural network architecture to identify Unmanned Aerial Vehicles (UAVs). Neural networks are capable of performing a large amount of parallel processing. Numerical simulation results of each type of architecture are shown. The results of each simulation have been validated using the real-time Hardware in the Loop (HIL) simulation technique for different sets of flight data.

Paper [24] provides an overview of the Critical and Rescue Operations using Wearables Wireless sensor networks (CROW^2). The main purpose of the article is to afford wireless communication and monitoring systems, and regarding this aspect, Wireless Sensor Networks (WSNs) have an important concern. Also, development of On-Body and Body-to-Body communications will allow beyond cross-layer communication architecture for wearable WSNs.

However, none of the previously mentioned solutions integrate multi-modal heat cameras, depth, toxicity, acoustic and video sensors, and, furthermore, are not designed specifically to improve first responder situational awareness. Our proposed solution is specifically tailored for primary responders to use it in an emergency situation, being based on our previously proposed secure e-health architecture [25], as we explain in the next section.

3 Platform Design

The purpose of this section is to analyze how to use the Parrot BEBOP 2 FPV drone to retrieve data from the environment. In order to collect the data from sensors mounted on the drone, the drone must go through predefined routes at certain time intervals.

The functional architecture containing all the components of the proposed platform can be seen in Fig. 1.

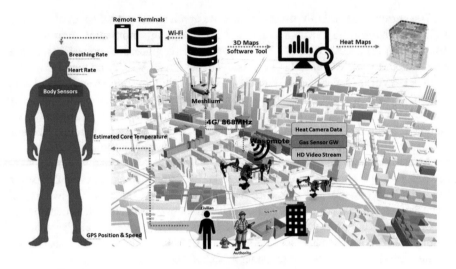

Fig. 1. Functional architecture of the monitoring platform

In our proposed approach, Parrot 2 FPV Drone flies over a certain perimeter at specific time intervals, e.g. 15 min. If an emergency situation is observed, using temperature and thermal sensors (provided by the Waspmote board) and a thermal imaging camera (which is installed additionally), the drone will quickly transfer the information regarding the environmental parameters and the people's condition from that area to the central monitoring and control unit, using the Meshlium gateway. Through the information provided by the Waspmote board and thermal camera it is possible to obtain the thermal map and fusion of other sensor data of that specific area, using a specialised 3D software tool from Pix4d.com. Also, the intervention authorities will be able to analyze the degree of danger of that perimeter by studying the thermal map and the vital parameters which will be received on mobile terminals, such as mobile phones and tablets.

The drone will also determine its GPS coordinates through its GPS receiver and use them to calculate the distance from the other sensors and georeferencing. Parrot Bebop 2 FPV offers a 3D viewable flight through FPV glasses, range up to 2 km, battery life of up to 25 min and full HD video capture capability. The drone has the option of "visual tracking" through Freeflight Pro, which is made possible by advanced visual recognition technology. The main uses are the location of people, as well as perimeter or environment research and analysis.

Waspmote is a sensor device that works with different protocols (ZigBee, Bluetooth, and GPRS) and capable of getting radio links of up to 12 km [26]. One of the main characteristics of it is low power consumption i.e. (9 mA–ON mode, 62 µA-sleep mode, 0.7 µA-hibernate mode). Different parameters that can be measured by the waspmotes are:

- Ambient temperature, relative humidity, atmospheric pressure, noise
- Radiations like ultraviolet, infrared, nuclear, solar, electromagnetic

- Gases such as carbon monoxide (CO), oxygen (O2), ozone (O3), nitrogen oxide (NO), nitrogen dioxide (NO2), sulfur dioxide (SO2), methane (CH4), particle matters (PM), etc.

The Parrot Bebop 2 FPV drone equipped with the integrated above-mentioned sensors in the Waspmote module is presented in Fig. 2. The Waspmote will be connected to a Libelium Meshlium gateway using LoRa.

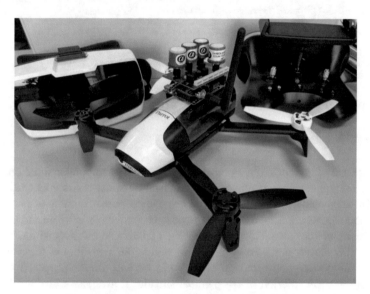

Fig. 2. Parrot Bebop 2 FPV drone with Waspmote sensors

For calibration and precision, an experiment was conducted to compare the GPS coordinates of Waspmote's GPS sensor and the drones GPS sensor. This was done in order to observe if there any differences between the GPS sensor outputs and the magnitude of these differences.

Figure 3 shows the two outputs of the Waspmote sensor and drone GPS sensor, located on the same ground. It is noted that the two graphs overlap in a large percentage, meaning that the results obtained with the Waspmote GPS sensor and those obtained with the drone GPS sensor are similar.

The potential usage of the two GPS sensors, one for the drone, and the other for the Waspmote sensor is to provide information on the condition of the drones (if available or not). They are used for various purposes, such as object tracking, drone recovery in case of emergency landing, smart parking, vehicle tracking, and many healthcare-specific applications. The sensors can be used simultaneously if the Waspmote GPS sensor is used to provide target positions, and the GPS sensor implemented on the drones' surface is used to track its location.

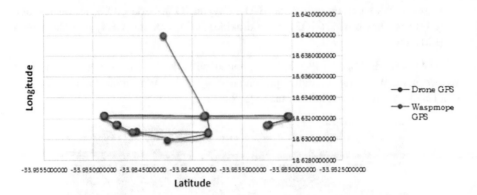

Fig. 3. Comparison of GPS data from Waspmote and drone integrated sensor

4   Conclusions

This paper shows an innovative approach to improve first responder situational aware-ness, by using multi-modal fusion of data from heat cameras, environment, acoustic and video sensors mounted on emergency first responders. Sensor communication technol-ogies such as LoRa are used in order to provide cost effective, reliable and low energy data acquisition. The base platform design for georeferenced data acquisition from drone mounted GPS sensors has been evaluated, as well as the modalities of transmitting this data to a 3D viewing platform. As future work, we will use the proposed system for developing a novel solution offering global situational awareness for both individual actors carrying out rescue operations and for the coordinating commanders.

Acknowledgments. This work has been supported in part by UEFISCDI Romania through projects 3DSafeguard, WINS@HI and ESTABLISH, and funded in part by European Union's Horizon 2020 research and innovation program under grant agreement No. 777996 (SealedGRID project).

References

1. Tu, D.H., Esten, I.G., Sujit, P.B., Tor, A.J.: Optimization of wireless sensor network and UAV data acquisition. J. Intell. Robot. Syst. **78**, 159–179 (2015)
2. Purvis, K., Astrom, K., Khammash, M.: Estimation and optimal configurations for localization using cooperative UAVs. IEEE Trans. Control Syst. Technol. **16**, 947–958 (2008)
3. Frew, E.W., Brown, T.X.: Networking issues for small unmanned aircraft systems. J. Int. Robot. Syst. **54**, 21–37 (2008)
4. Ruz, J.J., Arevalo, O., Pajares, G., Cruz, J.M.: UAV trajectory planning for static and dynamic environments, Ch. 27, pp. 581–600. InTech (2009)
5. Flushing, E.F., Gambardella, L., Caro, G.D.: Search and rescue using mixed swarms of heterogeneous agents: modeling, simulation, and planning, IDSIA/USI-SUPSI, Technical report (2012)

6. Bas, V., Huub, N., Geert, B., Bart, C.: Drone technology: types, payloads, applications, frequency spectrum issues and future developments (2016)
7. Everaerts, J.: The use of Unmanned Aerial Vehicles (UAVs) for remote sensing and mapping
8. Maria, J.: Civil drones in the European Union, European Parliamentary Research Service (2015)
9. Silvana, P.: Swiss military drones and the border space: a critical study of the surveillance exercised by border guards, Institute of Geography, University of Neuchâtel, Neuchâtel, Switzerland (2017)
10. Tom, M.: Institute of Geography, University of Neuchâtel, Neuchâtel, Switzerland (2016)
11. Paolo, T., Massimo, S., Giacomo, D.: Towards smart farming and sustainable agriculture with drones. In: IE (Intelligent Environments) (2015)
12. Pederi, Y.A., Cheporniuk, H.S.: Unmanned aerial vehicles and new technological methods of monitoring and crop protection in precision agriculture. In: IEEE International Conference Actual Problems of Unmanned Aerial Vehicles Developments (APUAVD) (2015)
13. HELIPSE: Border surveillance drones (UAV VTOL) (2016)
14. LIBELIUM: The first Smart Vineyard in Lebanon chooses Libelium's technology to face the climate change. http://www.libelium.com/the-first-smart-vineyard-in-lebanon-chooses-libeliums-technology-to-face-the-climate-change/
15. Kaur, G., Joshi, S., Singh, G.: Food sustainability using wireless sensors networks: Waspmote and Meshlium. Int. J. Comput. Sci. Inf. Technol. **5**(3), 4466–4468 (2014)
16. Zarco-Tejada, P.J., Suárez, L., Fereres, E., Berni, J.A.J.: Thermal and narrowband multispectral remote sensing for vegetation monitoring from an unmanned aerial vehicle. IEEE Trans. Geosci. Remote Sens. **47**(3), 722–738 (2009)
17. Watts, A.C., Ambrosia, V.G., Hinkley, E.A.: Unmanned aircraft systems in remote sensing and scientific research: classification and considerations of use. Remote Sens. **4**(6), 1671–1692 (2012)
18. Aden, S.T., Bialas, J.P., Champion, Z., Levin, E., McCarty, J.L.: Low cost infrared and near infrared sensors for UAVs. In: The International Archives of the Photogrammetry, Remote Sensing and Spatial Information Sciences, vol. XL-1 (2014)
19. Yufu, Q., Guirong, Z., Zhaofan, Z., Ziyue, L., Jiansen, M.: Active multimodal sensor system for target recognition and tracking. Sensors **17**(7), 1518 (2017). Open Access Journal
20. Jouni, R., Joakim, R., Peter, S.: Accurate and reliable soldier and first responder indoor positioning: multisensor systems and cooperative localization. IEEE Wirel. Commun. **18**, 10–18 (2011)
21. Francesco, N., Fabio, R.: UAV for 3D mapping applications: a review. Appl. Geomat. **6**, 1–15 (2013)
22. Sonya, A.H., Mac, J.M., Peter, F., David, I., Patti, J.C.: Emergency management: exploring hard and soft data fusion modeling with unmanned aerial systems and non-governmental human intelligence mediums (2016)
23. Vishwas, R.P., Sreenatha, G.A.: Comparison of real-time online and offline neural network models for a UAV. In: Proceedings of International Joint Conference on Neural Networks (2007)
24. Elyes, B.H., Muhammad, M.A., Mickael, M., Denis, B., D'Errico, R.: Wearable Body-to-Body Networks for Critical and Rescue Operations – The CROW^2 Project (2014)
25. Suciu, G., Suciu, V., Martian, A., Craciunescu, R., Vulpe, A., Marcu, I., Halunga, S., Fratu, O.: Big data, internet of things and cloud convergence–an architecture for secure e-health applications. J. Med. Syst. **39**(11), 141–145 (2015)
26. LIBELIUM: Sensor boards. http://www.libelium.com/products/waspmote/sensors/

Check for
updates

Cyber-Attacks – The Impact Over Airports Security and Prevention Modalities

George Suciu[1,2], Andrei Scheianu[1], Alexandru Vulpe[2],
Ioana Petre[1(✉)], and Victor Suciu[1]

[1] R&D Department, BEIA Consult International, Peroni 16, 041386 Bucharest, Romania
`{george,andrei.scheianu,ioana.petre,victor.suciu}@beia.ro`
[2] Telecommunication Department, University Politehnica of Bucharest, Bd. Iuliu Maniu 1-3,
061071 Bucharest, Romania
`{george.suciu,alex.vulpe}@radio.pub.ro`

Abstract. In recent years, we have witnessed an improvement of the systems developed for the prevention of cyber-attacks meant to harm the airports security, due to the increasing number of the sophisticated malicious attacks. The majority of government organizations depends on advanced systems consisting of inter-connected computing machines, used to provide public services. This connectivity and online services have many advantages for users, but also, they become vulnerable in case of a cyber-attack. One of the most targeted systems is the one used in the aviation sector, which is also vulnerable in front of physical threats. Since Sept 11th, 2001, an improvement of terrorist tactics was observed, reason why the security of this sector represents a critical consideration. This paper presents a summary of the ongoing research activities taken to improve the security level of airports and proposes a scalable solution for the integration of several security tools, services and fields.

Keywords: Airport · Security · Cyber-attack · Vulnerability

1 Introduction

Despite all the actions taken to improve the security of the international airports, the aviation sector is still a critical issue to be resolved as soon as possible. The terrorist attack that took place in the United States of America on September 11th, 2001 was the beginning of a worldwide attacks series, such as liquid explosions that occurred in 2006, body bombs (2009 and attempts occurring even today), attempts to destroy aircrafts by using improvised explosive devices (which were commonly seen in 2010), as well as the attacks that took place in Moscow, Sofia and Bulgaria [1]. These attacks have shown that, regardless of the methods used to prevent terrorist activities, attackers technique has continued to develop in direct proportion to the development of the security mechanism implemented by the international authorities responsible of aviation security.

The systems implemented in airports are the result of numerous research activities undertaken to prevent potential terrorist attempts. The main purpose of these systems is

© Springer International Publishing AG, part of Springer Nature 2018
Á. Rocha et al. (Eds.): WorldCIST'18 2018, AISC 747, pp. 154–162, 2018.
https://doi.org/10.1007/978-3-319-77700-9_16

to monitor the people inside airports by making an analysis based on their behavior, facial expressions and psychological profiles.

Opinions regarding the prevention of terrorist attacks through behavioral analysis differ. Many researchers believe that the practices used to detect the harmful trends to public safety have the potential to help the process of capturing terrorists and to provide an increased level of security and confidence, while others consider that the implementation of new security mechanisms can affect the public safety due to the provision of new technologies that can be exploited by cyber-attackers.

After many research activities in the field of cyber security and its influence on critical infrastructures, it was found that many developed countries believe that air transport security is a critical issue even for the future years. For example, Singapore has stated that civil aviation is a crucial element of the "Critical Infocomm Infrastructure Protection" program, intended to the CSP (Cyber Security plan) for 2018 [2]. For example, taking into account the cyber-attacks that took place in 2014, CNAIPIC (National Anti-crime Computer Centre for the Protection of Critical Infrastructures) reported 1151 such attacks, of which 161 were directed to the web protection sector and 50 of them to DDoS (Distributed Denial of Service) in Italy [3]. As time passed by, a significant increase of ICT (Information and Communication Technology) has been observed in civil aviation, particularly in the area of construction and development of air transport means and in the implementation of the communication methods and navigation equipment, but in addition to the benefits available, they have also introduced a set of issues related to cyber-security.

The main contribution of this paper is the development of a novel architecture for preventing cyber-attacks on critical infrastructures such as airports as well as two case studies built on top of this architecture. Also, prevention measures are given to be used as recommendation by interested stakeholders such as airport security staff, and airport network administrators.

The article consists of three other sections, as follows: Sect. 2 of this paper provides a general description of the activities undertaken by other agencies to address the cyber-security issues concerns, while Sect. 3 presents the proposed platform, consisting in a scalable solution developed to integrate several security tools, fields and services. The conclusions of the paper can be seen in Sect. 4.

2 Related Work

In this section will be presented the efforts undertaken internationally to improve the methods used to prevent the global attempts to breach the regulations related to the safety of airports headquarters and privacy of the online services provided to passengers.

Within the events of ICAO (International Civil Aviation Organization), ICCAIA (International Coordinating Council of Aerospace Industries Associations) has argued that the new technologies developed to ensure cyber-security are vulnerable to cyber-attacks. Therefore, it was necessary to create a research team responsible for the monitoring, management and implementation of the practices and procedures required to ensure the cyber-security. In 2013, IFALPA (International Federation of Air Line Pilots'

Associations) published the report [3] about the threat of cyber-attacks on aircrafts, airports and various online service delivery systems for travelers, with the main purpose of detailing the problems encountered in securing navigation data, which may affect the safety of airports if they are not adequately protected.

CANSO (Civil Air Navigation Services Organization) established in 2014 a research group, whose work led to the creation of a study called "Cyber Security and Risk Assessment Guide" [4]. The basic idea of the study was to improve the implementing provisions of the sectoral methods between the ANSP (Air Navigation Service Provider) providers.

Also, 2014 was the moment when several associations such as ICAO, CANSO, ICCAIA and IATA (International Air Transport Association) signed the CACSA (Civil Aviation Cyber Security Action) plan, which forced them to unite their efforts to effectively mitigate the cyber-attacks [3]. This partnership has contributed to promoting the development of a much more mature culture regarding the real risk of cyber-attacks and how to protect against them.

However, the global view is that the aviation sector for civil transport requires the improvement of the technologies and means of ensuring airports security. A difficulty in this regard is the fact that many of these technologies are also used in other industrial areas, which makes them even more vulnerable to potential attackers. According to the studies conducted in recent years [5, 6], it has been observed that the number of cyber-attacks directed against civil aviation has increased. Depending on the target system, it is possible to exploit a large number of vulnerabilities, on a case-by-case basis.

Scientific approaches have been taken to modelling attacks for airport security analysis. For instance, authors in [7] have developed a dynamic event-based model of the airport as a Cyber-Physical System (CPS) that predicts the impact of blackouts or other cyber-attacks on airport performance and can simulate the propagation of external stimulation within the airport network. A security QoS profiling for airport network systems was performed in [8]. Authors used Stateful Packet Inspection based on OpenFlow Application Centric Infrastructure for securing critical network assets. Other approaches have also been taken, including taking into consideration a hybrid approach of port and airport critical infrastructure security [9].

Furthermore, the following systems described could pose a serious issue to human safety and national security when they are compromised. The most targeted systems when talking about cyber-attacks are: international airport computer systems, in-flight aircraft control systems and air traffic management systems [10].

Internet is frequently used in airports to make simple operation such as informational exchange and messaging. Another important target could be represented by SCADA (Supervisory Control and Data Acquisition) systems that most often have control over ventilation systems, luggage transportation, etc. [11, 12]. Because most of the mobile applications needs an Internet connection to work, the airport security system will be endangered with repercussions on airport operations. This fact will make them become potential targets of cyber-attacks by a large number of physical operations (such as digital camera, USB drivers, computers, etc.) and virtual operations (trojans, DoS, phishing, etc.).

The impact of WiFi's introduction is considerable from the perspective of several security experts. Spanish expert Hugo Teso has demonstrated in [13] at the EASA (European Aviation Safety Agency) conference how to make an airplane crash by controlling the aircraft control system with a software mobile phone application created by himself. Ruben Santamarta demonstrated in [14] by using a WiFi-accessible flight entertainment system how it is possible to interfere with satellite communications, and therefore with a flight navigation system. Also, in 2015, Chris Roberts, succeeded to do the same particular thing, and according to an FBI (Federal Bureau of Investigation) documentation, Chris only typed in the command CLB (climb) and the plane's engine responded to the prompt [15].

US Department of Transportation Inspector General have made a report in 2009 in which the weaknesses and inefficiencies in the FAA's ATM systems were blamed when talking about the control procedures and the capability to identify the unlawful intrusions. GAO (Government Accountability Office) published a much more relevant report [2] in January 2015 and according to the report one of the deficiencies of this systems was the lack of cyber risk management, which is why there is no precise definition of the roles and responsibilities within the FAA (Federal Aviation Administration).

3   Airports Cyber-Attacks Prevention Methodology

In this section we present the most effective ways to analyze if a communication system is vulnerable or not to cyber-attacks by simulating these situations. Such a simulation can be performed on an Internet network that was developed specially to meet airport-specific requirements and services. Analyzing the results generated by the simulated attack, the vulnerabilities of the system could be analyzed in detail and more effective solutions to combat them could be developed. In the following section, we describe the procedure to be applied in order to obtain the desired results.

3.1   Use Case: Airports DDoS Attacks Prevention

Event: Let's assume that a group of attackers wants to blackmail a medium-to-large sized companies number by threatening them with a DDoS attack. Attackers have found that an airport company is the ideal target for such a cyber-attack, because it depends on Internet connection and, also controls substantial financial resources. Generally, for initiating a cybernetic-attack it's necessary to identify the IP addresses owned by the airport in question, which are not difficult to obtain, and some emitting sources which will be used to begin the attack. The first targets are the mobile terminals or infected networks that can be easily controlled.

Result: If the airport management does not agree to pay the amount required by the attackers the attack will be initiated, and the airport's Internet connection will be compromised. When it is accepted that the airport is indeed cyber-attacked, it is necessary to start investigations to resolve the critical situation. In Fig. 1 can be seen the functional architecture of the network, which contains all the airport network components. In order to exchange information between stakeholders, it is necessary to have an Internet connection. When the Internet connection is interrupted, the partners won't receive any more updated information from the Airport Operation Plan (AOP) sector, nor the Network Operation Plan (NOP) as the network will not be able to sustain a corresponding dataflow.

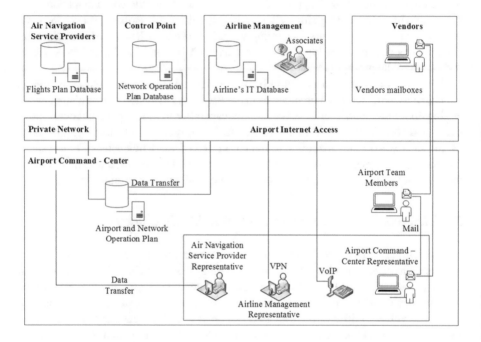

Fig. 1. The functional architecture of the network

Prevention measure - The following methods can be used [16]:

1. **Volumetric protection:** is one of the methods used by the ISPs (Internet Service Providers) to prevent and identify potential DDoS attacks. Also, the providers are capable to filter all the requests received from different emitting sources. In this way, ISPs can decrease the traffic that seems abnormal and grant only normal requests to the airport IP addresses. Certain IP addresses can be blacklisted if needed.
2. An alternative way used in cyber-attacks prevention is **the development of a second Internet connection** and the allocation of a different IP range, which will be used only in emergency situations.
3. To improve the efficiency of the protection mechanism, it is necessary to implement **high performance hardware devices** which have specialized network interfaces. The role of these devices is to permanently monitor the logging activities and existing traffic to specific IP addresses from the airport. In case of an attack detection, the traffic will be redirected to the cleaning equipment.

The protection method used in this research activity is **blackholing**. Specifically, the attacked IP addresses are automatically notified to the top-level providers, who will block the entire traffic to them. A procedure considered very important at this stage is to minimize the human factor from the traffic monitoring process. When receiving a phone call about a cyber-attack, the airport's representatives will be welcomed by an intelligent speech recognition system that will trigger the protection mechanisms, as it can be seen in Fig. 2.

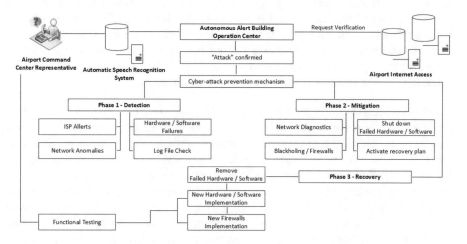

Fig. 2. Cyber-attack prevention mechanism functionality

Detection is based on unsupervised learning for anomalies detection. Unsupervised learning enables the identification or detection of behavioral anomalies at logical level. The detection method is implemented using a combination of clustering algorithms in the first phase (unsupervised learning), followed by the supervised learning phase (after discovering anomalies that suggest security issues, incidents, etc.).

3.2   Use Case: Blended Attack

Event: Let's assume that a group of attackers wants to disrupt the normal operation of an airport information flow, without the rapidly detection of the intrusion attempt. To achieve the desired goal, it has been chosen to use a blended attack involving the simultaneous initiation of multiple attacks, one of them having the main role of being quickly detected to create a diversion. In this sense, a DDoS attack similar to the one presented in the previous scenario will be initiated, which will affect the airport network performance. Meanwhile, to a limited number of aircrafts will be sent messages that will contain minor changes of the flight routes, which were specially designed to prevent them from being detected. The engineers and airport security officers will be distracted by the remediation of the effects caused by the first fake cyber-attack, while the real attack will not be noticed. Since the effects of the first attack can be easily resolved, the staff responsible with the provisioning of the airport security will have the impression that the situation is under control and that the communication network is safe, which is totally wrong.

Result: Undetected attacks will result in the change of the flight plans, such as aircrafts take-offs or landings delays. For example, if an airplane will arrive at its destination before the scheduled time, the staff responsible with the provisioning of the safe landing support of the airplane will have to come at the workstation earlier. If the attack will cause the aircraft to be delayed, the workers will be forced to wait for its arrival. This waste of time could be spent managing other flights, as needed. Maintaining in the airport space perimeter of an aircraft that has arrived earlier, until a landing strip is released, will cause a delay in the take-off of the other planes within the airport headquarters. This will also influence the aircrafts cleaning and fueling process, as well as the time required to load the luggage in the planes carriers. From an international point of view, the emergence of such a problem will affect the world-wide air traffic.

Prevention measure: The airport security staff will find out hard that the initial cyber-attack was not the real one. To avoid this, it is advisable to educate the authorized airport security staff about the possibility of masking a massive cyber-attack through a bait-attack. Also, in order to minimize the risks of occurrence of these situations, certain hardware or software components can be implemented in the network infrastructure to continuously monitor the operation of the airport equipment (e.g. Building Management Systems, Closed-Circuit Television, Communications and Flight Information Display Systems, etc.). Another way to prevent these blended attacks is to analyze similar attacks and develop and integrate more efficient security algorithms to monitor air traffic autonomously and to quickly respond if a suspect number of delays is detected.

3.3   Discussions

Ensuring a high level of security of the airport premises represents a critical aspect that must be solved as quickly as possible. The normal functionality of the airport operations centers can be easily disrupted if the availability and integrity of the security systems can be compromised. Considering the previously mentioned use-cases, the main cyber-problems that can affect the airport headquarters are:

- old-generation communications systems whose security mechanisms can be easily penetrated. In case of such equipment, ensuring a high-quality airport network is almost impossible;
- the use of customer services that are provided by third party companies. This event involves the existence of an increased number of users who can access the core infrastructure, which can affect the security of the network;
- even if most of the systems from an airport have been properly secured, it is necessary to ensure a certain level of security for all equipment within the network. For example, if airport access points are not reliable, the network security can be easily compromised. Furthermore, Internet of Things (IoT) platforms are being deployed in airports, but have several security issues [17].

If current problems will not be resolved soon, the prevention of terrorist attempts will be more and more difficult to achieve. This will have a major impact on international air transport, which will affect in time the public's confidence and the overall international security.

4   Conclusions

Considering that more and more organizations prefer the implementation of cloud-based functional models, thus supporting the technological evolution, it is clear that this phenomenon will influence the technologies currently used for airport headquarters protection. The number of cyber-attacks is rising and their nature is constantly changing. As a result, we presented how vital it is to adopt the implementation of superior security technologies and mechanisms that can effectively combat the existing threats. As future work we envision building a simulator for such kind of cyber-attacks.

Acknowledgments. This paper has been supported in part by UEFISCDI Romania through projects ALADIN, ToR-SIM and ODSI, and funded in part by European Union's Horizon 2020 research and innovation program under grant agreement No. 777996 (SealedGRID project).

References

1. Bernard, L.: Emerging threats from cyber security in aviation – challenges and mitigations. J. Aviat. Manag. (2014)
2. Tommaso, D.Z., Fabrizio, A., Federica, C.: The Defence of Civilian Air Traffic Systems from Cyber Threats. IAI (2016). http://www.iai.it/sites/default/files/iai1523e.pdf
3. IFALPA: Cyber threats: who controls your aircraft? IFALPA Position Papers, No. 14POS03 (2013). http://www.ifalpa.org/store/14POS03%20-%20Cyber%20threats.pdf
4. CANSO: Cyber Security and Risk Assessment Guide (2014). https://www.canso.org/node/753
5. CERT-UK: Common Cyber Attacks: Reducing the Impact (2015). https://www.gov.uk/government/uploads/system/uploads/attachment_data/file/400106/Common_Cyber_Attacks-Reducing_The_Impact.pdf
6. Accenture: Midyear cybersecurity risk review forecast and remediations (2017). https://www.accenture.com/t20171010T121722Z__w__/us-en/_acnmedia/PDF-63/Accenture-Cyber-Threatscape-Report.pdf
7. Koch, T., Möller, D.P.F., Deutschmann, A., Milbredt, O.: Model-based airport security analysis in case of blackouts or cyber-attacks. In: 2017 IEEE International Conference on Electro Information Technology (EIT), Lincoln, NE, pp. 143–148 (2017). https://doi.org/10.1109/eit.2017.8053346
8. Ugwoke, F.N., Okafor, K.C., Chijindu, V.C.: Security QoS profiling against cyber terrorism in airport network systems. In: 2015 International Conference on Cyberspace (CYBER-Abuja), Abuja, pp. 241–251 (2015). https://doi.org/10.1109/cyber-abuja.2015.7360516
9. Chiappetta, A., Cuozzo, G.: Critical infrastructure protection: beyond the hybrid port and airport firmware security cybersecurity applications on transport. In: 2017 5th IEEE International Conference on Models and Technologies for Intelligent Transportation Systems (MT-ITS), Naples, pp. 206–211 (2017). https://doi.org/10.1109/mtits.2017.8005666
10. ICAO: Aviation Unites on Cyber Threat (2014). http://www.icao.int/Newsroom/Pages/aviation-unites-on-cyberthreat.aspx

11. ENISA: Securing Smart Airports (2016). https://www.enisa.europa.eu/publications/securing-smart-airports/at_download/fullReport
12. Khalid, A.F.: An Analysis of Airports Cyber-Security (2016). http://www.caeaccess.org/research/volume4/number7/fakeeh-2016-cae-652129.pdf
13. Marcel, R., Gerald, T.: Cyber-Attack Warning: Could Hackers Bring Down a Plane? Spiegel Online International (2015). http://spon.de/aevsu
14. McAllister, N.: No, you CAN'T hijack a plane with an Android app. The Register (2013). http://goo.gl/news/0PmU
15. Kim, Z.: Feds Say That Banned Researcher Commandeered a Plane. Wired (2015). http://www.wired.com/?p=1782748
16. Olivier, D., Olivier, R., Eric, V., Chris, J., Matt, S., Piotr, S., Veronika, P.: Cyber-security application for SESAR OFA 05.01.01 - Final report (2016). http://www.sesarju.eu/sites/default/files/documents/news/Addressing_airport_cyber-security_Full_0.pdf
17. Arseni, S.C., Halunga, S., Fratu, O., Vulpe, A., Suciu, G.: Analysis of the security solutions implemented in current Internet of Things platforms. In: IEEE Grid, Cloud & High Performance Computing in Science (ROLCG), pp. 1–4 (2015)

Healthcare Information Systems
Interoperability, Security and Efficiency

Forecast in the Pharmaceutical Area – Statistic Models vs Deep Learning

Raquel Ferreira[1], Martinho Braga[2], and Victor Alves[3(✉)]

[1] Department of Informatics, University of Minho, Braga, Portugal
raquel.mferreira@hotmail.com
[2] Tlantic Portugal SI, Porto, Portugal
martinho.braga@tlantic.com
[3] Algoritmi Centre, University of Minho, Braga, Portugal
valves@di.uminho.pt

Abstract. The main goal of this work was to evaluate the application of statistical and connectionist models for the problem of pharmacy sales forecasting. Since R is one of the most used software environment for statistical computation, we used the functions presented in its forecast package. These functions allowed for the construction of models that were then compared with the models developed using Deep Learning algorithms. The Deep Learning architecture was constructed using Long Short-Term Memory layers. It is very common to use statistical models in time series forecasting, namely the ARIMA model, however, with the arising of Deep Learning models our challenge was to compare the performance of these two approaches applied to pharmacy sales. The experiments studied, showed that for the used dataset, even a quickly developed LSTM model, outperformed the long used R forecasting package ARIMA model. This model will allow the optimization of stock levels, consequently the reduction of stock costs, possibly increase the sales and the optimization of human resources in a pharmacy.

Keywords: ARIMA · Deep Learning · Forecast · LSTM · Pharmacy sales

1 Introduction

In the pharmaceutical retailing industry, it's essential to ensure there is an adequate stock of medications and supplies to serve the needs of pharmacy clients, so a careful inventory management can increase its profitability [1]. The constitution of stocks aims to combat stock disruption, the eventualities related with consumption, unforeseen circumstances associated with the delivery and serve as a regulator between deliveries and uses that are made at different rates. The manager of a pharmacy has as main tasks study the stock level and its required quantity, the amount to order and what is the best time to place the order. The objective is avoid an excess of stock and, at the same time, ensure that a stock rupture does not occur [1, 2]. It's important to highlight that drug shortage can disrupt the healing processes of patients and produce negative effects on public health.

© Springer International Publishing AG, part of Springer Nature 2018
Á. Rocha et al. (Eds.): WorldCIST'18 2018, AISC 747, pp. 165–175, 2018.
https://doi.org/10.1007/978-3-319-77700-9_17

Forecasting future demands can increase operational efficiency, effectiveness, and flexibility, improving customer services, and reducing both costs and problems. [1, 2]. In this context, the concept of Business Intelligence emerges, which brings a wide range of technologies that intend to extract from raw data, patterns, and trends that can be used to improve decision making in an organization, such as forecasting. These techniques represent a shift from tactical thinking to strategic thinking [3].

Forecasting is required in many situations, and whatever the circumstances or time horizons involved it is considered an important aid to effective and efficient planning. The predictability of an event depends on how well we understand the factors that contribute to it, how much data is available and whether the forecasts can affect what we are trying to forecast [4, 5]. When satisfied these three conditions, forecasts can be highly accurate. However, a forecasting model may in some cases not add any value. Anything that is observed sequentially over time is considered a time series and it aims to estimate how the sequence of observations will continue in the future. Forecasting time series commonly extrapolate seasonal trends and patterns, and all auxiliary information such as marketing initiatives, competitor activity, changing economic conditions can subsequently be identified by analyzing the data [4, 5].

A time series can be broken down into four elements:

1. Cycles C_t: Variations in the series that although periodic do not have any known periodicity.
2. Trend T_t: Increase or decrease in the behavior of the series over time.
3. Seasonality S_t: Cycles or repetitive patterns that occur over time.
4. Noise ε_t: Variations that occur along the time series without possible explanation.

2 Methods and Related Word

The experiments for the preparation of this study were based on a statistical and a deep learning approach.

Statistical models
With statistics, we can study, analyze and collect information for making predictions and conclusions about the future. R is a free software environment for statistical computation and graphics widely used by scientists, companies, mathematicians and analysts which choose to incorporate this language into their business logic since it offers a better view of all the information they are dealing with, does not require the development of customized tools and is very performant [4, 6, 7].

It has available a forecasting package, called *forecast* that was used in this study. The chosen integrated development environment for working with this software was *Rstudio*.

Deep learning models
Deep Learning (DL) is a subfield of Machine Learning (ML) where the system can discover how to learn, generate and transform features from data on its own, exploring neural networks with more than one layer of nonlinear processing [8, 9]. It is about learning multiple levels of representation and abstraction that help to make sense of data [9].

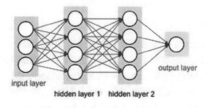

Fig. 1. Neural net connections [10].

Neural Networks (NN) consist in many processors called neurons simply connected, each producing a sequence of real-valued activations. Input neurons get activated through sensors perceiving the environment, while others get activated through weighted connections from previously active neurons [8]. There are basically three types of layers (input, hidden and output) and depending on the problem we want to model they can have more than one hidden layer (Fig. 1).

Time series usually contain temporal dependencies that make two otherwise identical time points belong to different classes or predict different behaviors. This feature usually increases the difficulty of analyzing and predicting them. Recurrent neural networks are a type of NN designed to deal with sequence dependency. A Long-Short Term Memory (LSTM) network is a recurrent neural network architecture that remembers data over arbitrary time intervals. It is a network optimized to learn from and act on temporal data, with indefinite or unknown temporal dimension. Its architecture allows persistence by giving the network a certain memory since it can be seen as multiple copies of the same network. It has been shown that LSTMs have the capability to predict various non-linear behaviors [11, 12].

In [13] the authors describe the use of a deep neural network architecture in the weather prediction. It uses multi-stacked LSTMs to map sequences of weather values of the same length. It was demonstrated that LSTMs are competitive with the traditional methods and can also be considered a better alternative to forecasting general weather conditions. Another study that has used LSTM is presented in [12] and it's about the human trajectory in crowded spaces. The authors argue that "the problem of trajectory prediction can be viewed as a sequence generation task, where we are interested in predicting the future trajectory of people based on their past positions". At the end, it was shown that LSTMs has the capability of predict various non-linear behaviors, such as a group of individuals moving together.

The programing language used for applying DL was Python as it is a general-purpose programming language, unlike R and Matlab, which are more oriented for numerical computing [14]. SciPy is a collection of mathematical algorithms and convenience functions built on the *NumPy* extension of Python, which enables the development of sophisticated programs and specialized applications [15]. *Scikit-learn* is a machine learning library, interoperating with Python numerical and scientific libraries such as *NumPy* and *SciPy*. Currently several numerical libraries can be used for developing deep learning models in Python, (e.g. *Theano, TensorFlow, Caffe*). In this work, we used *TensorFlow*, an open source software library for numerical computation using data flow graphs. It was originally developed for conducting machine learning and deep neural networks research [16]. On top of *TensorFlow* we used *Keras*, a high-level neural networks API, written in Python, that supports both convolutional networks and recurrent networks, as well as combinations of the two [17, 18].

3   Experiments

The dataset used contains 779569 records which correspond to a two years sales history of a set of pharmaceutical products and the first processing stage was the choice of the attributes that would be used to train the models, as well as the identification of the distinct categories of products present. The attributes that were selected include the date, the hour of the day and the quantity sold. All values were aggregated per day resulting in a dataset represented only by two attributes: date and quantity, resulting in a dataset with 733 records. Next, seven more attributes were added, one for each day of the week. For each row, the day of the week was identified and the correspondent attribute filled with one (1) leaving all the others filled with zeros (0). For the case oriented to statistics, if all variables were identified this would lead to a multicollinearity error, which occurs when at least two highly correlated predictors are assessed simultaneously in a regression model. So, for these models only six attributes were created since the algorithms themselves are capable of identifying the day that is missing. The identification of the months underwent a similar treatment: twelve new attributes were added to the result dataset, referring to the months of the year, and then populated with binary variables (for the statistical models, only from February to December).

Attributes for significant date events that could influence sales behavior were also added (Table 1). These events included flu season, holidays, bathing season, holiday season (weeks and days before and after), the carnival season, the Easter season, New Year's Day, Christmas Day, among others and were also identified through binary variables. It should be noted that these events are very important factors in identifying periods of seasonality and eventual peaks of sales, which would be more difficult to identify and explain without them. At this step, the dataset consisted of 733 records and 67 attributes. Outliers above 95% and below 5% for which we didn't have an explanation through the selected events were also removed.

Table 1. Attributes of the used dataset.

Attribute	Description	Characteristic
Date	Date of the sell (YYYY-MM-DD)	Linear
Quantity **(predictor)**	Quantity sold in a day	Integer
Days of the week	Monday to sunday (for R: one day less)	Binary
Months of the year	January to december (for R: one month less)	Binary
Events matrix	Days that could influence sales behavior	Binary

A subset of the currently available dataset was treated as future data to then compare forecasts with actual results. The models were developed using a set of data designated by training and then it was applied to the set designated by test, allowing to infer about the credibility and degree of confidence of the model [19]:

- x_train - training dataset (90% of the initial sample)
- y_train - training dataset variable to be predicted ($mean = 4428.13; SD = 1348.396$)

- x_test - test dataset (remaining 10% of the initial sample)
- y_test - test dataset variable to be predicted ($mean = 4106.836; SD = 1788.803$)

Model development
Both experiments followed the same model construction plan (Fig. 2).

Fig. 2. Forecast model development.

Experiment using statistical models
It was decided to use the *forecast* package that includes a large set of models capable of forecasting a univariate time series. Based on literature research, the group of models that are most commonly used and have good results are: Holt-Winters, NNETAR, TSLM and ARIMA [4]. However, since ARIMA was described as having the best results, it will be the model described and assessed in this study. Auto Regressive (AR) with Integrated Moving Average (IMA) (ARIMA) models aim to describe the autocorrelations present in the data. These models are defined by three parameters: order of the AR part (p), number of differentiations needed to obtain a stationary time series (d), order of the IMA part (q). The reason for this model to be one of the most used and the one that usually presents the best results, is due to the fact that it combines these two components, AR and IMA, which enable the creation of models much more robust and complex. The *auto.arima()* function presented in the R *forecast* package automatically chooses these three parameters, making the ARIMA model building an easy task. If the series is not stationary, the function itself differentiates the values [4, 20]. Since the vector constructed with several date events is likely to influence sales behavior it can be incorporate when building the model. This vector is represented in the code below by the variable *x_train*. To analyze the effect of this vector in the precision of the forecast, this experiment was also applied without it (model 2):

```
#Model 1 - auto.arima() function call with the events vector
M1=auto.arima(y_train,xreg=x_train[,-1],max.p=10,
max.P=10,max.q=10,max.Q=10)
#Model 2 - auto.arima() function call without the events vector
M2=auto.arima(y_train,max.p=10,max.P=10,max.q=10,
max.Q=10)
```

Experiment using Deep Learning

ML algorithms generally perform better if the time series has a consistent scale or distribution, which is achieved by normalization or standardization. Since the dataset consists mostly of binary variables and DL uses NN, this leads to choosing normalization rather than standardization because NNs assign different weights to the input values, so the dataset must be all on the same scale [21]. At the end, it was necessary to reverse this normalization so the error could be calculated on the same scale as the variable to be predicted [21]. For the structure of the LSTM networks it was necessary to determine how many layers would be the optimal, as well as the number of nodes and typology of each one. Using few neurons, the network could be unable to model complex data, resulting in *underfitting*. This situation can be recognized when the error is too great, both in the training and the test cases. If the number of neurons is too high for the dataset, it could result in *overfitting,* leading to loss of its predictive capability (the error in the training data decreases while in the test data it stagnates or starts to rise (Fig. 3)).

The final optimized architecture is presented in Fig. 4. Using *Keras* it is not necessary to define the first input layer because it is automatically created when the intermediate layers are defined. The following layers are two pairs of LSTM and Dropout layers. Finally, we included two Multi Perceptron Layers (Dense layers) with a Dropout layer in the middle. The LSTM layers appear in first place since they are the ones with memory capabilities (both old and recent).

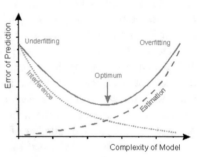

Fig. 3. Underfitting vs overfitting [22].

Layer (type)	Output Shape	Param #
lstm_1 (LSTM)	(None, 30, 128)	82944
dropout_1 (Dropout)	(None, 30, 128)	0
lstm_2 (LSTM)	(None, 64)	49408
dropout_2 (Dropout)	(None, 64)	0
dense_1 (Dense)	(None, 32)	2080
dropout_3 (Dropout)	(None, 32)	0
dense_2 (Dense)	(None, 1)	33

Total params: 134,465
Trainable params: 134,465
Non-trainable params: 0

Fig. 4. Architecture of the final model.

The activation function implements the decision regarding the output limits by combining the input values with their weights [10]. Since the case being studied was a regression problem of positive values, the chosen activation function for the *dense* layers was *ReLU*. It sets all negative values to zero and the rest remain as they are. In our first approaches in the design of the network architecture we verified the occurrence of overfitting. Generally, it can be reduced by adding more cases to the training set, reducing the complexity of the network architecture or adding *dropout* between the layers [10]. *Dropout* corresponds to erasing parts of the neural network allowing the network to generalize and depends on the problem and the results achieved in the training. Due to the fact that the size of the dataset was fixed, the number of nodes in the second LSTM layer was reduced and, consequently, in the first *Dense* layer also. Three *dropout* layers between the other layers were also needed.

An *epoch* consists of one full training cycle on the training cases. The *batch size* is the number of training cases that are used in each *epoch*. The number of *epochs* and *batch size* were increased until the appearance of overfitting. The best results were achieved with 1300 *epochs* and a *batch size* of 600.

4   Results

The presented results were obtained after a cycle process of parameter fine tuning with the goal of obtaining satisfactory results (Fig. 5).

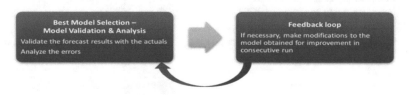

Fig. 5. Model fine-tuning cycle.

Statistical Models
The best model was selected using the Root Mean Squared Error (RMSE) analysis and the correlation between the real and predicted values. The discrepancy of the error between the two *auto.arima()* models highlights the relevance of the *x_train* vector in the predictive capability of the model. RMSE is commonly used to measure accuracy since it measures the mean magnitude of the errors in a set of predictions and the average magnitude of the error itself. This is probably the most easily interpreted error since it's presented in the units of the variable that is being predicted and depends on its range of values [23]. The correlation, as the name indicates, aims to measure the degree of relationship between variables, in this case, between *y_test* values and the predictions made.

The model that presented the best results was the *auto.arima()* using the *x_train* vector, which obtain 1352,951 for the RMSE and reached 66% in the correlation test (Table 2). The obtained values for the *p*, *d* and *q* parameters were 3, 1 and 3, respectively, which means that the time series was not considered stationary by the model and it had to differentiate the series once.

The *noise* (residuals) relating to predicted values should be similar to a white noise, i.e., independently of the fraction of time that is observed, it should look like floating points with no meaning. If the residuals don't exhibit a white noise-like behavior it is an indication that additional improvements can be made to the forecast model. Through the AutoCorrelation function (*acf*) graph, the histogram (check normal distribution) and applying the *Box-Pierce* and *Ljung-Box* it was possible to verify if the residuals were similar to a *white noise*. The *acf* is a measure of the correlation between the observations of a time series that are separated by k (lags) units of time $(y_t \, e \, y_{t-k})$. It is possible to have 1 or 2 significant correlations in higher order lags, however they can be due to random errors [24]. In the *Box-Pierce* test, if the *p-value* is less than 0.05 indicates that the series could be non-stationary [4].

Finally, it was concluded that the model successfully captured all the information present in the data since all these tests reported that there was no correlation present in the residuals.

Deep Learning

The DL analysis was done during the construction of the model. Like the loop presented in Fig. 5, the parameters were changed several times until they were finally adjusted to the model. Loss graphs were analyzed, showing the error that occurred for the training and test data during the training periods (*epochs*).

Table 2. RMSE and correlation average measure

	ARIMA	AVG (LSTM)
RMSE	1352.95	1220.55
Correlation	0.65	0.69

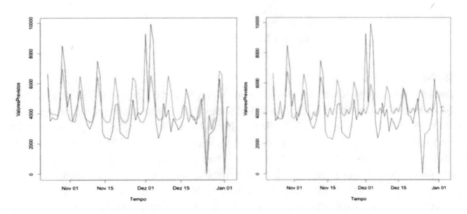

Fig. 6. Forecast graphs using the modeling function *auto.arima()* (with the events vector - left graph; without the events vector – right graph).

The validation of the model was also done using the RMSE and the correlation value. Since the initialization of the network is random it causes a variation in these results, so the model was applied to the training data seven times and then an average of its results was made. The maximum learning occurred at 750 *epochs* and 600 *batch size*.

The results of the two approaches can be observed and compared in Table 2 and Figs. 6 and 7. The RMSE of the ARIMA model and the average of the DL model when compared with the mean of the *y_test* values (4106.836) are

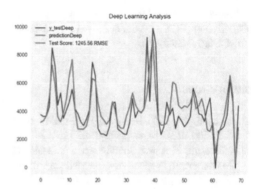

Fig. 7. Forecast graphs using the DL approach.

32,94% and 29,72%, respectively. In the graph of the ARIMA model at the right (without the events vector), the model is doing mostly an average of the values, being the predicted values near the middle of the graph. On the other ARIMA graph, the effect of using the events vector, can be verified by the two low values in the end. The LSTM networks also could identify this "outliers" with the corresponded events, showing the capacity of learning by these networks. In general, both the models could follow the behavior of the *y_test* series.

5   Conclusions

After comparing the results obtained by the two experiments it was possible to affirm that the deep learning approach achieved better results than the usually used traditional statistical model ARIMA, both in the RMSE and the correlation value. In both models, when the obtained *noise* series were analyzed, it was concluded that they captured the maximum information present in the data, which means that the models were correct and considered accurate.

DL in general and LSTM in particularly is still a growing technique but it is one that is increasingly emphasized by its success in several areas and this study has demonstrated the potential for deep learning in the forecasting area. Each experiment must be analyzed separated and all datasets have different behaviors, translating into different accuracies of precision by the models, however when this study was realized another three very different datasets were also analyzed and DL presented better results for them also.

An interesting analysis would be to study the introduction of a new product in the market, since it does not have a set of historical data to train and construct a model. Eventually using clustering, we could define to which group/category of products it belongs, and do a study from there. Another interesting approach would be to aggregate daily sales per week to verify if this could achieve less error. First it would be necessary to find the correspondent percentage of sales for each day of the week and then divide the weekly forecast by the respective percentage.

Acknowledgments. This work has been supported by COMPETE: POCI-01-0145-FEDER-007043 and FCT – Fundação para a Ciência e Tecnologia within the Project Scope: UID/CEC/00319/2013.

References

1. Yousefi, N., Alibabaei, A.: Information flow in the pharmaceutical supply chain. Iran. J. Pharm. Res. **14**(4), 1299–1303 (2015)
2. Microsoft: Business Intelligence for Healthcare : The new prescription for boosting Cost Management, Productivity and Medical Outcomes. Business Intelligence for Healthcare: The New Prescription for Boosting Cost Management, Productivity and Medical Outcomes. BusinessWe, February 2009
3. Ashrafi, N., Kelleher, L., Kuilboer, J.-P.: The impact of business intelligence on healthcare delivery in the USA. Interdiscip. J. Inf. **9**(9), 117–130 (2014)
4. Hyndman, R.J., Athanasopoulos, G.: Forecasting: Principles and Practice, pp. 46–51. OTexts, Australia (2014)
5. Shumway, R.H., Stoffer, D.S.: Time Series Analysis and Its Applications, 3rd edn. Springer, Heidelberg (2017)
6. Ohri, A.: Why every business analyst needs to learn R? (2012). http://analyticstraining.com/2012/why-every-business-analyst-needs-to-learn-r/. Accessed 01 June 2017
7. Verma, E.: A Quick Guide to R Programming Language for Business Analytics (2015). https://www.simplilearn.com/r-programming-language-business-analytics-quick-guide-article. Accessed 01 June 2017
8. Schmidhuber, J.: Deep learning in neural networks: an overview. Neural Netw. **61**, 85–117 (2015)
9. Deng, L., Yu, D., Deep Learning : Methods and Applications (2013)
10. Russakovsky, O.: Convolutional Neural Networks for Visual Recognition (2015)
11. Olah, C.: Understanding LSTM Networks (2015)
12. Alahi, K., Goel, V., Ramanathan, A., Robicquet, Fei-Fei, L., Savarese, S.: Social LSTM : Human Trajectory Prediction in Crowded Spaces (2014)
13. Zaytar, M.A., El Amrani, C.: Sequence to sequence weather forecasting with long short-term memory recurrent neural networks. Int. J. Comput. Appl. **143**(11), 975–8887 (2016)
14. May, M.: An Overview of Python Deep Learning Frameworks (2017)
15. Community, S.: SciPy Reference Guide (2015)
16. TensorFlow: https://www.tensorflow.org/. Accessed 01 Nov 2017
17. Keras: https://keras.io/. Accessed 01 Nov 2017
18. Brownlee, J.: Deep Learning With Python. https://machinelearningmastery.com/deep-learning-with-python/. Accessed 18 June 2017
19. Steinberg, D.: Why Data Scientists Split Data into Train and Test (2014). http://info.salford-systems.com/blog/bid/337783/Why-Data-Scientists-Split-Data-into-Train-and-Test. Accessed 11 June 2017
20. Hyndman, R.J.: New in forecast 4.0 (2012). https://robjhyndman.com/hyndsight/forecast4/. Accessed 12 June 2017
21. Brownlee, J.: How to Normalize and Standardize Time Series Data in Python (2016). http://machinelearningmastery.com/normalize-standardize-time-series-data-python/. Accessed 18 June 2017

22. Dieterle, D.F.: Overfitting, Underfitting and Model Complexity (2016)
23. Wesner, J.: MAE and RMSE — Which Metric is Better? (2016). https://medium.com/human-in-a-machine-world/mae-and-rmse-which-metric-is-better-e60ac3bde13d. Accessed 15 June 2017
24. Mukaka, M.M.: Statistics corner: a guide to appropriate use of correlation coefficient in medical research. Malawi Med. J. **24**(3), 69–71 (2012)

Data Science Analysis of HealthCare Complaints

Carlos Correia[1], Filipe Portela[1(✉)], Manuel Filipe Santos[1], and Álvaro Silva[2]

[1] Algoritmi Research Center, University of Minho, Guimarães, Portugal
{cfp,mfs}@dsi.uminho.pt
[2] ERS - Entidade Reguladora da Saúde, Porto, Portugal

Abstract. Nowadays, any health-related issue is always a very sensitive issue in the society as it interferes directly in the people well-being. In this sense, in order to improve the quality of health services, a good quality management of complaints is essential. Due to the volume of complaints, there is a need to explore Data Science models in order to automate internal quality complaints processes. Thus, the main objective of this article is to improve the quality of the health claims analysis process, as well as the knowledge analysis at the level of information systems applied to referred health. In this article, it is observable the development of data treatment in two stages: loading the data to an auxiliary database and processing them through the Extract, Transform and Load (ETL) process. With the data warehouse created, the Online Analytical Processing (OLAP) cube was developed that was later interconnected in Power BI enabling the creation and analysis of dashboards. The various models studied showed somehow a poor quality of the data that supports them. In this sense, with the application of the filters, it was possible to obtain a more detailed temporal perception, as the height of the year in which there is more affluence of registered complaints. Thus, we can find in this study the main analysis of paper complaints and online complaints. For paper complaints, a total of 234 records of the selected period is well-known for the "Unknown" valence affluence with 72.67% of the registrations. With regard to online complaints, a total of 42 records of the selected period is notorious for the following incidence: Typification "Other subjects" with 19.05% of registrations; State "Inserted" with 90.48% of registrations; Ignorance "Unknown" with 95.24% of registrations; Typology "Complaint" with 69.05% of registrations.

Keywords: Data Science · Knowledge discovery in database
Health information systems · Business Intelligence
Quality of healthcare complaints

1 Introduction

Nowadays, a good management of information quality by organizations is most important. Through a consistent systematization of the data, it is possible to create standards that can later be able to be molded giving answers to the needs of the client, allowing greater access and perception of the information generated. With the advancement of technologies, it is possible to design standards using Data Science (DS) techniques. Information systems have been playing a key role in health as they enable better organization and

© Springer International Publishing AG, part of Springer Nature 2018
Á. Rocha et al. (Eds.): WorldCIST'18 2018, AISC 747, pp. 176–185, 2018.
https://doi.org/10.1007/978-3-319-77700-9_18

quality of information. All information recorded in hospital settings may be essential in problems with future users. With the advancement of technologies, it is possible to improve the internal processes of high importance of the hospital centers. Nowadays, data quality is one of the main concerns of health entities. In this sense, this study is based on the exploration of Data Science techniques in order to understand how the processes of complaint analysis can be improved, in the quality. In the development of this study, techniques such as Business Intelligence (BI) and Big Data were still considered and applied. In this way, BI was applied with the purpose of data analysis and processing and demonstration of results, such as dashboards. On the other hand, Big Data was applied in terms of the amount of data recorded and their processing in real time.

After the introduction, in the second section, a presentation of the work plan was made. In the third section, the development of the solution is presented, that is, a description of the data handling as well as the development of dashboard analysis. This article ends with the discussion of the results and a section with the conclusion and future work.

2 Background

As stated before, this article aims to offer a study about the quality of health claims, identifying possible improvements in the quality of its analysis process using Data Science.

2.1 Complaints Management Entity

Since the 90 s, Portugal has undergone some changes in the main basis of life, Health. As reform initiatives were varied, always in the sense of improvement essentially hospital management [1]. According to Reis [2], healthcare providers establishments have been subject to changes in management, in search of efficiency and quality of the services provided. In order to improve more and more regulation and monitoring of healthcare establishments, a creation of the Health Regulatory Entity (ERS) emerges in 2003. This is a public and independent entity, endowed to serve the users essentially at security levels, quality and rights [3]. The ERS faced serious problems shortly after its creation, for this new entity was a bad entity by many people. According to Moreira [4], "an ERS won between the test of time and the test of legitimacy, only for the intransigent, atavistic and sectarian opposition of the doctors order." The ERS seeks to access its efficiency in the regulation of health in two fundamental aspects, the economic and social, being the economic aspect is essentially with the control of prices, production and the market, since the social aspect essentially deals with the control of compliance of users rights [5].

2.2 Quality of Information in Health

The quality of information is one of the foundations for the survival and greater competitiveness of organizations. Still, quality is increasingly demanded and equally important in the everyday life of any human being.

The role of information quality in health has been growing exponentially over the years. The quality of information, as well as the availability of quality, grows parallel to the technological advances, thus obliging the adaptation of the healthcare providers establishments. In this way, the availability of information in the largest communication network, the internet, became essential.

According to Berner [6], a good analysis of the content of the information consulted by the user is essential in order to confirm the veracity of the same, in addition, the user must evaluate if the information is effectively useful for him, this because the majority of the available information are generalized, namely aspects such as symptoms of some type of diseases that can vary from user to user.

With regard to the quality of the complaints register, it is considered by many to be an essential component of health systems, as important as health promotion strategies. A good record of complaints can contribute effectively to the improvement of the health system, allowing the identification of occasional failures or trends. In addition to improving the quality of the service provided, effectively dealing with complaints can improve internal communication between health - care professionals, not to mention the increase of trust and satisfaction of the user [7]. Nowadays it is fundamental in critical areas to have intelligent systems that are able to support the decision making process giving important information in the right moment [8].

2.3   Related Work

An early study was focused on the visualization of the information contained in the complaints [9, 10]. The only common aspect in this research study with Oliveira [10] is the data and the typology of the same. Thus, following this study, the data available are related to online complaints and paper complaints. The data provided are the result of complaints made by users of health care providers, who somehow felt the need to manifest themselves. Contrary to the name itself, these users do not only resort to complaints to express negative opinions, but also serve as a means of communication to share positive opinions with the body responsible for receiving and handling these opinions, the ERS. This data also results from the aggregation of all the opinions shared by these users through complaint books (paper complaints) and complaints portal (online complaints). ERS, like any other entity responsible for aggregation of information, is totally impartial to the information contained in the complaints, and it is perfectly normal to find useless information without possibility of analysis. The opposite is also apparent in such complaints, that is, constructive criticism, positive testimony, and even acknowledgment of good practice by healthcare providers.

3   Solution Development

This point reflects the practical development of the study and is divided into five sections, including data preparation, the Extract, Transform and Load (ETL) process, the creation of the Online Analytical Processing (OLAP) cube, the introduction to Power BI with the development and analysis of dashboards. In the first section you can identify all stages of data preparation. In the following section we present the entire ETL process as well as the treatment to which the data were subjected. In the following section you can see the process of creating the OLAP cube, responsible for making data available for analysis. Finally, the process of creating dashboards as well as their analysis is presented.

3.1 Data Analysis

The process of data preparation is fundamental for success in the development of the study, and it is important to avoid any gaps that may exist. In this way, it is important to point out that the data provided appeared as unstructured data, with too many errors and with some lack of information. For a better perception of the treatment performed, a multidimensional model was developed, according to Fig. 1.

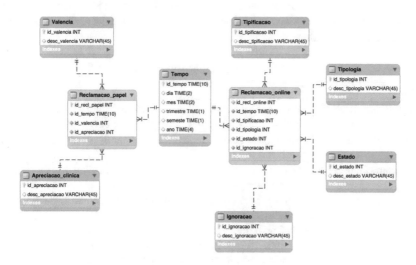

Fig. 1. Multidimensional model

As can be seen in Fig. 1, the tables of facts "Paper Complaint" (Reclamacao_papel) and "Online Complaint" (Reclamacao_online) are constituted only by id's coming from the dimensions that support them, being "Valence" (Valencia) and "Appreciation_clinical" (Apreciacao_clinica) for paper complaints and "Typification" (Tipificacao), "Typology" (Tipologia), "State" (Estado) and "Ignoration" (Ignoracao) for online complaints. There is also a dimension common to the two tables of facts, the dimension "Time" (Tempo), endowed with different types of granularity.

The tables directly related to the "Paper Complaint" (Reclamacao_papel) and "Online Complaint" (Reclamacao_online) fact tables are composed of id's and their descriptions for which there is a relationship to the claims. All other descriptions that do not meet these parameters are not considered.

After the development of the multidimensional model, it was identified the need to create an auxiliary database so that it was possible to establish the association of the data present in the non-structured data tables "ers_complaints" (ers_reclamacoes) and "ers_complaintsonline" (ers_reclamacoesonline) with the necessary tables to model.

First, we identified the data of direct correspondence, that is, data that were already in separate tables. The result of this first identification is the tables "Valence" (Valencia), "State" (Estado) and "Typification" (Tipificacao). Through a query it was then possible to insert the desired data (id and description) into a database created in SQL Server previously prepared with its data structuring. It should be emphasized, once again, that only the data that intersect were considered in this study.

3.2 Data Preparation

The ETL process is the process responsible for extracting data from various sources, cleaning, optimizing and inserting the same data in a destination database, also known as DW (data warehouse) [11]. In this phase of the project through the SQL Server Integration Services (SSIS) module of visual studio data tools was organized the entire process of data transformation and loading.

Table 1. Others changes in data

Table	Change
Appreciation_clinical (Apreciação_clinica)	–Placing the description "Not considered" for data with id 0; –Placing the "Considered" description for data with id 1.
Typology (Tipologia)	–Placement of the description "Other" for data with the description "Null" and "–1".
Ignoration (Ignoração)	–Placing the description "Unknown" for data with the description "Null" and ""; –Placing the description "Out of scope" for data with the description "Unsubstantiated"; –Placing of the "Anonymous" description for data with the description "Anonymous", "Anonymous Claim", "Anonymous, with no material to analyze" and "Anonymous Denunciation; internal inspection request"; –Placement of the description "Repetition" for data with the description "repetition".
Typifications (Tipificação)	–Placing the description "Other Subjects" for data with the description "Other", "Other", "Other topics" and "Null". –If there was no correspondence for typifications with id "Null" and "–1", id 296 was assigned to this data, referring to the description "Other Matters".
Valence (Valência)	–If there was no correspondence for valences with id less than 3 and greater than 44, id 80 was assigned to this data, referring to the description "Unknown".

This process can be divided into five phases, such as database creation, data processing, data loading, processing of facts and loading of facts.

In a first phase, a query was created (Cria DW_Tese) responsible for the creation of all the necessary tables for the study. This query resulted from the previous analysis of the data, guaranteeing that there were no problems of inconsistency in the loading of data, that is, that there were agreement of types, namely size and accuracy of the field as well as the corresponding type (numeric or text). The time table, also developed through a query, is responsible for loading all the dates needed for the study, based on the time interval considered in the given dataset. With this phase completed, it was then necessary to process the data previously loaded and organized in an auxiliary database. At this stage, the data not subject to treatment were the data regarding valences and states. All others have undergone the necessary changes for this development, then presented in Table 1 for better understanding.

After processing the data, safeguarding all hypotheses of error, it was possible to process this model by inserting 12255 records for paper complaints and 1799 records for online complaints. With the handling and data loading process developed, with a total of 14054 complaints records, conditions for cube development are met in the SQL Server Analysis Services (SSAS) module of visual studio data tools.

3.3 Cube Creation

The OLAP cube, considered by Berson and Smith [12] a data structure capable of providing a rapid analysis of the data through multidimensional views, allows the identification of patterns and trends in the data. These systems are very common in multidimensional data models such as that used in research project, data warehouse. In this stage, the OLAP cube was developed, which will later be considered in the data exploration through dashboards in the Power BI tool. In the development of the cube, a link was created to the previously created data warehouse. The representative multidimensional model is exactly the same as the multidimensional model considered in the previous steps. With the established connections it was necessary to develop the necessary hierarchies, and the most important in this process is the hierarchy of the time dimension, since all other dimensions need only the description of the data in question. In this sense, the hierarchy considered for the time dimension is: Year (Ano); Semester (Semestre); Quarter (Trimestre); Month (Mês) and Day (Dia).

4   Dashboards and Discussion of Results

At this point, dashboards have been created and exploited, evidencing in some way a poor quality of the data that supports them. In this sense, with the application of the filters it is possible to obtain a more detailed time perception, such as the height of the year in which there is more affluence of recorded complaints. This study focused essentially on the analysis of the year, semester, quarter, month and day with more complaints registered. Then we can find the main analysis of a total of five elaborate.

- Paper complaints

   The analysis developed for the year 2014 in which there are records of the full year, has a total of 11013 complaints, 322 of which relate to the selected period, as can be seen in Fig. 2. In this period, the "Unknown" valence (valencia) prevails with 72.67% of the registrations with respect to 234 registrations. The next valence (valencia), no less important is "Orthopedics" with 4.04% which relates to 13 registrations.

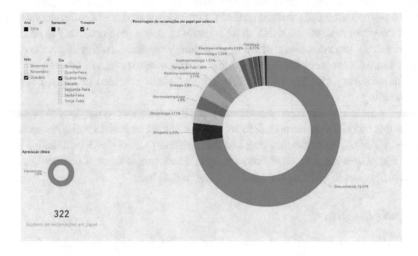

Fig. 2. Analysis of paper complaints in 2014

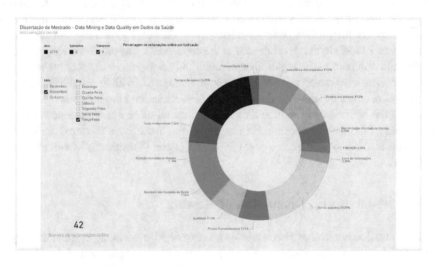

Fig. 3. Analysis of online complaints in 2014

Regarding the analysis of paper claims previously presented, based on the season of the year (Autumn/Winter) we can consider that the influx of complaints may be due to the onset of cold and consequent of the appearance of the first flu. According to the DGS influenza is a seasonal disease that manifests mainly during the winter (Freitas, 2015). However, they are abstract conclusions, since the great affluence of the registers presents unknown valence. In these cases, consideration should be given to the possibility of extending the valence by giving more specificity to this topic, thus reducing the high number of inconspicuous registrations.

• Online complaints

The analysis developed for the year 2014 in which there are full-year records has a total of 1511 complaints, of which 42 relate to the selected period, as can be seen in Figs. 3 and 4. In this period prevails Table 2:

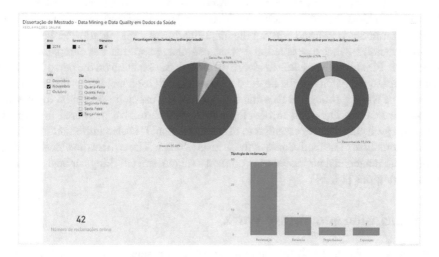

Fig. 4. Analysis of online complaints in 2014 (continuation)

Table 2. Observations in dashboards data

Table	Observations
Typification (Tipificação)	–"Other business" (Outros assuntos) with 19.05% of registrations (8 registrations); –"Waiting times" (Tempos de espera) with 14.29% of the registers (6 registers).
State (Estado)	–"Inserted" (Inserida) with 90.48% of the registers (38 registers).
Ignoration (Ignoração)	–"Unknown" (Desconhecida) with 95.24% of the records (40 records).
Typology (Tipologia)	–"Complaint" (Reclamação) with 29 records.
Typification (Tipificação)	–"Other business" (Outros assuntos) with 19.05% of registrations (8 registrations); –"Waiting times" (Tempos de espera) with 14.29% of the registers (6 registers).

Regarding the analysis of the online complaints presented above, based on the season of the year (Winter) we can consider, as in paper complaints, that the influx of complaints may be due to the cold and consequent of the strong occurrence of influenza. Regardless of the reason for the high number of complaints at this time of year, we can observe that most of the registrations are of the complaint type, which alone is already cause for alarm because something does not please the majority of users. On the other hand, we have easily identified that most complaints have the status entered, that is, it is accepted and analyzed. However, should these complaints already entered and reviewed be ignored, they must present a reason. In this analysis we can also identify that the vast majority of reasons for ignorance are unknown. Before this case we can draw two conclusions, we can consider that the data do not correspond and are incorrect or we can consider lack of professionalism at the moment of analysis. If we are faced with the second case and the complaint is ignored without the identification of a reason, the user has no way of understanding what has failed and what can change so that the complaint sent is successful. Finally, it is possible to identify a clear dissatisfaction on the part of the users with regard to the quality of the health care provided. Subsequent to this dissatisfaction are also the waiting times, the fees and the rights of the users. It is essential to take into account the above facts and to develop plans with corrective measures in a timely manner, modifying aspects essential to user satisfaction. It is also important to draw up reports to analyze the effectiveness of the measures taken. The artifact developed is ready to be implemented in a pervasive environment using internet of things or business intelligence solutions [13, 14].

5   Conclusion and Future Work

With the study finished, it is time to make an evaluation regarding the accomplishment of the initially defined objectives. Together with the objectives, it is important to reflect on the scientific question posed, which seeks answers to the problem presented by the Health Regulatory Entity: *How can Data Science improve the quality of the health claims process?*

In this sense, in response to the scientific question, we can affirm that through the research carried out Data Science can help to improve the process of analysis of the quality of complaints, because based on the reports developed it is possible to observe a strong incidence of data without relevant information, showing the poor quality of the complaints received. Thus, it is necessary to rethink the restructuring of the existing complaint model, influencing the process of specification of the themes by the user. Attained the main objective of this study, resulting in automatic and real-time reporting of the data, based on all the processing and processing of the initial dataset. After the end of the investigation and as a form of evaluation of the fulfillment of the objectives, we can say that the results obtained were satisfactory.

After completing this study, there are some points to be developed in the future, highlighting among them:

- Addition of study attributes to the already developed model, extending the exploration of indicators through the development of dashboards;
- Development and application of regression and classification models;
- Development of the same study, but based on updated ERS data.

Acknowledgement. This work has been supported by COMPETE: POCI-01-0145-FEDER-007043 and FCT – Fundação para a Ciência e Tecnologia within the Project Scope: UID/CEC/00319/2013

References

1. Simões, J.: As parcerias público-privadas no sector da saúde em Portugal. Revista Portuguesa de Saúde Pública **4**, 79–90 (2004)
2. Ribeiro, V.: O que é Data Warehouse? (2011). https://vivianeribeiro1.wordpress.com/2011/03/30/o-que-e-data-warehouse/
3. Anabela, L.: Entre o direito a reclamar e o direito à saúde. Serviço social em gabinetes do cidadão, do SNS (2014)
4. Moreira, V.: Regulação dos serviços da saúde - público (2011)
5. Almeida, L.: A criação da Entidade Reguladora da Saúde em Portugal | Comunicação Por conta e Risco (2010). https://porcontaerisco.wordpress.com/2010/12/03/a-criacao-da-entidade-reguladora-da-saude-em-portugal/#comments. Accessed 18 Oct 2016
6. Berner, E.S.: Clinical decision support systems: theory and practice. Springer, New York (1999)
7. Vasconcelos Parra, R.: Reclamações no setor público da saúde (2014)
8. Portela, F., Gago, P., Santos, M.F., Machado, J., Abelha, A., Silva, Á., Rua, F.: Implementing a pervasive real-time intelligent system for tracking critical events with intensive care patients. Int. J. Healthc. Inf. Syst. Inform. **8**(4), 1–16 (2013)
9. Oliveira, A., Portela, F., Santos, M.F., Neves, J.: Towards an intelligent system for monitoring health complaints, pp. 639–649. Springer, Cham (2016)
10. Oliveira, A., Portela, F., Machado, J., Abelha, A., Neves, J.M., Vaz, S., Silva, A., Santos, M.F.: Towards an ontology for health complaints management. In: KMIS 2015 - International Conference on Knowledge Management and Information Sharing, Lisbon, Portugal, pp. 174–181. SciTePress (2015). ISBN978-989-758-158-8
11. Negash, S.: Business intelligence. Commun. Assoc. Inf. **13**(15), 177–195 (2004)
12. Berson, A., Smith, S.J.: Data Warehousing, Data Mining, and OLAP. McGraw-Hill, New York (1997)
13. Guarda, T., Augusto, M.F., Barrionuevo, O., Pinto, F.M.: Internet of things in pervasive healthcare systems. In: Machado, J., Abelha, A., Santos, M., Portela, F. (eds.) Next-Generation Mobile and Pervasive Healthcare Solutions, pp. 22–31. IGI Global, Hershey (2018). https://doi.org/10.4018/978-1-5225-2851-7.ch002
14. Pereira, A., Portela, F., Santos, M.F., Abelha, A., Machado, J.: Pervasive business intelligence: a new trend in critical healthcare. In: Procedia Computer Science - ICTH 2016 - International Conference on Current and Future Trends of Information and Communication Technologies in Healthcare, vol. 98, pp. 362–367. Elsevier (2016). ISSN1877-0509

Check for updates

An Overview of Big Data Architectures in Healthcare

Hugo Torres, Filipe Portela[(✉)], and Manuel Filipe Santos

Algoritmi Research Center, University of Minho, Guimarães, Portugal
{cfp,mfs}@dsi.uminho.pt

Abstract. It is proven that Big Data is related to an increase in efficiency and effectiveness in many areas. Although many studies have been conducted trying to prove the value of Big Data in healthcare/medicine, few practical advances have been made. In this project, an analysis and a comparison were made of the existing Big Data technologies applied in healthcare. We analyzed a Big Data solution developed for the INTCare project, a Hadoop-based solution proposed for the Maharaja Yeshwatrao hospital located in India and a solution that uses Apache Spark. The three solutions mentioned above are based on open source technology. The IBM PureData Solution for Healthcare Analytics solution used at Seattle's Children's Hospital and the Cisco Connected Health Solutions and Services solution are part of the proprietary solutions analyzed.

Keywords: Big Data · INTCare · HealthCare

1 Introduction

Since the invention of computers, large volumes of data are generated at a surprising rate [1]. Up to 2003, 5 exabytes of data were created by the human being; currently, this amount is created in 2 days [2]. This is how Big Data began to reveal its true potential in dealing with large volumes of data from various sources and generated at high speeds. The health industry generates huge amounts of data, though most of it is stored in non-digital format. Nowadays, the trend is to digitize most of the information [3]. According to Feldman et al. [4], the increase in the volume of data in the health industry comes, not only from the creation of new forms of data (three-dimensional images, biometric sensor readings, and others), but also from the transformation of existing data, such as radiology images, DNA sequence data and other, to digital format. Given the noticeable delay in the adoption of Big Data technologies by the health industry, it is necessary to identify the challenges and potential use of Big Data in this industry and to identify cases of adoption of Big Data technologies in hospitals/healthcare clinics. To ease the adoption of Big Data technologies in the health industry, this study aims to identify, analyze, filter and compare the solutions identified.

This paper is divided into six sections: Introduction; Background; Methods and Tools; Case Study; Discussion; Conclusion and Future Work. The second section presents the challenges and potential of Big Data in the healthcare industry. In Sect. 3, the methods and tools utilized for this project are presented and described. In Sect. 4 the

© Springer International Publishing AG, part of Springer Nature 2018
Á. Rocha et al. (Eds.): WorldCIST'18 2018, AISC 747, pp. 186–194, 2018.
https://doi.org/10.1007/978-3-319-77700-9_19

case study is presented, the various solutions found and the comparison between the filtered solutions. In Sect. 5 the results are analyzed in the project context. In Sect. 6, a summary of this paper is given, describing the main conclusions. Finally, in Sect. 7 a short description of the future work is presented.

2   Background

2.1   Big Data

"Big data refers to datasets whose size is beyond the ability of typical database software tools to capture, store, manage, and analyze." Manyika et al. [8].

According to Zikopoulos et al. [5] the three characteristics that define Big Data are: Volume, Variety and Velocity. Hurwitz et al. [6] state that the three V's quoted above is too simplistic and proposed a fourth V, veracity. In the literature, there are some authors who add a fifth V, value [7]. Therefore, Big Data can be characterized by: (a) Volume – refers to the large amounts of data generated that grows exponentially and comes from a variety of sources; (b) Variety - According to Zikopoulos et al. [5], society invests a large part of its time with structured data (representing 20% of the total volume of data generated), but the challenge lies in the remaining 80% which are semi-structured or unstructured; (c) Velocity - The speed at which the data is generated. Today, many of the data that is generated has an "expiration date", that is, they are only relevant to organizations if they are analyzed almost in real time [5]; (d) Value - This characteristic is related to the economic value of the data; (e) Veracity - This characteristic is related to the quality of the data.

According to Feldman et al. [4], the increase in the volume of data in the health industry comes, not only from the creation of new forms of data (three-dimensional images, biometric sensor readings, among others), but also from the digitization of existing data, such as radiology images, DNA sequence data, among others.

The McKinsey Global Institute conducted a study to understand the potential of Big Data in five areas, one of which was the health area in the United States. Despite the importance of this sector in the country's economy (representing 17% of GDP), it is still possible to notice a delay in the adoption of Big Data in relation to other industries [8].

3   Big Data Architectures Used in Healthcare

3.1   Big Data Architecture for the INTCare Project

INTCare is a research project developed at the Intensive Care Unit (ICU) of the Centro Hospitalar do Porto, which, at an early stage, was designed to develop an intelligent system to predict organ failure and its effects on users (Portela et al. [10]). The INTCare project uses a continuous flow and real-time data collection, from several sensors and monitoring devices, which generates a volume of data from 50 to 500 Terabytes [9]. According to Portela et al. [10], in 2009 the excessive amount of medical records on paper or manually entered into the database became apparent. After a set of studies aimed at identifying the gaps in the ICU information system, it was possible to develop a new solution based on intelligent systems capable of performing automatic tasks such as data acquisition and processing [10]. Nowadays, INTCare is a Pervasive Intelligent Decision Support System (PIDSS) that acts automatically and in real-time, in order to provide new information to decision-making entities in the ICU, i.e. physicians and nurses [10–12]. The Data Management subsystem of this architecture relies on Apache Kafka and Apache Storm for streaming data processing. For operational data processing it relies on Apache Phoenix and Apache HBase. The processing of analytical data is ensured by Apache Hive. Security, administration and operations are ensured by the following Hadoop subprojects: Apache Sqoop; Apache Knox; Apache Ranger; Apache Flume; and Apache Oozie.

3.2   Big Data Architecture with Apache Spark

Liu et al. [13], proposed an architecture for a Big Data processing tool, designed for the health area, that includes Apache Spark.

The "Data Storage" layer of this architecture, has the Hadoop Distributed File System, and Apache HBase. The "Data Processing" layer contains Apache Spark and Spark Streaming. Apache Hive and Spark SQL, are a part of the "Access to data" layer. Finally, the "Analytics and Business Intelligence" layer contains the MLib, GraphX and SparkR tools.

3.3   Big Data Architecture Designed for the Maharaja Yeshwatrao Hospital

Ojha and Mathur [14] proposed a Big Data architecture to address the needs of the Maharaja Yeshwatrao hospital located in Indore, India, which is considered, by the authors, as the largest public hospital in central India. Ojha and Mathur [14] state that the hospital generates large volumes of data based on the number of citizens who attend it daily. With the implementation of Big Data technologies, the authors intend to improve the quality of patients life, especially those in need, since long waiting times have a negative impact on the poorest patients because it forces citizens to lose working days and consequently lose a portion of their salary.

Using Big Data and data analysis tools, it will be possible to store the data the Maharaja Yeshwatrao Hospital generates and, therefore, health professionals will be able to find new knowledge, hidden patterns and trends which may result in improved treatments, reduction of readmissions, reduction of expenses, and others Ojha and Mathur [14].

The "Data Storage" layer is composed of the HDFS as well as Apache HBase. The "Data Processing" layer contains the Hadoop MapReduce. Apache Hive, Apache Pig and Apache Avro are part of the "Data Access" layer. The "Management" layer consists of the Apache Zookeeper and Apache Chukwa.

3.4   IBM PureData Solution for Healthcare Analytics

This architecture is comprised of the IBM PureData Solution for Healthcare Analytics solution that is being used at Seattle's Children's Hospital to improve diagnostic and patient care capabilities [15].

IBM PureData Solution for Healthcare Analytics is a solution developed by IBM that integrates various technologies to meet the Big Data needs of a healthcare organization. This solution has the following components: IBM Cognos Business Intelligence - business Intelligence suite; IBM PureData System - a highly scalable system that relies on servers, databases, storage, and others; IBM Healthcare Provider Data Model - set of data models and business solution models; IBM InfoSphere Information Server for DataWarehouse - a data integration platform that supports the capture, integration and transformation of large volumes of structured or unstructured data [16].

3.5   Cisco Connected Health Solutions and Services

The infrastructure developed by Cisco Systems, Inc., integrates multiple services into a single solution that can meet "all" the needs of a healthcare organization.

On the official Cisco Systems, Inc. website, we can see the various applications and the various services they offer. The services are divided into 6 categories: Personalized service to the user; Remote assistance and collaboration; Simplify clinical workflows; Increase efficiency in the workplace; Connect the research and development department with the production; Enable security and compliance. This solution was presented to the scientific community by Nambiar et al. [17].

4   Benchmarking

In this chapter we will compare the three selected architectures, more specifically, their components (from the "Data processing" layer). The architectures have been selected based on the type of their license, only open source solutions will be compared.

As it can be seen from Table 1, in the "Data Storage" layer, all solutions consist of the HDFS and Apache HBase tools.

In the "Data Processing" layer, the Big Data Architecture with Apache Spark consists of Apache Spark and Spark Streaming, the Big Data Architecture designed for the Maharaja Yeshwatrao Hospital has the Hadoop MapReduce, and the Big Data Architecture for the INTCare Project consists of Apache Kafka, Apache Storm and Apache Phoenix.

In the "Management" layer the Big Data Architecture with Apache Spark does not present any tool, the Big Data Architecture designed for the Maharaja Yeshwatrao Hospital has Apache Zookeeper and Apache Chukwa, the Big Data Architecture for the INTCare Project has Apache Oozie and Apache Flume tools.

Table 1. Comparative table of the selected architectures

Layer	Big Data Architecture with Apache Spark	Big Data Architecture designed for the Maharaja Yeshwatrao Hospital	Big Data Architecture for the INTCare Project
Data Storage	HDFS	HDFS	HDFS
	Apache HBase	Apache HBase	Apache HBase
Data Processing	Apache Spark	Hadoop MapReduce	Apache Kafka
	Spark Streaming		Apache Storm
			Apache Phoenix
Management	No information	Apache Zookeeper	Apache Flume
		Apache Chukwa	Apache Oozie
Data Access	Apache Hive	Apache Hive	Apache Hive
		Apache Pig	Apache Sqoop
		Apache Avro	
Analytical and Business Intelligence	Spark SQL	No information	Knowledge Management subsystem of the INTCare project
	MLib		
	GraphX		
	SparkR		
Security	No information	No information	Apache Knox
			Apache Ranger

The "Data Access" layer has Apache Hive present in all architectures, but the Big Data Architecture designed for the Maharaja Yeshwatrao Hospital also includes Apache Pig and Apache Avro and the Big Data Architecture for the INTCare Project also includes Apache Sqoop.

The Apache Spark Architecture presents the Spark SQL, MLib, GraphX and SparkR tools for the "Analytical and Business Intelligence" layer, while the Big Data Architecture for the INTCare Project has the knowledge management subsystem developed for the INTCare project and the Big Data Architecture designed for the Maharaja Yeshwatrao Hospital does not present any tool.

Finally, in the "Security" layer only the Big Data Architecture for the INTCare Project presents tools, which are Apache Knox and Apache Ranger.

4.1 Comparison Between Hadoop MapReduce and Apache Spark

In this subchapter, the differences between the two data processing frameworks, Hadoop MapReduce and Apache Spark, will be presented. Afterwards, two experiments will be presented comparing the performance of the two frameworks in several scenarios.

Table 2 shows the main differences between MapReduce and Apache Spark.

Table 2. Main differences between Hadoop MapReduce and Apache Spark [18].

Hadoop MapReduce	Apache Spark
Stores data on disk	Stores the data in memory. The data is first stored in memory and then processed
Computing based on disk memory, partial use of RAM (Random Access Memory)	Computing based on RAM memory, partial use of disk memory
Fault tolerance is achieved through replication	Fault tolerance is achieved through RDDs
Difficult to process and analyze data in real time	Can be used to modify data in real time
Inefficient for applications that need to constantly reuse the same dataset	Stores the dataset in RAM for efficient reuse

Shi et al. [19] conducted an experiment to compare the performance between the two frameworks. The experiment consisted of the execution of several workloads (WordCount, Sort, K-means) that simulated the real-world use of these frameworks.

Apache Spark performed better in the execution of WordCount. For a 1 GB input, Apache Spark was 34 s faster, for 40 GB it was 110 s faster and finally, for 200 GB Apache Spark was 398 s faster at executing the task.

When executing Sort, for a 1 GB input, Apache Spark performed the task in less time with a difference of 3 s compared to Hadoop MapReduce, but the same was not visible for an input of 100 and 500 GB, where MapReduce executed the task with a difference of 1.5 m and 20 m respectively.

For both the first and subsequent iterations of the k-means execution, Apache Spark presents shorter execution times, to emphasize the fact that the difference in time is accentuated in the following iterations due to the caching mechanism of Apache Spark.

Gu and Li [20] conducted an experiment to compare the performance of Hadoop MapReduce and Apache Spark in performing iterative tasks. PageRank was the algorithm chosen for the experiment.

Runtimes varied depending on the size of the dataset. For small datasets (between 1 and 11 MB) Apache Spark was 25–40 times faster to complete the tasks. For datasets with sizes between 40 MB and 89 MB Apache Spark was about 10–15 times faster than MapReduce. For datasets whose size is comprehended between 200 and 600 MB Apache Spark was between 3 to 5 times faster than MapReduce. When the dataset size exceeds 1 GB, MapReduce performed better than Apache Spark, and for some cases, Apache Spark failed during the execution while MapReduce concluded the task.

4.2   Comparison Between Apache Spark and Apache Storm

The experience conducted by Lu et al. [21] consisted of the execution of 7 workloads (Identity, Sample, Projection, Grep, Wordcount, DistinctCount, Statistics) to simulate various scenarios of the real use of these frameworks. After analyzing the experience that compares Apache Spark Streaming to Apache Storm, it is possible to observe that Apache Spark Streaming has better throughput values (average number of processed records per second), but the same does not happen in the latency (average of the intervals from the arrival of each record until the end of processing it) values, where Apache Storm presents better values, except for the values obtained in the execution of

WordCount workload. As for the ability to handle data, Apache Storm presents worse results compared to Apache Spark Streaming.

5   Discussion

Based on this study, it is possible to conclude that there is not much scientific documentation about the implementation of Big Data technologies in hospitals/health clinics. It is also possible to conclude that the approval of the scientific community can help to overcome some of the challenges that are presented to the adoption of Big Data technologies in the health area.

Given the results obtained in the analyzed experiments, it is possible to conclude that:

- The Big Data Architecture for the INTCare Project is best suited to handle streaming data. This solution combines Apache Kafka and Apache Storm to handle data from bedside monitors (vital signs, ventilation and others) [9];
- The Maharaja Yeshwatrao Architecture is best suited to handle large volumes of data, although Hadoop MapReduce performs poorly against Apache Spark in most of the tests presented in subchapter 6.1, it has been able to handle large volumes of data;
- The Big Data Architecture with Apache Spark is a hybrid solution as it has proven capable of handling streaming and batch data. However, the performance of Apache Spark is very dependent on the configuration of the cluster.

Although it was not possible to make a direct performance comparison of all the solutions chosen for benchmarking, it is possible to conclude that the most appropriate architecture for a healthcare organization is the Big Data Architecture for the INTCare Project. This solution presents in detail all the components and how they will interact with the system where they are inserted and, more importantly, it is the only solution that presents components in the "Security" layer (as it can be seen in Table 1). Since security is one of the challenges to implementing Big Data in healthcare, it is considered necessary to integrate tools that ensure data and system security in general.

6   Conclusion and Future Work

The realization of this project made it possible to understand the state of implementation of Big Data technologies in healthcare, it is potential and the main challenges. The research of applications used or designed to be implemented in hospitals/health clinics has proved to be the most challenging task of this project, due to the scarcity of literature regarding the implementation of Big Data solutions in healthcare. The research of experiments carried out on applications similar to those chosen for comparison allowed to evaluate the performance of the applications in several scenarios, therefore, it was possible to perceive the strengths and weaknesses of the chosen solutions. The research of Big Data technologies used in healthcare, revealed the variety of solutions to be explored, and showed that there is no ideal solution that can satisfy all the needs. Still there are some areas to be explored in the future, among them which include the research of Big Data solutions similar to those presented that have not yet been presented to the scientific community and the execution of practical tests on the tools presented with real data.

Acknowledgements. This work has been supported by COMPETE: POCI-01-0145-FEDER-007043 and FCT – Fundação para a Ciência e Tecnologia within the Project Scope: UID/CEC/00319/2013.

References

1. Yaqoob, I., et al.: Big data: from beginning to future. Int. J. Inf. Manag. **36**(6), 1231–1247 (2016)
2. Sagiroglu, S., Sinanc, D.: Big data: a review. In: 2013 International Conference on Collaboration Technologies and Systems, pp. 42–47 (2013)
3. Raghupathi, W., Raghupathi, V.: Big data analytics in healthcare: promise and potential. Heal. Inf. Sci. Syst. **2**, 3 (2014)
4. Feldman, B., Martin, E.M., Skotnes, T.: Big data in healthcare - hype and hope. Dr. Bonnie 360 degree (bus. Dev. Digit. Heal. **2013**(1), 122–125 (2012)
5. Zikopoulos, P., Eaton, C., DeRoos, D., Deutsch, T., Lapis, G.: Understanding Big Data: Analytics for Enterprise Class Hadoop and Streaming Data. McGraw-Hill, New York (2012)
6. Hurwitz, J., Nugent, A., Halper, D.F., Kaufman, M.: Big Data for Dummies. John Wiley & Sons Inc., Hoboken (2013)
7. Taurion, C.: Big Data (2013)
8. Manyika, J., et al.: Big data: the next frontier for innovation, competition, and productivity. McKinsey Glob. Inst., p. 156, June 2011
9. Gonçalves, A., Portela, F., Santos, M.F.: Towards of a real-time big data architecture to intensive care. In: Procedia Computer Science - ICTH 2017 - International Conference on Current and Future Trends of Information and Communication Technologies in Healthcare, pp. 585–590. Elsevier (2017). ISSN 1877-0509
10. Portela, F., Santos, M.F., Machado, J., Abelha, A., Silva, Á., Rua, F.: Pervasive and intelligent decision support in intensive medicine – the complete picture (2014)
11. Guarda, T., Augusto, M.F., Barrionuevo, O., Pinto, F.M.: Internet of Things in pervasive healthcare systems. In: Next-Generation Mobile and Pervasive Healthcare Solutions, pp. 22–31 (2018)
12. Guarda, T., Orozco, W., Augusto, M.F., Morillo, G., Navarrete, S.A., Pinto, F.M.: Penetration testing on virtual environments. In: Proceedings of the 4th International Conference on Information and Network Security, ICINS 2016, pp. 9–12 (2016)
13. Liu, W., Li, Q., Cai, Y., Li, Y., Li, X.: A prototype of healthcare big data processing system based on spark, no. Bmei, pp. 516–520 (2015)
14. Ojha, M., Mathur, K.: Proposed application of big data analytics in healthcare at Maharaja Yeshwantrao hospital. In: 2016 3rd MEC International Conference on Big Data and Smart City, ICBDSC 2016, pp. 40–46 (2016)
15. Krishnan, S.M.: Application of analytics to big data in healthcare. In: Proceedings of the 32nd Southern Biomedical Engineering Conference, SBEC 2016, pp. 156–157 (2016)
16. IBM, IBM PureData Solution for Healthcare Analytics (2013)
17. Nambiar, R., Sethi, A., Bhardwaj, R., Vargheeseh, R.: A look at challenges and opportunities of big data analytics in healthcare, pp. 17–22 (2013)

18. Verma, A., Mansuri, A.H., Jain, N.: Big data management processing with Hadoop MapReduce and spark technology: a comparison. In: 2016 Symposium Colossal Data Analysis Networking, CDAN 2016 (2016)

19. Shi, J., et al.: Clash of the titans: MapReduce vs. spark for large scale data analytics. Proc. VLDB Endow. **3**, 2110–2121 (2015)

20. Gu, L., Li, H.: Memory or time: performance evaluation for iterative operation on hadoop and spark. In: Proceedings of the 2013 IEEE International Conference on High Performance Computing and Communications, HPCC 2013 and 2013 IEEE International Conference on Embedded and Ubiquitous Computing, EUC 2013, pp. 721–727 (2014)

21. Lu, R., Wu, G., Xie, B., Hu, J.: Stream bench: towards benchmarking modern distributed stream computing frameworks. In: Proceedings of the 2014 IEEE/ACM 7th International Conference on Utility and Cloud Computing, UCC 2014, pp. 69–78 (2014)

Mobile Collaborative Augmented Reality and Business Intelligence: A System to Support Elderly People's Self-care

Marisa Esteves, Filipe Miranda, José Machado[✉], and António Abelha

Department of Informatics, Algoritmi Research Center, University of Minho, Braga, Portugal
{id6884,id6883}@alunos.uminho.pt, {jmac,abelha}@di.uminho.pt

Abstract. The ageing of the population increases the number of elderly people dependent in self-care. Thus, being dependent in a home context is a fact that deserves attention from social support entities integrated into the community, such as nursing homes. In this sense, this study is aimed at elderly dependent people in self-care, their caregivers, and members of nursing teams, and emerged to ensure predominantly the continuity of care of patients from Portuguese nursing homes and to strengthen the communication strategies between the different elements of the target audience. Therefore, at this stage of the project, the design of a preliminary archetype of a mobile collaborative augmented reality and business intelligence system is proposed, which main objectives are to accompany, teach, and share information between its users. It will be a reinforcement, that is, a way to promote and complete the knowledge and skills to deal with patients' health.

Keywords: Health Information and Communication Technology · Telenursing
Mobile health · Collaborative learning · Augmented reality · Business intelligence
Data warehousing · Nursing homes · Elderly people · Self-care

1   Introduction

In recent decades, a noticeable demographic change has been felt worldwide: in many countries, official statistics data show that the population is aging [1–4]. This situation poses several challenges to our society, and the topics of independence and mobility for the elderly become increasingly a critical situation [5], raising the need of ensuring multidisciplinary nursing teams, but also solutions that promote the potential of each dependent's autonomy.

To deal with those challenges, researchers have suggested applying health information and communication technology (ICT), particularly for the remote assistance of patients, which has been advocated as a concept that can radically transform and improve the delivery of healthcare [6–10].

In this paper, the design of a preliminary archetype of a mobile collaborative augmented reality (AR) and business intelligence (BI) system to support elderly people's self-care is proposed, which will be sustained by the Design Science Research (DSR)

© Springer International Publishing AG, part of Springer Nature 2018
Á. Rocha et al. (Eds.): WorldCIST'18 2018, AISC 747, pp. 195–204, 2018.
https://doi.org/10.1007/978-3-319-77700-9_20

methodology. The novelty of this project lies in trying to fulfil the current lack of communication strategies to support the remote assistance by Portuguese nursing homes (case studies) to the elderly and their caregivers using emerging technologies. Therefore, we believe that rooting technological innovation into elders' home care is an answer to support their self-care, but also to prevent their dependence, and, thus, support elders' independence in a home context.

Regarding the structure of this document, in Sect. 2, the state of the art related to the research area of this project is described. Thereafter, Sect. 3 – "Selected Technologies and their Main Advantages" – presents a brief description of each technology chosen to develop the proposed system, as well as their main advantages. Then, Sect. 4 presents the results already achieved regarding the design of a preliminary archetype of a mobile collaborative AR and BI system to support elderly people's self-care, and they are subsequently discussed in the same section. In Sect. 5, the conclusion and future work conclude briefly this paper.

2   State of the Art

In recent years, big software/hardware companies committed their time and money into the development of augmented reality (AR) software. Solutions for the enhancement of reality are now available in small mobile devices, such as smartphones and tablets, and gaming consoles. According to Mota et al. in [11], the AR technology refers to the inclusion of virtual elements in views of the physical world creating a mixed reality in real time. On the same line of thought, Carlson et al. in "Augmented Reality Integrated Simulation Education in Healthcare" [12], states that AR is a software technology that allows a virtual 2D or 3D computer-generated image to be overlaid onto a real environment.

AR is being used in diverse fields since maintenance to healthcare with the purpose to facilitate people's life or increase knowledge on a subject. Even though this new emerging trend is in its early stages of development, it presents a wide variety of learning domains, and a considerable amount of literature. Bacca et al. in "Augmented Reality Trend in Education: A Systematic Review of Research and Applications" states that "We are only beginning to understand effective instructional designs for this emerging technology.", naturally referring to AR [13].

In developed countries, evidence shows a growth of elderly mobile users, and therefore the possible trends of using AR solutions to support them [14, 15]. Contrary to some common beliefs, elderly people are aware of the importance and benefits of modern technologies, showing a keen interest in learning and using advancing technologies, and a good acceptance of multimedia applications such as videoconferencing and making online calls [15, 16].

Nonetheless, there is currently a lack of potential solutions using AR for addressing older population's requirements and experience, despite its enormous potential for promoting learning in healthcare [5, 14, 17, 18]. Most of those related studies assessed reported initial prototypes, and they do not integrate clinical competencies to ensure patients' safety [17, 18].

In the healthcare industry, it also has been pointed out the use of business intelligence (BI) to obtain useful real-time knowledge to improve the decision-making process [19–22]. It refers to the process of collecting, transforming, organizing, analysing, and distributing data from various external sources of information to improve the decision-making process [19]. Solutions that resort to BI include the application of several processes including the construction of data warehouses for structuring data to facilitate its analysis – data warehousing; the extract, transform, load (ETL) process that handles the extraction, clean-up, normalization, and loading of data; and, finally, the visualization, analysis, and interpretation of the information represented by data [20, 21].

3   Selected Technologies and Their Main Advantages

This section presents a brief description of each technology chosen to develop the proposed system, as well as their main advantages, which justify the choice made concerning their use.

3.1   MySQL Relational Database Management System

Every web application must be supported by a good relational model and relational database management system (RDMS), the so-called "back-end" or the model component of the model-view-controller (MVC) view.

MySQL is a free-to-use open source RDMS maintained by Oracle Corporation since 2010. They are a few features that make this RDMS a perfect match for this project. The website "DevOps.com" [23] indicates that data security, on-demand scalability, high performance due to the storage-engine framework, and flexibility of the free open source are some of the features that most contribute to the success of MySQL.

In this project, the free open source and high performance components influenced a lot this decision as well as past successful experiences with it. On the other hand, when dealing with healthcare data, data security is a subject that matters a lot. Therefore, the MySQL RDMS presents a good alternative.

3.2   ReactJS JavaScript Library

ReactJS is a JavaScript library for the development of interfaces maintained by Facebook in collaboration with Instagram [24]. Created in 2013, the main feature of this library is the possibility of creating large web applications that can change over time without reloading the whole page. In the model-view-controller (MVC) pattern, ReactJS corresponds to the view (V), so it is responsible to process user interfaces.

The renowned blog "PTC – PRO-TEK Consulting" points out that the advantages of using ReactJS to develop a web application are [25]:

1. Efficiency – enormous flexibility and amazing gain in performance;
2. Makes writing code in JavaScript easier by using a special syntax called JSX, which allows to mix HTML with JavaScript;
3. Has "out-of-the-box" developer tools;

4. Awesome for search engine optimization (SEO);
5. Easy to test.

Combining the above advantages with the fact that the user interface code is readable and maintainable [25], choosing this library for the development of a web application was an easy choice.

3.3 Vuforia Augmented Reality SDK

There are two types of augmented reality (AR) applications: the marker-based supported by image recognition; and the location-based that uses the global positioning system (GPS), accelerometers, and digital compasses to establish location and create AR objects. The first scenario is simpler, once it uses the camera of the device to detect a certain pattern or marker (QR codes or images), and afterwards overlays the digital information.

The first step in the development of an AR application is the choice of the software development kit (SDK). According to the website "UpWork", several criteria (9) must be considered, namely type of licence, supported platforms, smart glasses support, Unity support, cloud recognition, on-device (local) recognition, 3D tracking, geolocation, and SLAM (simultaneous, localization, and mapping) [26].

A list of popular AR SDKs can be assembled from the web pages [26, 27], which are compared in Table 1, where "F" stands for free, "FOS" for free and open source, "C" for commercial, "A" for Android, "W" for Windows, "UWP" for Universal Windows Platform, and "L" for Linux.

Table 1. Comparison between the most popular augmented reality SDKs where "F" stands for free, "FOS" for free and open source, "C" for commercial, "A" for Android, "W" for Windows, "UWP" for Universal Windows Platform, and "L" for Linux [26, 27]

	Vuforia	Wikitude	EasyAR	Kudan	ARToolKit	Maxst	ARKit	XZIMG
Licence	F, C	C	F, C	F, C	FOS	F, C	F	F, C
Supported platforms	A, iOS, UWP	A, iOS	A, iOS, UWP, macOS	A, iOS	A, iOS, L, W, macOS	A, iOS, W, macOS	iOS	A, iOS, W
Smart glasses support	Yes	Yes	No	No	Yes	Yes	Yes	No
Unity support	Yes	Yes	Yes	Yes	Yes	Yes	Yes	Yes
Cloud recognition	Yes	Yes	Yes	No	No	No	Yes	No
3D recognition	Yes	Yes	Yes	Yes	No	Yes	Yes	No
Geolocation	Yes	Yes	No	No	Yes	No	Yes	No
SLAM	No	Yes	Yes	Yes	No	Yes	Yes	No

From the analysis of Table 1, SDKs with only commercial releases are no good for the development of this project, so Wikitude is out. Another important point is the future potential adaptation of the AR application to smart glasses, which therefore eliminates the SDKs EasyAR, Kudan, and XZMIG. On the other hand, the solution was designed in a first stage for Android devices – the world's most popular mobile operating system, which leaves ARKit behind. Between the Vuforia and Maxst SDKs, only the first one uses geolocation, which is an even more important aspect than SLAM since it is essential for location-based applications. Another important factor is the capability of Vuforia to store the markers on the cloud unlike ARToolKit. Additionally, the new release of Unity 2017.2 integrates the Vuforia engine, making it even easier to create AR solutions. Therefore, the best choice is the Vuforia SDK.

As stated previously, Vuforia is an AR SDK for mobile devices that enables the development of AR applications. In short, it uses Computer Vision to recognize and track planar images such as ImageTargets and VuMarks, and 3D objects, in real time [28]. Therefore, it supports a myriad of 2D and 3D target types.

VuMarks is the next generation bar code, delivering AR experiences on any object, and allowing freedom for a customized design. When compared with the standard ImageTargets, both have the same basis, and are recognized and tracked by the Vuforia SDK. Nonetheless, the advantages of VuMarks are [29]:

- Capability of presenting millions of uniquely instances of a VuMark;
- Capability of encoding a variety of data formats;
- Possibility of differentiating among identical looking products based on their instance ID.

Some of their use cases comprehend the identification of parts and equipment, as well as precisely register service and operations instructions to the areas and surfaces they pertain to [29]. In accordance with the objectives of this study, the VuMark solution is ideal and easy to implement.

3.4 Unity3D Game Engine and IDE

According to Pietro Polsinelli, Unity3D is a cross-platform game engine with a built-in integrated development environment (IDE) developed by Unity Technologies [30]. It supports the development of applications for web plugins, desktop platforms, consoles, and mobile devices.

Introduced in 2005 by Unity Technologies, Unity3D allows a workflow based on the separation of concerns. Therefore, developers, game designers, graphical designers, modellers, and audio guys can all develop their work individually with Unity.

Since the beginning of game development, Unity3D positioned itself as the best game engine and IDE for augmented reality (AR) development. When comparing Unity with other popular game engines like Unreal Engine 4, it is possible to recognise that it is a better choice for novice developers, a large set of components and IDE extensions are available via the Asset Store, and that it also has a good relationship with Android development and the Vuforia SDK. Although it is more limited in terms of graphic capabilities, Unity has more documentation available (and of quality), and it is programmer and designer friendly in counterpoint with Unreal Engine 4 which is only designer friendly [31]. Thereby, the choice of Unity3D for the development of an AR application was clear.

4   Results and Discussion

The different elements of the target audience of the mobile collaborative augmented reality (AR) and business intelligence (BI) system include elderly dependent people in self-care, their caregivers, and members of nursing teams from Portuguese nursing homes (case studies). Each user type will have different roles and permissions assigned.

A preliminary version of its architecture is described in the next Sect. 4.1, which is followed by a short discussion (Sect. 4.2).

4.1   Preliminary Architecture of the Mobile Collaborative Augmented Reality and Business Intelligence System

The system will encompass two main distinct components – a Web application and a mobile application.

The Web application will primarily serve as a means of communication between the users of the application, especially among professionals and their patients and caregivers, the presentation of clinical and performance indicators by applying BI, and patient and resource management, among other functionalities. It will be developed with the JavaScript library ReactJS in JSX (a syntax extension to JavaScript), and supported by MySQL relational database management system (RDMS) as the backend database. The sharing of data between the data warehouse and the Web application will be performed through RESTful Web APIs in PHP.

A schematic representation of the Web application is illustrated in Fig. 1.

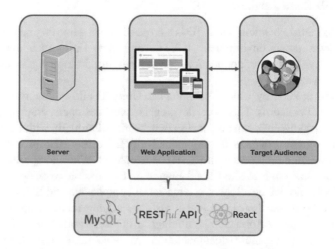

Fig. 1. Schematic representation of the Web application.

On the other hand, the mobile application will essentially incorporate nursing services supported by AR, and it will be developed using the Unity3D game engine and IDE, and Vuforia Augmented Reality SDK, in the programming language C#. In a first stage, the solution will be mainly designed for Android devices since Android is currently the world's most popular mobile operating system. Unity enables the creation and export of ".apk" files, which is the package file format used by the Android operating system for the installation of mobile apps. Nonetheless, Unity solutions can be deployed to other platforms such as iOS and UWP.

A schematic representation of the mobile application is exemplified in Fig. 2.

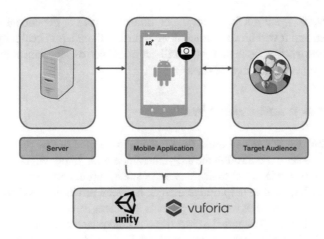

Fig. 2. Schematic representation of the mobile application.

Therefore, this system will include three main technologies into its design and development, namely data warehousing to organise the data related to patients, their caregivers, and professionals from the nursing homes; AR to support patients and their caregivers in some self-care services; and BI to create clinical and performance indicators to assist professionals in their decision-making process by using the clinical records available to present helpful information.

The competition of its design will involve the realization of semi-structured interviews with professionals from the nursing homes (in person or via Skype) to gather and analyse their feedback concerning the services that will be sustained by AR and the indicators that will be created, but also other functionalities that could be integrated into the system. Hence, it is intended to integrate clinical competencies from the nursing homes to ensure patients' safety. On the other hand, it is as well projected to collect and analyse the opinion of patients and their caregivers via online questionnaires (remotely) created with Google Forms.

4.2 Discussion

In a hope to improve the lives of elders and their self-care, this solution will be used to prevent problems, e.g. through BI, which presents the potential to improve the quality of healthcare delivery and reduce the costs and waste associated, enhance experiences using emerging technologies such as AR, and accompany and assist the elderly and their caregivers. Through the AR technology, it is envisioned to assist patients and their caregivers in medication administration, as well as to carry out a better management in home support services, including hygiene and cleaning tasks, and the feeding and mobility of patients.

With this solution, it is anticipated to strengthen the communication between the different elements of the target audience to support the current lack of remote assistance by nursing teams. It will thus also facilitate the access to nursing services by rural populations.

Therefore, the system will ensure the continuity of care of dependent patients in a home context, and it will be a reinforcement to deal with patients' health by improving elders' lifestyle quality. Nonetheless, it will act in a sense of supplementary, and never to replace completely nursing services.

5   Conclusion and Future Work

Finally, we believe that the solution designed and presented in this paper will support elderly people's self-care through the aid of nursing homes, but it will also enable elders to live more independently at their homes. As stated throughout this manuscript, it is predominantly intended to help bridge the gap between professionals from Portuguese nursing homes (case studies) and their patients and caregivers.

Therefore, this study intended to identify a problem, and motivate its resolution. A solution and its objectives were subsequently defined, and a preliminary version of its architecture was presented, which included the selection of the most appropriate technologies to develop it.

Regarding future work, it is expected to begin the development of the solution already in the beginning of the next year, and to conclude its design till the end of 2018, which will include the definition of the data warehouse, the BI indicators to create, and the services to be supported by AR.

The next phases of this project will continue to follow rigorously the phases of the Design Science Research (DSR) approach since it is intended to construct and evaluate a useful and rigorous health information and communication technology (ICT) solution to support elderly people's self-care through the aid of nursing homes.

Acknowledgments. This work has been supported by Compete POCI-01-0145-FEDER-007043 and FCT - *Fundação para a Ciência e Tecnologia* within the Project Scope UID/CEC/00319/2013.

References

1. Boll, F., Brune, P.: Online support for the elderly – why service and social network platforms should be integrated. Procedia Comput. Sci. **98**, 395–400 (2016)
2. Macis, S., Loi, D., Raffo, L.: The HEREiAM tele-social-care platform for collaborative management of independent living. In: 2016 International Conference on Collaboration Technologies and Systems (CTS), pp. 506–510. IEEE (2016)
3. Kuo, M.-H., Wang, S.-L., Chen, W.-T.: Using information and mobile technology improved elderly home care services. Health Policy Technol. **5**, 131–142 (2016)

4. Sendra, S., Granell, E., Lloret, J., Rodrigues, J.J.P.C.: Smart collaborative mobile system for taking care of disabled and elderly people. Mob. Netw. Appl. **19**, 287–302 (2014)
5. Kurz, D., Fedosov, A., Diewald, S., Guttier, J., Geilhof, B., Heuberger, M.: Towards mobile augmented reality for the elderly. In: IEEE International Symposium on Mixed and Augmented Reality (ISMAR), pp. 275–276. IEEE (2014)
6. Zhang, N.J., Seblega, B., Wan, T., Unruh, L., Agiro, A., Miao, L.: Health information technology adoption in U.S. acute care hospitals. J. Med. Syst. **37**, 1–9 (2013)
7. Cresswell, K.M., Sheikh, A.: Health information technology in hospitals: current issues and future trends. Futur. Hosp. J. **2**, 50–56 (2015)
8. Bardhan, I.R., Thouin, M.F.: Health information technology and its impact on the quality and cost of healthcare delivery. Decis. Support Syst. **55**, 438–449 (2013)
9. Lee, J., McCullough, J., Town, R.: The impact of health information technology on hospital productivity. Rand J. Econ. **44**, 545–568 (2013)
10. Buntin, M.B., Burke, M.F., Hoaglin, M.C., Blumenthal, D.: The benefits of health information technology: a review of the recent literature shows predominantly positive results. Health Aff. **30**, 464–471 (2011)
11. Mota, J.M., Ruiz-Rube, I., Dodero, J.M., Arnedillo-Sánchez, I.: Augmented reality mobile app development for all. Comput. Electr. Eng. 1–11 (2017)
12. Carlson, K.J., Gagnon, D.J.: Augmented reality integrated simulation education in healthcare. Clin. Simul. Nurs. **12**, 123–127 (2016)
13. Bacca, J., Fabregat, R., Baldiris, S., Graf, S.: Kinshuk: augmented reality trends in education: a systematic review of research and applications. Educ. Technol. Soc. **17**, 133–149 (2014)
14. Liang, S.: Research proposal on reviewing augmented reality applications for supporting ageing population. Procedia Manuf. **3**, 219–226 (2015)
15. Page, T.: Touchscreen mobile devices and older adults: a usability study. Int. J. Hum. Factors Ergon. **3**, 65–85 (2014)
16. Saracchini, R., Catalina-Ortega, C., Bordoni, L.: A mobile augmented reality assistive technology for the elderly. Comunicar **23**, 65–74 (2015)
17. Zhu, E., Hadadgar, A., Masiello, I., Zary, N.: Augmented reality in healthcare education: an integrative review. PeerJ. **2**, e469 (2014)
18. Akçayır, M., Akçayır, G.: Advantages and challenges associated with augmented reality for education: a systematic review of the literature. Educ. Res. Rev. **20**, 1–11 (2017)
19. Mettler, T., Vimarlund, V.: Understanding business intelligence in the context of healthcare. Health Inform. J. **15**, 254–264 (2009)
20. Hočevar, B., Jaklič, J.: Assessing benefits of business intelligence systems – a case study. Management **15**, 87–119 (2010)
21. Bonney, W.: Applicability of business intelligence in electronic health record. Procedia Soc. Behav. Sci. **73**, 257–262 (2013)
22. Brandão, A., Pereira, E., Esteves, M., Portela, F., Santos, M., Abelha, A., Machado, J.: A benchmarking analysis of open-source business intelligence tools in healthcare environments. Information **7**, 57 (2016)
23. McLean, M.: 8 Advantages of Using MySQL. https://devops.com/8-advantages-using-mysql/
24. Facebook Inc.: React - A JavaScript Library for Building User Interfaces. https://reactjs.org/
25. PTC PRO-TEK Consulting: Advantages & Disadvantages of ReactJS. http://www.pro-tekconsulting.com/blog/advantages-disadvantages-of-react-js/
26. Btyksin, G.: Best Tools for Building Augmented Reality Mobile Apps. https://www.upwork.com/hiring/for-clients/building-augmented-reality-mobile-apps/

27. SocialCompare: Augmented Reality SDK Comparison. http://socialcompare.com/en/comparison/augmented-reality-sdks

28. PTC Inc.: Vuforia Developer Library - Getting Started. https://library.vuforia.com

29. PTC Inc.: Vuforia Developer Library – VuMark. https://library.vuforia.com/articles/Training/VuMark

30. Polsinelli, P.: Why is Unity so Popular for Videogame Development? https://designagame.eu/2013/12/unity-popular-videogame-development/

31. Eisenberg, A.: Unity or Unreal – Which Is the Best VR Gaming Platform? https://appreal-vr.com/blog/unity-or-unreal-best-vr-gaming-platforms/

New Approach to an openEHR Introduction in a Portuguese Healthcare Facility

Daniela Oliveira[1], Ana Coimbra[1], Filipe Miranda[1], Nuno Abreu[2],
Pedro Leuschner[2], José Machado[1(✉)], and António Abelha[1]

[1] Algoritmi Research Center, University of Minho, Braga, Portugal
jmac@di.uminho.pt
[2] Centro Hospitalar do Porto, Porto, Portugal

Abstract. Implementing a new EHR data system is not easy, as the systems already in place and user mentality are very difficult to change. The openEHR architecture introduces a new way of organizing clinical information using archetypes and templates. The present paper focuses on the initial steps of the implementation of an openEHR based EHR in a Portuguese major HealthCare provider. The system comprises operational templates creation through the creation of a validation mechanism and after that storage, a platform for data generation dynamically constructed from templates and an interoperability mechanism through the implementation of an HL7 V3/CDA message system.

Keywords: EHR · openEHR · Archetypes · Operational templates
HL7 V3/CDA · SNOMED CT · Interoperability

1 Introduction

An Electronic Health Record (EHR) storages a large amount of medical data, data that must be available throughout the lifetime of a patient. Besides the effort and cost the solution must protect information when data loss occurs and at the same time be persistent and reliable across the years. The problem is often not the quantity of available data, instead the major issue is the fact that most of the information is made up to free text serving for nothing more than registering and consulting information.

Since 2004, the openEHR foundation has published a series of design specifications for semantically interoperable and future-proof EHR systems. The main feature of the openEHR design is the separation between clinical concerns and technical design, the so called, two-level modelling [1]. The first level, Reference Model (RM), represents the technical concerns (information structure and data types). The second level of the model handles the clinical domains (representation of communication of the semantics) [2]. This enables the construction of stable EHR systems without specific clinical content necessary in different fields [1].

The use of archetyping in openEHR enables new relationships between information and models. An archetype stands for a computable expression of a domain in the form of structured constraint statements, so openEHR archetypes are based on the openEHR Reference Model [2]. These can be composed into larger structures called the templates [2].

© Springer International Publishing AG, part of Springer Nature 2018
Á. Rocha et al. (Eds.): WorldCIST'18 2018, AISC 747, pp. 205–211, 2018.
https://doi.org/10.1007/978-3-319-77700-9_21

The purpose of the present document is to demonstrate the initial steps taken in the implementation process of an openEHR based EHR in a Portuguese major healthcare unit. The solution composes the creation of templates (modification of archetypes and translation), a system for validation and storage of the previous and the subsequent creation of web forms with basis on that operational templates. The system will also feature the generation and storage of information and exchange of information using HL7 Version 3 guidelines and HL7 V3 CDA. HL7 International specifies several flexible standards, guidelines, and methodologies by which various healthcare systems can communicate with each other. Such guidelines or data standards are a set of rules that allow information to be shared and processed in a uniform and consistent manner. These data standards are meant to allow healthcare organizations to easily share clinical information. Theoretically, this ability to exchange information should help to minimize the tendency for medical care to be geographically isolated and highly variable.

After the introduction, section two is entitled "Background" and is based on an intensive review of the literature on the theme, as well as, in the opinion of several authors, using them as motivations and strengths of the present work. Subsequently, the main section is presented as "System overview" and initially describes the proposed system in a general way, and then each component is described in detail. In the section "Discussion", through the SWOT analysis, a primary evaluation of the described system was carried out and, in the last section "Conclusion and Future Work", the main conclusions of the work accomplished, as well as, future steps are mentioned.

2 Background

Structuring large amount of information is not easy, as noted by several failed attempts reported in the past. According to Rector in "Clinical Terminology: Why is it so hard?" one of main reasons is that structuring requires standardization of the structured element that usually is implemented in a top-down manner [3]. The openEHR architecture presents a new and interesting approach that empowers the local user to decide how they want to structure the EHR [3].

Although reported in various informatic research projects with considerable success, reports from real large scale implementations are still scarce [4]. This statement highlights the importance and difficulty of this kind of projects. Ellingsen et al. reported the first efforts to implement large-scale EHR in Western hospitals (in Norway) conforming to the openEHR architecture [4], offering an insight into the first two years of the process, and the socio-technical challenges they met along the way. Wollersheim et al. in "Archetype-based electronic health records: a literature review and evaluation of their applicability to health data interoperability and access" presents an overview of the current archetype literature relevant to Health Information Managers [5]. In this paper, different developments are presented, concerning different settings of healthcare: elder patient care, complementary and alternative medicine, nursing, discharge summary or even security concerns focused on the security of MEDIS. The same work also presents a distribution of works by country, with Australia assuming a leading role in the openEHR implementation.

One of these Australian works belongs to Murat Gok, entitled "Introducing an openEHR-Based Electronic Health Record System in a Hospital", provides a road map for future implementations of an openEHR system in the Austin Hospital Emergency Department, Melbourne, including history, architecture and relationships to other standards. These relations play a central role on the present work, R. Qamar and A. Rector presented in "Semantic Mapping of Clinical Data to Biomedical Terminologies to Facilitate Data Interoperability" interesting points of view on this subject, stating that interoperability of clinical systems requires integration of data models, such as HL7 messages or openEHR archetypes, with terminologies such as SNOMED-CT or ICD-X [6].

A key to the success of an EHR system is inevitably its usability. If the Graphical User Interface (GUI) of a system is not intuitive and appealing there will be greater resistance from healthcare professionals, leading to under-use or misuse of the system [7]. Template4EHR (http://template4ehr.azurewebsites.net/) and EhrScape Framework (https://www.ehrscape.com/) are examples of tools that dynamically generate GUI from archetypes for Health applications [8].

3 System Overview

The purposed system consists in the implementation of an openEHR based EHR in a Portuguese major healthcare unit and it can be divided into three major components, as shown in Fig. 1. The first, the information workflow, focuses on the creation and edition of archetypes and templates. The second component, data generation, uses the operational templates previously created and generates dynamical web forms. The last part of the system, HL7 Binding, uses HL7 V3 guidelines and HL7 V3 CDA to generate, storage and exchange the information. An overview of each of these components is presented below.

3.1 Information Workflow

On the clinical domain, concepts can be organized through the archetypes, i.e., sets of independent data, which may be more or less specialized, and composed or decomposed for a new use. Blood pressure, glucose and diagnosis are some examples of the archetypes. These are written in Archetype Definition Language (ADL). The workflow of information is illustrated in Fig. 2.

When new data is inserted in a specific case, such as clinical reports or specific messages, is the system requires the use of templates, previously defined and composed of various archetypes. In this case, Medical Observations template combine three independent archetypes, but the Diabetic Checkup use only two. In this new approach, operational templates (OPT) will be used because they have a flexible structure that facilitates the implementation processes, using extensible markup language (XML) or java object notation (JSON). With respect to this, the chosen strategy to save the OPTs in a specific database shall encounter consistency and reasoning. Furthermore, these OPTs can be compiled and transformed into specific artifacts, ready to be used by software developers, messaging systems implementers or data managers [9].

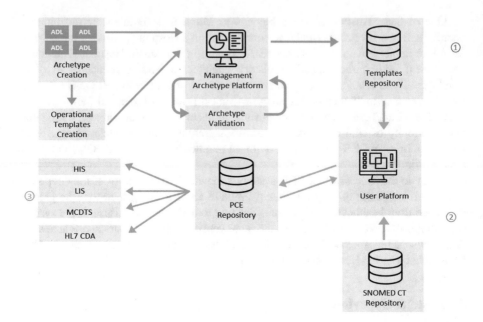

Fig. 1. Archetypes system overview with HL7 binding.

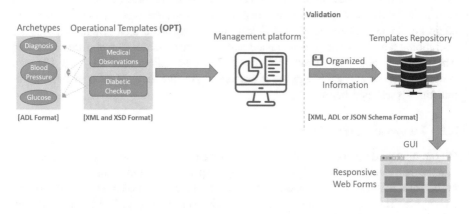

Fig. 2. Workflow of the information in the proposed system.

The proposed system will contain a multi-function support platform that allows one to submit new templates or consult existing ones. When a new template is submitted, it must be approved manually or automatically. In manual mode, a designated professional will accept or not the submitted templates, respecting the governance rules. After this validation, the template will be saved in the repository with a compatible format for the next processes, as aforementioned. In this step, the information conjugated from different HISs generates structured knowledge (Concepts and Definitions), ready to be use and, posteriorly, to create dynamic and responsive web forms. These web forms constitute the User Platform presented in Fig. 1.

3.2   Data Generation

The second part of the system consists in the dynamic generation of GUI from Operational Templates.

A web service will be responsible for reading the operational templates, extract the data and generate the GUIs, creating a webform platform.

It will use two repositories: one for storage of the OPTs (the label and definition of what is being clinically observed, and another for storage of the data inserted by healthcare professionals (the values, or results of the observation).

Semantic interoperability is ensured by binding the SNOMED CT terminology and the informational structures, represented by archetypes.

3.3   HL7 Binding

Health Level Seven version 3 (HL7 V3) defines standards for messages that are exchanged in the healthcare workflow. Relying on an object-oriented principle, where all messages are derived from the Reference Information Model (RIM) that together with data types and vocabularies are serialized in XML syntax defined by the Model Interchange Format (MIF). Version 3 of HL7 uses XML, in an evolution from v2, where "|" was used to separate the fields of a message. Although both openEHR and HL7 V3 rely on reference models, they are different, so they must be mapped to allow the conversion of information from one architecture to the other. An HL7 V3 message, as all messages, begins with a trigger event. Figure 3 represents the approach intended for the health unit.

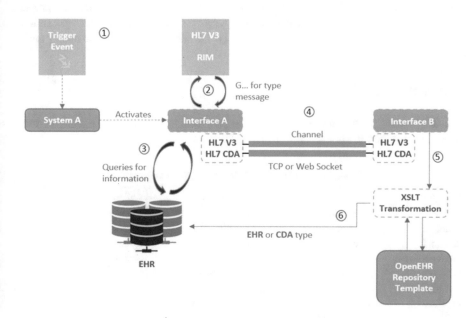

Fig. 3. Interoperability approach for openEHR based EHR.

The trigger event (1) deploys a chain reaction that leads to the share of information or documents from one health system to another. The creation of an interface for TCP or Web sockets will enable the byte stream across the internet. After the trigger event, the interface must assemble the HL7 message v3 or CDA using the EHR in openEHR format, through query methodologies like ADL (2, 3). A XML-like document is created, converted to byte stream and sent to the interface B. This point will require a XSLT transformation in order to convert the XML associated with the HL7 v3 or CDA message to openEHR compatible content. This is based on the templates stored in a openEHR repository, designed specifically for that purpose (5). After that step, the information is ready for storage in openEHR-validated archetypes or templates.

3.4   System Communication: Multi-agent Systems

As the volume of clinical data is very high, it was necessary to build consistent and organized data models. In this case, each repository of the system will be updated through multi-agent systems (MAS) and, consequently, the *front-end* applications will be synchronized. Being multi-agent architectures a field of research of distributed artificial intelligence, this technology is intrinsically connected to the concepts that define a distributed architecture, while being distinct in the definition of an agent versus the properties of the general middle-wares of many others distributed architectures. A MAS is a computer system with several autonomous agents that that interact with each other to perform certain tasks. Its main characteristics are the autonomous capacity to make decisions, as well as the capacity to interact with other agents through social interaction protocols, reaching the desired levels of coordination and cooperation [10].

Each health unity already aggregates information under the same roof, through the Agency for Integration, Diffusion and Archive of Medical Information (AIDA) system. AIDA is an agent-based platform with the purpose of ensuring the interoperability among HIS, i.e., is an example of a MAS [11]. On other hand, different health units communicate through hl7 v2 agents.

4   Discussion

In order to analyze the pros and cons of the proposed system a SWOT analysis was made, as described in this section (Table 1).

Table 1. SWOT analysis for the proposed system.

Strengths	Weakness	Opportunities	Threats
Better structure of data and interoperability	Connection to the hospital intranet is required	Modernization and organizational development	User resistance of adopting a new system by healthcare professionals
User-friendly and intuitive interface	Manual creation of archetypes by health professionals	Economic benefit of using an open source solution (openEHR)	–

5 Conclusion and Future Work

Verifying the SWOT analysis, it is possible to conclude that the proposed system is expected to grant significant added value for the Portuguese healthcare unit where it is to be implemented. For a better understanding of the benefits of this new system it is important to design and perform a study that compares this new system with the one previously implemented in the healthcare unit.

In the future could be added to the system a decision support tool with the aim to help the healthcare professionals for example, when a healthcare professional fills the diagnosis field in the form the symptoms field should be automatically filled in.

Acknowledgments. This work has been supported by Compete POCI-01-0145-FEDER-007043 and FCT - Fundação para a Ciência e Tecnologia within the Project Scope UID/CEC/00319/2013.

References

1. Beale, T., Heard, S.: Archetype definitions and principles. In: Ocean Informatics, p. 15 (2007)
2. openEHR Foundation: openEHR - Design Principles. In: Design Principles (2017)
3. Rector, A.L.: Clinical terminology: why is it so hard? Methods Inf. Med. **38**(4–5), 239–252 (1999)
4. Ellingsen, G., Christensen, B., Silsand, L.: Developing large-scale electronic patient records conforming to the openEHR architecture. Procedia Technol. **16**(2212), 1281–1286 (2014)
5. Wollersheim, D., Sari, A., Rahayu, W.: Archetype-based electronic health records: a literature review and evaluation of their applicability to health data interoperability and access. Health Inf. Manag. J. **38**(2), 7–17 (2009)
6. Sciences, P., Qamar, R., Rector, A.: Semantic mapping of clinical model data to biomedical terminologies to facilitate. In: HealthCare Computing Conference 2007, p. 257 (2007)
7. Schuler, T., Garde, S., Heard, S., Beale, T.: Towards automatically generating graphical user interfaces from openEHR archetypes. Stud. Health Technol. Inform. **124**, 221–226 (2006)
8. De Araujo, A.M.C., Times, V.C., Da Silva, M.U., Alves, D.S., De Santana, S.H.C.: Template4EHR: building dynamically GUIs for the electronic health records using archetypes. In: Proceedings of the 16th IEEE International Conferences on Computer and Information Technology (CIT 2016), pp. 26–33 (2017). 6th International Symposium on Cloud Service Computing (IEEE SC2 2016), 2016 International Symposium on Security Privacy in Social Networks
9. Oliveira, R.: Transformation of clinical data from HL7 messages to openEHR compositions (2016)
10. Oliveira, D., Duarte, J., Abelha, A., Machado, J.: Step towards interoperability in nursing practice. Int. J. Pub. Health Manag. Ethics (IJPHME) **3**(1), 26–37 (2018)
11. Peixoto, H., Santos, M., Abelha, A., Machado, J.: Intelligence in interoperability with AIDA. In: 20th International Symposium on Methodologies for Intelligent Systems, World Intelligence Congress, Macau. LNCS, vol. 766. Springer, Heidelberg (2012)

Intelligent and Collaborative Decision Support Systems for Improving Manufacturing Processes

Computer Aided Data Acquisition and Analysis for Value Stream Mapping

Krzysztof Żywicki, Adam Hamrol, and Paulina Rewers[✉]

Faculty of Mechanical Engineering and Management,
Poznan University of Technology, Piotrowo 3, 60-965 Poznan, Poland
{krzysztof.zywicki,adam.hamrol,
paulina.rewers}@put.poznan.pl

Abstract. The article presents original computer software designed to assist in data acquisition and analysis for the purpose of value stream mapping for one of tools of lean manufacturing concept. The AnaPro program uses the collected data about manufacturing processes to analyse product demand, material supplies or manufacturing processes, which is then used as a basis for creating current state maps. Furthermore, the program allows for preparing variants of future state maps through an analysis of turnover classification, product demand and methods of production flow control. The article describes functions of AnaPro with regard to value stream mapping and a practical example of the software application is provided.

Keywords: Value stream mapping · AnaPro software
Analysis of manufacturing processes · Lean manufacturing

1 Introduction

Dynamic changes in economy along with continuously increasing requirements of clients force production companies wanting to maintain their position on the market to seek for solutions to facilitate manufacturing processes and to cut down production costs [1, 2]. When performing subsequent operations the value of manufactured products is generated, thus generating a value stream. It is important to ensure that the values adding up to a product price are acceptable to clients. Therefore, improving manufacturing processes is growing increasingly significant.

The Lean Manufacturing concept (LM) [3] makes pursuing that goal possible. The LM tools and methods are applied at companies to improve manufacturing processes by eliminating all sorts of waste. The Lean Management traces its roots back to the best manufacturing practices from the end of the 19th and the first half of the 20th century [4, 5]. The prototype for the Lean Manufacturing concept was the Toyota Production System developed after WWII as a combination of various methods, techniques and principles, including, inter alia, J. Juran's management by quality, and solutions developed by Japanese practitioners: T. Ohno, E. Toyoda, S. Shingo [6].

To meet the growing requirements of clients who wish to receive products that are of good quality and cheap, in shortest time possible, the manufacturing process must be properly organized and subject to an ongoing improvement. There are various tools and

© Springer International Publishing AG, part of Springer Nature 2018
Á. Rocha et al. (Eds.): WorldCIST'18 2018, AISC 747, pp. 215–224, 2018.
https://doi.org/10.1007/978-3-319-77700-9_22

methods of analysing manufacturing processes and supporting its improvement. These aim at identification and then elimination of waste resulting, inter alia, from [7, 8]:

– maintaining excessive stocks of raw materials, goods in process, or finished goods,
– manufacturing more goods than demanded by the client (overproduction),
– unnecessary transfers of operators and material,
– unnecessary waiting of goods in the process,
– failure to utilize the personnel's potential.

A basic tool for analysing production flow in the manufacturing system is value stream mapping – VSM [9]. Having analysed the current state, future state maps are developed, which contain adequate process improvement variants. Production may be improved at different levels. If a workplace is improved, methods such as 5S [10], SMED (Single-Minute Exchange of Die) [11, 12], standardized work [13] or poka-yoke [14] may be applied. On the other hand, material flow may be facilitated with use of just in time [15], Kanban [16] and levelled production [17–19] methods.

2 Value Stream Mapping

VSM is a graphic method of presenting material and information flow in the manu-facturing system. The map shows all tasks performed under the process, from purchase of raw materials to delivery of finished products to the client. Analysis of the map enables to identify the waste occurring in the manufacturing process, and to set directions for further actions in order to eliminate such waste [20].

The main stages of the mapping include [21]:

– developing a current state map which will enable to analyse the manufacturing processes in current manufacturing system conditions;
– designing modifications and developing a future state map that will serve as a basis for drawing up an implementation plan.

The information necessary for developing the current state map should be gathered directly on the factory floor, in the place where the actual work is performed, in the so-called gemba. The data which must be identified and then placed on the map relate to:

– method of information flow: between individual stage of production, between the client and the company, and between the supplier and the company;
– requirements of the clients and order specifications;
– material delivery specifications and specifications of raw materials used in production;
– inventory levels of raw materials, goods in process, finished goods;
– manufacturing process: stages of technological process and transport between individual workstations;
– time of individual operations and changeovers, number of employees, level of rejects, and others.

A timeline is placed at the end. It shows a total time during which a given material stays at a company (from delivery as raw material, to shipment as finished goods) and value adding time. The information required to describe the timeline is read out from the map.

The analysis of the data on the map is of key importance for developing the future state maps. At that stage, the previously collected data are structured and analysed. The analysis aims at identifying the waste occurring during production. The following data can be analysed at that stage: lead times of individual orders, inventory levels of raw materials, goods in process, finished goods, times of operations, changeovers, waiting times, availability and output of workstations, indicators: EPE (Every Part Every e.g. hours), OEE (Overall Equipment Effectiveness), etc. others.

Identification, collection and further analysis of the data is the more difficult the more different types of products a given company manufactures. Each family product (a family is defined as a group of products which are technologically alike) should be analysed separately, which is often connected with a great number of current state maps, and thus a larger volume of data to be analysed.

Having analysed the current state, the future state maps are developed. Mapping the future state consists in designing modifications and drawing up a plan of implementations. The major objective to be pursued while introducing modifications in the manufacturing system is mostly to shorten the goods throughput time. Reducing the throughput time might be achieved in various ways, starting from shortening the time of certain operations or changeovers, to work standardization and maintaining workplace order, to changing the production flow method. Any developed modifications should ensure the most effective production flow. This is possible inter alia by developing and analysing several variants of production flow organization.

Developing a few future state maps will provide an opportunity for evaluating the benefits and disadvantages of individual modifications. In this case it is also connected with the complexity of analysing large volumes of data concerning manufacturing processes.

It is not easy to select a proper variant, especially with such a high number of variables. In such a situation it is useful to apply decision support methods which, based on a certain knowledge and information about potential consequences, help find the best solution [22]. In times of violent development of information technologies, computer systems start to play a significant role in decision processes. They are used in situations when large volumes of data have to be processed to make a decision, or when a specific nature of decision requires application of complex computational models. Having selected the variant of improvement, the suggested solutions are implemented. Depending on the assumed objective, adequate Lean Manufacturing tools and methods can be applied.

Value stream mapping is mostly a graphic method of reflecting the flow of materials and information under a manufacturing system. There are some IT solutions that only support graphic presentation of value stream with use of standard graphic icons. However, the graphic form is not as important from the point of view of improvement actions as the fact whether the data on the map are reliable and will allow to make reasonable decisions. The AnaPro software described in this article is a solution to identify and analyse the data used for diagnosing the current value stream and for suggesting some improvement solutions.

3 Functional Structure of AnaPro

AnaPro is made up of several modules used for creating and managing data bases, which contain the necessary algorithms to perform analytical calculations to assist in developing the current and future state maps (Fig. 1). The program databases include:

- workstations – defining workstations which make up a manufacturing system,
- products – specification of products and their structure,
- technology – information about the course of technological processes performed to manufacture the products in the scope of workstations and in the scope of their potential variants, taking into account the alternative structures of the process (the so-called technology version and different variants of workstations used for performance of a given technological operation (the so-called alternative workstations),
- production orders – information about lead times,
- manufacturing processes – information about specific parameters of manufacturing processes, e.g.: cycle time, changeover time,
- demand for products – data on the quantities and the dates of product sales,
- material supplies – data on the quantities and dates of production material supplies.

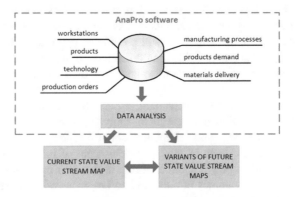

Fig. 1. Functional structure of AnaPro: own work.

The program supports the analysis of the main areas connected with material flow in the value stream: demand for products, delivery of materials, manufacturing processes, which are used for preparing a current state map and variants of the value stream future state.

4 Analysis of the Value Stream Current State

4.1 Separating Product Families

The basic stage before the value stream mapping is separating product families, which are then analysed in the context of the production flow organization. It is all the more important since the analysis of material flow for product families allows for unequivocal picturing of all the relations between different production resources.

Product family separation module is the program's functionality responsible for performing an analysis which results in a matrix of specified product families. Technological similarity of products, expressed by the use of the same workstations while manufacturing the products (Fig. 2), was assumed as a criterion for separating product families.

Fig. 2. Diagram of the process of separating product families: own work.

4.2 Analysis of Product Requirements

Product requirements is analysed by determining basic parameters connected with requirement for products. It is performed on the basis of information specified in the database "Product Requirements". The demand parameters determined in the program include: average monthly sales, average daily sales, average sales frequency, minimum shipment batch, and maximum shipment batch.

4.3 Analysis of Material Deliveries

As far as the analysis of material deliveries in the program is concerned, it is performed to determine some basic characteristics connected with: average monthly deliveries, average daily deliveries, average frequencies of deliveries, minimum delivery batches and maximum delivery batches.

4.4 Analysis of Manufacturing Processes

This module of AnaPro analyses the data on manufacturing processes collected in real conditions. Based on such data, the following information is analysed (Fig. 3): cycle

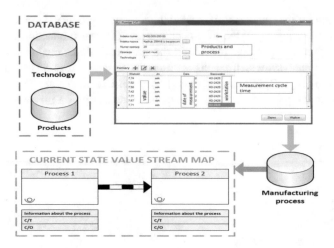

Fig. 3. Diagram of the process of manufacturing process measurement: own work.

times of subsequent operations (C/T), changeover times (C/O), lead times and work-in-process inventory levels.

Thanks to the above information it is possible to perform a detailed illustration of material flow in manufacturing processes.

5   Analysis of Data for Value Stream Improvement Actions

The objective which should be pursued when introducing changes into a manufacturing process is above all to shorten the time required for the product to pass through the process. This is possible by eliminating tasks and operations which do not add value. Mapping of the future state consists in designing changes and developing a plan of implementations to ensure that the production rate is adjusted to the rate of order placement by the clients and reduction of inventory levels. Therefore, developing a few future state maps will provide an opportunity for evaluating the benefits and disadvantages of individual modifications.

5.1   Classification of Products Requirements Turnover

AnaPro enables to classify goods pursuant to two methods in which the criterion of division is related to products requirements quantities. The first method divides products into three categories: fast-moving (A), medium-moving (B) and slow-moving (C). The second method of classification is the so-called Glenday Sieve, where products is divided into four categories (red, yellow, green and blue) (Fig. 4).

Such a classification provides information required to select methods connected with development of variants of lean production flow.

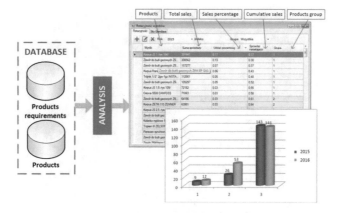

Fig. 4. Diagram of the process of products requirements turnover: own work.

5.2   Methods of Production Flow Control

The program enables to perform basic calculations connected with the following assumptions of the lean manufacturing production flow control: a supermarket pull system and a sequential pull system – FIFO lanes.

Furthermore, it indicates a possibility of combining more operations of manufacturing processes into a continuous flow on the basis of comparison of cycle times (C/T) and takt times (T/T). Calculations concerning the pull system come down to determining a specific inventory level in the supermarket (rotating stock, buffer stock, safety stock) or the maximum quantity of materials (FIFO lane) between subsequent stages in the process.

6   An Example of Application of AnaPro for Value Stream Mapping

AnaPro was used for value stream mapping in a company manufacturing plumbing and gas fittings. 15 product families were separated during the analysis, which were then subject to current state mapping and value stream improvement actions.

On the basis of the received system data and the data collected directly during the manufacturing processes, analyses were performed in AnaPro to develop a current state value stream map for each product family.

The following production effectiveness coefficients were determined for such maps: throughput time, inventory level and productivity (of operators, machines and equipment). Furthermore, the areas of material flow and information flow were indicated which were of critical importance for the obtained effectiveness coefficients.

Different variants of production flow control models in the form of future state value stream maps were developed under the improvement actions (Fig. 5). These actions were performed on the basis of analyses of product demand and production control methods prepared in AnaPro. As a result of the analyses a model of production

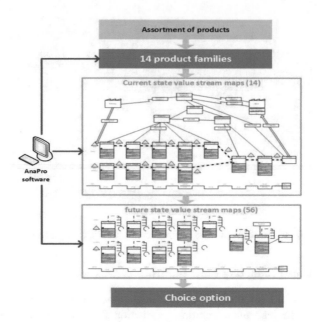

Fig. 5. Example of map the current and future state: own work.

planning and control for a given production family was selected. The suggested solutions resulted in enhancing the parameters of production flow, which was reflected by shortening throughput times and reduction of inventory levels in the production system.

7   Conclusions

To develop the current and future state maps many analyses need to be performed to investigate functioning of production system, such as, for example: product demand, material deliveries, diversity of products, etc. The AnaPro software was developed to automate the analyses. It provides assistance in the production data acquisition and analysis as well as during sales. The program makes it possible to shorten the time required for analyses and to eliminate calculation errors. This allows for more effective development of current and future state maps.

The exemplary use of AnaPro for value stream mapping in a production company presented in this article contributed to more effective implementation of the improvement process. The study shows that development of current state maps, and consequently future state maps, with use of AnaPro resulted in shortening the throughput time by an average of 40%. Thanks to this, it was possible to manufacture more products (Fig. 6).

The AnaPro software is in its development phase. Another step in the work is to develop support modules for introducing levelled production and other modules used for analysis of the production data.

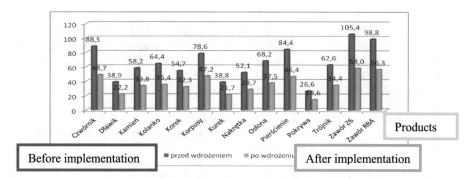

Fig. 6. The obtained results of shortening lead times for the analysed product families: own work.

References

1. Zawadzki, P., Żywicki, K.: Smart product design and production control for effective mass customization in the Industry 4.0 concept. Manag. Prod. Eng. Rev. **7**(3), 105–112 (2016)
2. Dostatni, E., Diakun, J., Grajewski, D., Wichniarek, R., Karwasz, A.: Functionality assessment of ecodesign support system. Manag. Prod. Eng. Rev. **6**(1), 10–15 (2015). https://doi.org/10.1515/mper-2015-0002
3. Trojanowska, J., Kolinski, A., Galusik, D., Varela, M.L.R., Machado, J.: A methodology of improvement of manufacturing productivity through increasing operational efficiency of the production process. In: Hamrol, A., Ciszak, O., Legutko, S., Jurczyk, M. (eds.) Advances in Manufacturing. Lecture Notes in Mechanical Engineering, pp. 23–32. Springer, Cham (2018)
4. Shah, R., Ward, P.: Lean manufacturing: context, practice bundles, and performance. J. Oper. Manag. **21**(2), 129–149 (2003)
5. Holweg, M.: The genealogy of lean production. J. Oper. Manag. **25**(2007), 420–437 (2007)
6. Liker, J.K.: The Toyota Way. McGraw-Hill, New York (2004)
7. Womack, J.P., Jones, D.T., Roos, D.: The Machine That Changed the World: The Triumph of Lean Production. Rawson Macmillan, New York (1990)
8. Araújo, A.F., Varela, M.L.R., Gomes, M.S., Barreto, R.C.C., Trojanowska, J.: Development of an intelligent and automated system for lean industrial production adding maximum productivity and efficiency in the production process. In: Hamrol, A., Ciszak, O., Legutko, S., Jurczyk, M. (eds.) Advances in Manufacturing. Lecture Notes in Mechanical Engineering, pp. 131–140. Springer, Cham (2018)
9. Rother, M., Shook, J.: Learning to See: Value Stream Mapping to Add Value and Eliminate Muda. The Lean Enterprise Institute, Cambridge (2003)
10. Labach Elaine, J.: Using standard work tools for process improvement. J. Bus. Case Stud. **6** (1), 39–47 (2010)
11. Chabowski, P., Rewers, P.: Impact of Changeovers on Production Flexibility, Logistics, vol. 4, Poznan (2015). (in Polish)
12. Trojanowska, J., Żywicki, K., Varela, M.L.R., Machado, J.: Improving production flexibility in an industrial company by shortening changeover time: a triple helix collaborative project. In: Peris-Ortiz, M., Ferreira, J., Farinha, L., Fernandes, N. (eds.) Multiple Helix Ecosystems for Sustainable Competitiveness. Innovation, Technology and Knowledge Management, pp. 133–146. Springer, Cham (2016)

13. Rewers, P., Trojanowska, J., Mandziuk, M.: Applications use standardized work purpose of increase the production capacity-case study. Res. Logist. Prod. **5**(2), 191–200 (2015)
14. Dudek-Burlikowska, M., Szewieczek, D.: The Poka-Yoke method as an improving quality tool of operations in the process. J. Achiev. Mater. Manuf. Eng. **36**(1), 95–102 (2009)
15. Ohno, T.: Toyota Production System. Productivity Press, Portland (1988)
16. Budynek, M., Celińska, E., Dybikowska, A., Kozak, M., Ratajczak, J., Urban, J., Materene, K.: Strategies of production control as tools of efficient management of production enterprises. Manag. Syst. Prod. Eng. **21**(1), 20–24 (2016)
17. Żywicki, K., Rewers, P., Bożek, M.: Data analysis in production levelling methodology. In: Rocha, Á., Correia, A., Adeli, H., Reis, L., Costanzo, S. (eds.) Recent Advances in Information Systems and Technologies. WorldCIST 2017. Advances in Intelligent Systems and Computing, vol 571, pp. 460-468. Springer, Cham (2017)
18. Rewers, P., Hamrol, A., Żywicki, K., Kulus, W., Bożek, M.: Production leveling as an effective method for production flow control—experience of polish enterprises. Procedia Eng. **182**, 619–626 (2017)
19. Rewers, P., Trojanowska, J., Diakun, J., Rocha, A., Reis, L.P.: A study of priority rules for a levelled production plan. In: Hamrol, A., Ciszak, O., Legutko, S., Jurczyk, M. (eds.) Advances in Manufacturing. Lecture Notes in Mechanical Engineering. Springer, Cham (2018)
20. Hines, P., Rich, N.: The seven value stream mapping tools. Int. J. Oper. Prod. Manag. **17**(1), 46–64 (1997)
21. Rewers, P., Chabowski, P., Trojanowska, J.: Methods of improvement of production processes - a literature review. In: Research and Development of Young Scientists in Poland - Technical and Engineering Sciences, vol. VII, pp. 85–92. Young Scientists, Poznan (2016). (in Polish)
22. Kujawińska, A., Rogalewicz, M., Diering, M., Piłacińska, M., Hamrol, A., Kochański, A.: Assessment of ductile iron casting process with the use of the DRSA method. J. Min. Metall. Sect. B Metall. **52**(1), 25–34 (2016)

An Approach to Buffer Allocation, in Parallel-Serial Manufacturing Systems Using the Simulation Method

Slawomir Kłos[(⊠)] and Justyna Patalas-Maliszewska

Faculty of Mechanical Engineering, University of Zielona Góra,
Licealna 9, 65-417 Zielona Gora, Poland
{s.klos, j.patalas}@iizp.uz.zgora.pl

Abstract. The buffer allocation problem (BAP) concerns real, discrete manufacturing systems that complete repetitive, multi-assortment productions. In the automotive industry, the allocation of the correct, buffer capacity is especially important in order to obtain an acceptable throughput and work-in-progress. The BAP is an NP-hard combinatorial optimisation problem. The methodology for the allocation of buffer capacity in a parallel-serial, manufacturing system is proposed, in the paper and deals with the compromise between high, system throughput and low-level work-in-progress. The methodology is based on the simulation method. In order to analyse the behaviour of the manufacturing system, Tecnomatix Plant Simulation Software is used. Simulation experiments are conducted for the different capacities of buffer allocation within a manufacturing system. To evaluate the results of the simulation, a system performance index is proposed.

Keywords: Buffer allocation problem · Computer simulation
Parallel-serial manufacturing system · Throughput · Work-in-progress

1 Introduction

The parallel-serial manufacturing system includes several (minimum of two) production lines where the machines and buffers from parallel lines are connected (see Fig. 1). Single production line includes serially connected machines which realise different technological operations. In a production line, between each neighbour pair of machines a buffer is allocated. The structure of parallel-serial manufacturing systems is often used for repetitive production where redundant manufacturing resources are required, in order to guarantee the reliability of the manufacturing process. Buffer allocation capacity directly impacts on the throughput and work-in-progress of a parallel-serial manufacturing system. In the paper, implementation of the simulation method, in order to find acceptable results for buffer allocation, is proposed. BAP (buffer allocation problem) is a well-known, NP-hard (NP-hardness, in computational complexity theory, is the defining property of a class of problems) combinatorial optimisation problem and many researchers are currently engaged in attempting to solve the issue. The authors of this paper propose using the simulation method to

© Springer International Publishing AG, part of Springer Nature 2018
Á. Rocha et al. (Eds.): WorldCIST'18 2018, AISC 747, pp. 225–235, 2018.
https://doi.org/10.1007/978-3-319-77700-9_23

analyse the allocation of buffer capacity in a parallel-serial manufacturing system on the average throughput and work-in-process of the systems.

1.1 Literature Overview

The Buffer Allocation Problem (BAP) is one of the most important questions facing a serial production designer. It is a combined, NP-hard, combinatorial, optimisation problem when designing production lines; the issue is studied by many scientists and theorists around the world.

Demir et al. proposed the tabu search algorithm, in order to find near-optimal buffer allocation plans for a serial production line with unreliable machines. To estimate the throughput of the line with a given specific buffer allocation, they used the analytical decomposition approximation method. They demonstrated the performance of the tabu search algorithm, based on existing benchmark problems [1]. Costa et al. proposed a novel, parallel, tabu search algorithm, equipped with a proper, adaptive, neighbourhood generation mechanism, in order to solve the primal buffer allocation problem. They implemented an evaluative method based on a specific algorithm, in order to simulate the behaviour of the system [2]. Nahas et al. describe a new, local search approach for solving the buffer allocation problem in unreliable production lines. The proposed approach allows the allocation plan to be calculated, in a computationally efficient manner, subject to a given amount of total buffers slots [3]. Sabuncuoglu et al. analysed the problem in order to characterise the optimal buffer allocation; specifically, they studied the cases with single and multiple bottleneck stations under various experimental conditions. They developed an efficient, heuristic procedure to allocate buffers to serial production lines, in order to maximise throughput [4]. Kose and Kilincci proposed a hybrid, approach-based simulation optimisation, in order to determine the buffer sizes required in open serial production lines for the maximisation of the average production rate of the system. The hybrid approach, using a genetic algorithm and simulated annealing, was used as a search tool to create candidate buffer sizes [5]. The critical literature overview in the area of buffer allocation and production line performance was done by Battini et al. [6]. The simulation methods are often used to improve support for a decision and increase the effectiveness of discrete manufacturing systems. Jithavech and Krishnan present a simulation-based method in order to develop an efficient layout design facility, albeit with uncertainty as to the demand for the product. The validation of the simulation approach, against analytical procedures, was detailed firstly and the methodology for a simulation based approach was then provided. Results from case studies showed that this procedure results in a risk reduction as high as 80% [7]. Trojanowska et al., proposed a tool for supporting decision making in job-shop scheduling. This tool, introduced in the article, enables scheduling based on the author's priority rule and allows maximum usage of the most loaded resource; this is known as the 'critical resource' and determines the efficiency of the production system [8]. Iassinovski et al. presented a structure, along with the components of a shared, decision-making system for complex, discrete systems and process control. The proposed model allows on-line simulations, state-graph search expert systems and other, decision making methods to be used [9]. Negahban and Smith provided a comprehensive review of discrete event simulations presented in publications which were published between 2002 and 2013 and are focussed particularly on the application of simulations, vis-à-vis manufacturing. The use of simulation methods, for the study of maintenance processes is included in the paper [10]. Kłos presents different simulation models for discrete manufacturing systems using Tecnomatix Plant Simulation [11].

1.2   The Buffer Allocation Problem

The buffer allocation problem is concerned with the allocation of a certain number of buffers, P, among the N-1 intermediate buffer locations of a production line, in order to achieve a specific objective [12]. A production line consists of machines working in sequence, separated by buffers where the machines are denoted as $M_1, M_2,..., M_N$ and the buffers as $B_1, B_2,..., B_{N-1}$. The classification of production lines can be described as: blocking, that is, those which block either before or after service; job transfer timing, such as, asynchronous, synchronous and continuous; production control mechanisms, such as, push and pull; workstation types and whether they are reliable or unreliable and career requirements, whether open or closed [13]. BAP can be formulated in three cases, depending on the function of the objective. In the first case, the main objective is maximisation of the throughput rate for a given, fixed number of buffers. The first BAP case is formulated as follows (1)–(3) [14]:

find

$$B = (B_1, B_2, \ldots, B_{N-1}) \tag{1}$$

so as to

$$\max \ f(B) \text{ or } \mathbf{min} f(\mathbf{B}) \tag{2}$$

subject to

$$\sum_{i=1}^{N-1} B_i = P \tag{3}$$

Where B represents a buffer size vector and f(B) represents the throughput rate of the production line, as a function of the size vector of the buffers and P is a fixed, non-negative integer, denoting the total buffer space available in the manufacturing system. The second BAP case is formulated as follows (4)–(6):

$$B = (B_1, B_2, \ldots, B_{N-1}) \tag{4}$$

so as to

$$\min \sum_{i=1}^{N-1} B_i \tag{5}$$

subject to

$$f(B) = f^* \tag{6}$$

where f^* is the desired throughput rate. The third **BAP** case is formulated as follows (7)–(10):

find

$$B = (B_1, B_2, \ldots, B_{N-1}) \tag{7}$$

so as to

$$\min Q(B) \tag{8}$$

subject to

$$f(B) = f^* \tag{9}$$

$$\min \sum_{i=1}^{N-1} B_i \leq P \tag{10}$$

where $Q(B)$ denotes the inventory of the average work-in-progress as a function of the size vector of the buffers and f^* is the desired throughput rate. This formulation of the problems expresses maximisation of the throughput rate for a fixed, given number of buffers which achieves the desired throughput rate, with the minimum total buffer size - or with the minimisation of the inventory of the average work-in-progress- which is subject to the constraints of total buffer size and the desired throughput rate. A model of the parallel-serial manufacturing system is described in the next section.

2   A Model of the Parallel-Serial Manufacturing System

The structure of a parallel, serial manufacturing system is presented in Fig. 1. The system includes N technological operations and M production lines.

Each production line includes N machines and N-1 intermediate buffers. Each machine is connected to neighbouring buffers and each buffer is connected to

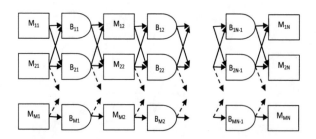

Fig. 1. The structure of a parallel-serial manufacturing system.

neighbouring machines at 2 or 3 connections. The elements are transferred between machines and buffers, using the Round-Robin priority rule [15]. If a buffer or machine is busy, then the next, empty buffer or machine is chosen. The model of the system was prepared using Tecnomatix Plant Simulation Software (Fig. 2). The model of the system includes 9 machines and 6 buffers. Three technological operations are undertaken in the system, firstly by machines M_{11}, M_{21}, M_{31}, secondly by machines M_{12}, M_{22}, M_{32} with the final operation being undertaken by machines M_{13}, M_{23} and M_{33}. This means that production in the system can be undertaken *via* several, alternative routes. For example, if an element is located, initially, on machine M_{11}, 6 alternative routes can be taken into consideration: (1) $M_{11}\text{->}M_{12}\text{->}M_{13}$; (2) $M_{11}\text{->}M_{22}\text{->}M_{23}$; (3) $M_{11}\text{->}M_{22}\text{->}M_{33}$; (4) $M_{11}\text{->}M_{32}\text{->}M_{23}$; (5) $M_{11}\text{->}M_{32}\text{->}M33$; (6) $M_{11}\text{->}M_{22}\text{->}M_{13}$. The processing and set-up times of manufacturing resources are determined as lognormal distribution. Lognormal distribution is a continuous distribution, where a random number has a natural logarithm which corresponds to normal distribution. The undertakings are non-negative, real numbers. The function of the density of the probability of lognormal distribution takes the following form:

$$f(x) = \frac{1}{\sigma_0 x \sqrt{2\pi}} \cdot exp\left[-\frac{(lnx - \mu_0)^2}{2\sigma_0^2}\right] \tag{11}$$

where the mean value of the lognormal distribution is

$$\mu = exp\left[\mu_0 + \frac{\sigma_0^2}{2}\right] \tag{12}$$

and variance σ^2 takes on the value of

$$\sigma^2 = exp\left(2\mu_0 + \sigma_0^2\right) \cdot \left(exp\left(\sigma_0^2\right) - 1\right) \tag{13}$$

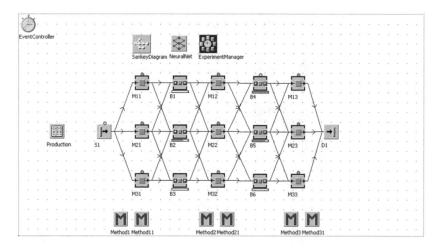

Fig. 2. The simulation model of a parallel-serial manufacturing system.

The simulation model of the parallel-serial manufacturing system, prepared using Tecnomatix Plan Simulation Software, is presented in Fig. 2. The machines are represented by icons M_{11}, M_{21}, M_{31}, M_{12}, M_{22}, M_{32}, M_{13}, M_{23}, M_{33} and buffers as B_1, B_2, ... B_6. S_1 represent input and D_1 output of the system. In table *Production* the production batch-sizes are defined. *SenkeyDiagram*, *NeuralNet* and *Experiment manager* are tools for the system behaviour analyse. The *Methods* define operation and setup times.

The processing and set-up times are defined on the basis of actual, automotive industry data.

2.1   The Parameters of the Model of a Manufacturing System

The set-up and processing times, defined for machines, in the model of a parallel-serial manufacturing system, are presented in the Table 1.

Table 1. Set-up and processing times defined for the machines

Setup times	μ	σ^2	Processing times	μ	σ^2
M_{11}	840	200	M_{11}	480	20
M_{12}	840	300	M_{12}	480	20
M_{13}	840	100	M_{13}	480	20
M_{21}	840	200	M_{21}	480	20
M_{22}	840	300	M_{22}	480	20
M_{23}	840	100	M_{23}	480	20
M_{31}	840	200	M_{31}	480	20
M_{32}	840	300	M_{32}	480	20
M_{33}	840	100	M_{33}	480	20

Table 2. The sizes of the production batches

Product	Size of production batches
A	100
B	300
C	80
D	120
A	60
B	200
C	150
D	80

The sizes of the production batches are presented in Table 2 and are executed in cyclical sequence. The availability of manufacturing resources is determined as 95%, for all machines defined for the model of manufacturing system.

2.2   The Input and Output Values of the Simulation Experiments

Thirty-three (33) different buffer allocation capacities are proposed as input values for the simulation experiments. The values of the capacities are presented in Table 3. The input values of the buffer capacity are selected, having been based on an analysis of 100 cases of randomly generated, simulation experiments where the buffer capacity ranged between 1 and 20 products. The hourly, throughput values of the system and the average lifespan of the products, are chosen as the output values of the experiments. The average lifespan of the products reflects the level of work-in-progress. For each simulation experiment, 3 observations are assigned. The results of the simulation experiments are presented in the next section.

Table 3. Buffer capacities as input values of the simulation experiments

	B_1	B_2	B_3	B_4	B_5	B_6		B_1	B_2	B_3	B_4	B_5	B_6		B_1	B_2	B_3	B_4	B_5	B_6
Exp01	1	1	1	1	1	1	Exp12	1	2	1	2	1	2	Exp23	20	2	2	2	20	2
Exp02	1	1	1	1	1	2	Exp13	10	1	1	1	1	10	Exp24	1	11	10	1	16	8
Exp03	1	1	1	2	1	1	Exp14	2	2	1	1	2	1	Exp25	5	5	5	5	5	5
Exp04	1	1	2	1	1	1	Exp15	2	2	1	1	2	2	Exp26	2	18	2	2	7	2
Exp05	1	1	1	1	2	1	Exp16	8	1	1	5	2	6	Exp27	20	20	20	20	20	20
Exp06	2	1	1	1	1	1	Exp17	2	2	2	2	2	2	Exp28	20	10	11	10	1	1
Exp07	1	1	2	2	1	1	Exp18	3	3	3	3	3	3	Exp29	18	11	20	1	1	9
Exp08	1	1	1	1	10	10	Exp19	1	10	1	10	1	1	Exp30	19	15	15	1	1	10
Exp09	1	2	1	1	1	1	Exp20	1	8	3	4	8	7	Exp31	15	15	15	1	1	15
Exp10	2	1	2	1	1	1	Exp21	1	8	5	10	13	9	Exp32	20	20	20	1	1	1
Exp11	2	2	2	1	1	1	Exp22	1	10	1	1	10	1	Exp33	1	20	1	1	20	1

3   The Results of the Simulation Experiments

The results of the simulation experiments are presented, in Figs. 3 and 4. The throughput, results chart shows that in the first 22 experiments, the throughput value ranges between 19,42 (Exp13) and 20,36 (Exp18) products per hour.

For experiments Exp23, Exp26 and Exp33, however, the value of the throughput is extremely low, at, respectively, 16,37; 17,33 and 16,99 products *per* hour. The critical reduction in system throughput for Exp23 and Exp33 shows that if buffer capacities B_1 and B_5 are relatively large, (20) and the remaining buffers are small (2), then the throughput of the system is low. Based on the above observation, the following thesis can be formulated: *Given, is a parallel-serial, manufacturing system with defined processing and set-up times, availability of manufacturing resources and production batch sizes. One buffer allocation capacity is responsible for a significant lowering of the performance of the system.* The average product lifespan is presented in Fig. 4.

Min-Max intervals with quartiles

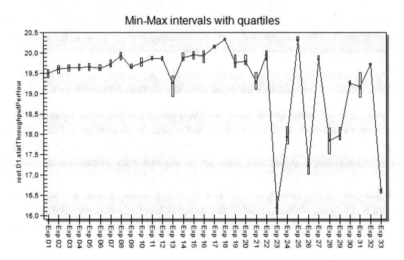

Fig. 3. The throughput value *per* hour, for the simulation experiments.

Min-Max intervals with quartiles

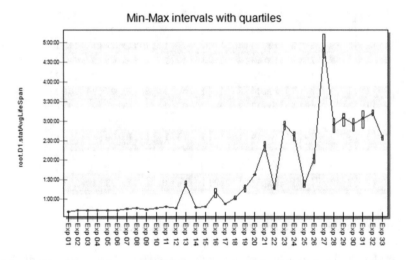

Fig. 4. The value of the average lifespan of products for the simulation experiments.

For the first 12 experiments, the average, product lifespan showed similar values, ranging between 0:41:01 (Exp01) and 0:48:36 (Exp11). For Exp13, the average lifespan showed a value of 1:16:13. The greatest value for the average lifespan of products was obtained for Exp27, with the average time that the system was in use being greater than 4 h -4:30:12; this gave the maximum, buffer capacity value.

The total effectiveness of a system should include both throughput and work-in-progress values, that is, the average, product lifespan. To evaluate the total effectiveness of a parallel-serial, manufacturing system, the following system performance rate θ is proposed:

$$\theta = \frac{f(B)}{\omega} \tag{14}$$

where f(B) – throughput of the system, ω – average product lifespan.

The results of the system performance rate are presented in Fig. 5. The greatest value of the index is obtained for Exp01 (θ = 685,5). The smallest values for the system performance rate are obtained for Exp23 (θ = 141,1) and Exp27 (θ = 105,8). Based on this analysis of the poorest throughput values, an additional 6 experiments were conducted. The results are presented in Table 4. For the buffer allocation capacity at experiment Exp'01, the smallest value of throughput is obtained (15,52). The system performance rate for the experiment is 112,08. The value is obtained on the basis of 100, random, simulation experiments. Very important result and novelty of the study is that there are allocation of buffer capacity for parallel-serial manufacturing system which extremely reduce the system performance rate (see experiments 23 and 27). The results of the simulation experiments can be implemented in multi-criteria decision support system for production planning [15].

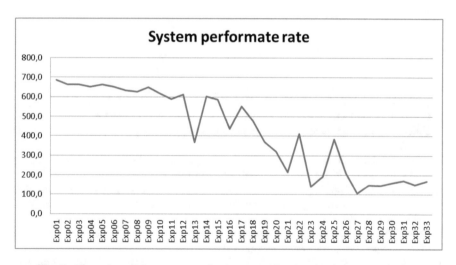

Fig. 5. The value of the system performance rate for the simulation experiments.

Table 4. Buffer allocation, resulting in the poorest throughput values

	B_1	B_2	B_3	B_4	B_5	B_6	f(B)	ω	θ
Exp'01	20	3	2	20	3	2	15,52	03:19:26,38	112,08
Exp'02	20	2	3	20	2	3	15,60	03:15:14,23	115,05
Exp'03	19	2	3	20	2	3	15,74	03:07:46,48	120,74
Exp'04	19	3	2	20	2	1	15,76	03:05:25,23	122,40
Exp'05	19	3	2	20	3	3	15,80	03:14:57,55	116,69
Exp'06	19	3	2	18	3	3	15,84	03:08:19,67	121,12

4 Conclusions

In the paper, the impact of the buffer on throughput and work-in-progress, in a parallel-serial manufacturing system, was analysed. A simulation model of a parallel-serial manufacturing system was prepared, using Tecnomatix Plant Simulation Software. On the basis of the research conducted, the following conclusion may be formulated:

- buffer allocation capacity has a significant impact on the throughput and work-in-progress of parallel-serial, manufacturing systems,
- one allocation of buffer capacity is responsible for a significant decrease in throughput, as also for an increase of work-in-progress,
- using the simulation method, buffer allocation capacity can be shown to be chiefly responsible in determining poor throughput and work-in- progress values.

To evaluate the efficiency of the parallel-serial system, the system performance rate is proposed. In future research, the impact of the efficiency of production and maintenance workers on the total efficiency of a parallel-serial, manufacturing system, will be analysed.

References

1. Demir, L., Tunali, S., Løkketangen, A.: A tabu search approach for buffer allocation in production lines with unreliable machines. Eng. Optim. **43**(2), 213–231 (2011)
2. Costa, A., Alfieri, A., Matta, A., Fichera, S.: A parallel tabu search for solving the primal buffer allocation problem in serial production systems. Comput. Oper. Res. **64**, 97–112 (2015)
3. Nahas, N., Ait-Kadi, D., Nourelfath, M.: A new approach for buffer allocation in unreliable production lines. Int. J. Prod. Econ. **103**, 873–881 (2006)
4. Sabuncuoglu, I., Erel, E.: Gocgun. Y. Analysis of serial production lines: characterization study and a new heuristic procedure for optimal buffer allocation. Int. J. Prod. Res. **44**(13), 2499–2523 (2006)
5. Kose, S.Y., Kilincci, O.: Hybrid approach for buffer allocation in open serial production lines. Comput. Oper. Res. **60**, 67–78 (2015)
6. Battini, D., Persona, A., Regattieri, A.: Buffer size design linked to reliability performance: a simulative study. Comput. Ind. Eng. **56**, 1633–1641 (2009)
7. Jithavech, I., Krishnan, K.: A simulation-based approach to the risk assessment of facility layout designs under stochastic product demands. Int. J. Adv. Manufact. Technol. **49**, 27–40 (2010)
8. Trojanowska, J., Varela, M.L.R., Machado, J.: The tool supporting decision making process in area of job-shop scheduling. In: Rocha, Á., Correia, A., Adeli, H., Reis, L., Costanzo, S. (eds.) Recent Advances in Information Systems and Technologies, WorldCIST 2017, Advances in Intelligent Systems and Computing, vol. 571, pp. 490–498 (2017)
9. Iassinovski, S., Artiba, A., Fagnart, C.: A generic production, rules-based system for on-line simulation, decision making and discrete process control. Int. J. Prod. Econ. **112**(1), 62–76 (2008)
10. Negahban, A., Smith, J.S.: Simulation for manufacturing system design and operation: literature review and analysis. J. Manufact. Syst. **33**, 241–261 (2014)

11. Klos, S.: The simulation of manufacturing systems with Tecnomatix Plant Simulation, Zielona Góra, Oficyna Wydaw, Uniwersytetu Zielonogórskiego (2017)
12. Demir, L., Tunali, S., Eliiyi, D.T.: The state of the art on the buffer allocation problem: a comprehensive survey. J. Intell. Manufact. **25**, 371–392 (2014)
13. Dallery, Y., Gershwin, S.B.: Manufacturing flow line systems: a review view of models and analytical results. In: Queueing Systems: Theory and Applications, vol. 12, Springer-Verlag (1992)
14. Rewers, P., Trojanowska, J., Diakun, J., Rocha, A., Reis, L.P.: A study of priority rules for a levelled production plan. In: Advances in Manufacturing, Lecture Notes in Mechanical Engineering, pp. 111–120. Springer (2018)
15. Kalinowski, K., Krenczyk, D., Paprocka, I., Kempa, W., Grabowik, C.: Multi-criteria evaluation methods in the production scheduling. In: IOP Conference Series: Materials Science and Engineering, vol. 145 (2016)

Check for updates

A Text Mining Based Supervised Learning Algorithm for Classification of Manufacturing Suppliers

V. K. Manupati[1], M. D. Akhtar[1], M. L. R. Varela[2(✉)], G. D. Putnik[2], J. Trojanowska[3], and J. Machado[2]

[1] Vellore Institute of Technology, Vellore, Tamilnadu, India
manupativijay@gmail.com, mdakhtar20067@gmail.com
[2] University of Minho, Guimarães, Portugal
{leonilde,putnikgd}@dps.uminho.pt,
jmachado@dem.uminho.pt
[3] Poznan University of Technology, Poznań, Poland
justyna.trojanowska@put.poznan.pl

Abstract. With the expeditious growth of unstructured massive data on the World Wide Web (WWW), more advanced tools, techniques, methods, and approaches for information organization and retrieval are desired. Text mining is one such approach to achieve the above mentioned demand. One of the main text mining applications is how to classify data presented by different industries into groups. In this paper, the classification of data into various groups based on the choice of the users at any given point of time is proposed. Here, a support vector machine (SVM) based classification algorithm is established to classify the text data into two broad categories of Manufacturing and Non-Manufacturing suppliers. Later, the performance of the proposed classifier was tested experimentally using most commonly used accuracy measures such as precision, recall, and F-measure. Results proved the efficiency of the proposed approach for classification of the texts.

Keywords: Text mining · Support vector machine · Classification algorithm

1 Introduction

Recent advancements in information and communication technology change the configuration of supply chains from their traditional services to decentralized/distributed services to meet the quick requirements of the customers. To meet the mentioned customer needs the suppliers have to be responsive, appropriate and technologically capable and competent. This forced companies to make themselves visible in online portals and websites or any other mode of global advertisement. Although, there is a solution available but it is a challenging task in the field of supply chain due to the participation of multi domain service providers (suppliers, manufacturers, distributers, etc.) with different functional requirements. Therefore, not only identifying the multiple service provider's requirements, integration of usability methods on semantic level is essential (Holzinger 2008).

© Springer International Publishing AG, part of Springer Nature 2018
Á. Rocha et al. (Eds.): WorldCIST'18 2018, AISC 747, pp. 236–244, 2018.
https://doi.org/10.1007/978-3-319-77700-9_24

This paper mainly focuses on distinguishing between manufacturing and non-manufacturing suppliers along with providing a strong foundation for developing a market. Enormous amount of work have been done regarding text mining and classification approach but limited work has been done on manufacturing supplier's context. The constraint followed in this paper is the unstructured data, involving varieties of suppliers, for classifying them into the above mentioned categories. The idea has been reflected in (Yazdizadeh et al. 2015). Various fundamental text mining tasks and techniques like text pre-processing, classification, clustering and the like have been incorporated in (Allahyari et al. 2017). A novel approach for enabling organization of textual documents along with extraction of thematic patterns in manufacturing corpora has been shown in (Shotorbani et al. 2016), through document clustering and topic modelling techniques. A method for extracting beneficial insights from drilling activities and then classifying them with the help of a Support Vector Machine has been exhibited in (Esmael et al. 2010). Various text mining classification techniques for classification of publicly available accident narratives have been elucidated in (Goh et al. 2017). Text mining techniques like entities recognition and topic hierarchies are applied to textual data, as shown in (Domingues et al. 2014), for automatic attainment of contextual data. The origins and motivation of building Manufacturing Corpus Version 1 (MCV1) using text mining techniques has been shown in (Liu et al. 2004). Machine learning strategies designed for leveraging positively labelled data using text mining techniques has been explained in (Yeganova et al. 2011). Three different text mining techniques for extracting various service aspects from the same online data has been examined in (Vencovsky et al. 2017). The general concepts behind text mining and comparison of its techniques, along with handful of text mining applications are discussed in (Dang et al. 2014). The use of text mining techniques for regulation of an ontology, representing a dynamic domain, has been shown in (Nora et al. 2009). Various text mining techniques have been used to analyse mobile learning based approaches, which have been detailed in (Salloum et al. 2018). The use of hybrid applications of text mining to classify textual data into two different classes has been explained in (Rahman et al. 2012). Manufacturing and non-manufacturing companies produce substantial quantity of narrative texts related to their products and services as a by-product of the processes. These narrative texts include companies profile, products catalogue, services detail etc. (Mark et al. 2000).

In this huge era of data related operations, the establishment of data mining techniques (DMTs) and prediction of the stability of the DMTs related applications in the field of production management at present as well as in future has been elucidated in (Cheng et al. 2017). Exploitation of work orders (WOs) and downtime data (DD), for precise extraction of failure time data after a text mining procedure is suggested, has been carried out in (Arif-Uz-Zaman et al. 2016). The effectiveness of text mining techniques had been evaluated, for executing and updating the proper technique selection by making use of a class of criteria, in (Hashimi et al. 2015). Regarding process industries, the practices carried out in their various units and the sustainability trends are recognized with the help of various text mining techniques in (Liew et al. 2014). A bottom-up approach related issue which is responsible for itemizing environmental sustainability indicators, by making use of their text-based objective information, and which plays an important role in investigating perceptions based on industrial and indicator uses has been expressed in (Park and Kremer 2017).

In this paper, classification of supplier families into manufacturing as well as non-manufacturing sectors based on the text narratives available in the online profiles has been investigated. In addition, it is necessary to improve the search process by improving the intelligence of the search tools to obtain the accurate and precise results. In this research study, in order to classify the mentioned sectors, the textual portfolios from their websites and internet gateways helps to categories and organize diverse documents. Here, we use the classification tools such as Naïve Bayes and Support Vector Machine (SVM) methods as classifier to classify the text. The extracted textual information from these classifiers are verified and validated with confusion matrix. Finally, the comparison study has been done between two classification algorithms with different accuracy measures.

Further details pertaining to the above problem is discussed and organized as follows: In Sect. 2 the detail problem description has been described. The proposed methodology for classifying the manufacturing and non-manufacturing sectors is described with an architecture in Sect. 3. Section 4 detailed the proposed flowchart with different modules for the classification of sectors. In Sect. 5 the experimentation with two different classification algorithms has been described. Finally, the conclusions have been drawn and future work has been defined in Sect. 6.

2 Problem Description

Modern industries are quick, responsive and competitive in nature due to the information and communication technology advancements. Still, identification of right supplier at right time for the right order is one of the challenging tasks in recent decentralized environments. In supply chain management philosophy "a chain is only as strong as its weakest link". Similarly, in the decentralized environment identification of a wrong supplier is more vulnerable for the entire project. Thus, in this research work, our aim is to identify and classify the appropriate suppliers for a specific order out of many different services available. Here, we considered that the data is unstructured having different variety of suppliers that need to be processed and categorized as manufacturing and non-manufacturing suppliers. Due to its ambiguity and complexity, and the amount of the shapeless data of different sources need to retrieved and categorized. Thus, there is a need of sophisticated technique and methodology to classify the data into its used form.

3 Proposed Methodology for Categorization of Services in Text Mining Environment

The classification of manufacturing and non-manufacturing services with text mining approach as an architecture is depicted in Fig. 1. The proposed architecture is divided into three major sections VIZ. input module, processing module, and output module. Beginning with input module, the data has been collected from weblogs and internet portals

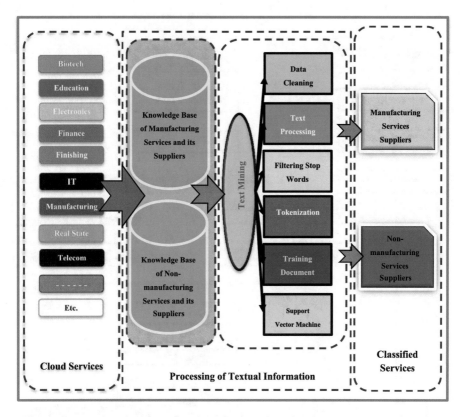

Fig. 1. Proposed methodology for categorization of services in text mining environment

which is not specific to one sector. Secondarily, the collected data has to be classified according to the decision maker requirement. Here, manufacturing as a case with pre-processing and processing using text mining approach the data has been classified. Finally, the filtered data is used for training, testing, and validation with the adopted classifier. In this paper, we used support vector machine as a classifier to get the final classified result (Manufacturing and Non-manufacturing services suppliers). The detailed step by step procedure of implementation of this architecture is detailed in next sections.

4 Proposed Flowchart to Classify the Suppliers

In this section, the developed flowchart has been clearly shown in Fig. 2. The step by step function of the flowchart and their description is as follows:

Step 1. Preparation of Training data

Datasets are collected from companies' websites and online gateways for training purpose. To build the training dataset, suppliers from both categories (i.e.

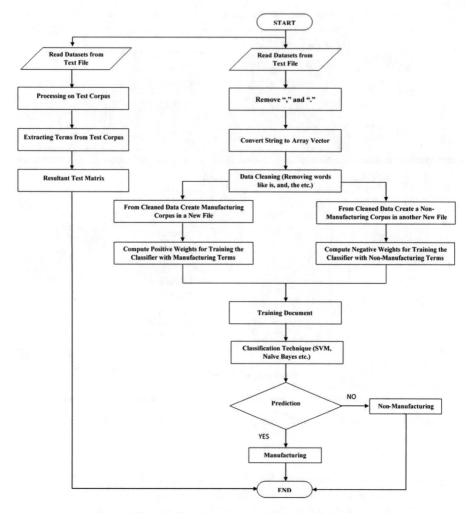

Fig. 2. Flowchart to classify the suppliers

manufacturing and non-manufacturing) are selected and their online profiles are converted into textual documents. These documents collectively form the training corpus. After going through a series of pre-processing steps, as described below, the corpus documents are categorized successfully. Read datasets obtained from sourcing portals or websites of various industrial suppliers.

Step 2. Pre-Processing of textual information

Preprocessing of unstructured data collected from weblogs and online portals were first converted from text files data to string format and store it in a vector array and then cleaned to remove unwanted information and words such as punctuations, commas, semi colons etc., delimiters ("," ".") which are present in the text files.

Step 3. Filtration of Stop words

This is a method of information retrieval which involves in deleting unnecessary words which occur too frequently. It helps in reducing the time and space complexity of the program. Stop words can be filtered by using simple string manipulation functions available in the string library.

Step 4. Tokenization

In this step, breaking a stream of text up into words, phrases, symbols, or other meaningful elements were done and it is called tokens. The list of tokens becomes input for further processing such as parsing or text mining. To analyze the data we create two text-corpus, one consisting of manufacturing terms and the other of non-manufacturing terms.

Step 5. Text Processing

The weights of the terms present in each corpus are computed based on the frequency of their occurrence. However, the assignment for both corpus is based on the sgn function. The function is defined below:

$$\text{sgn}(f) = \begin{cases} +f, & \text{if} \quad \in \quad manufacturing \\ -f, & \text{if} \quad \in \quad non-manufacturing \end{cases} \tag{1}$$

Step 6. Training and testing of Textual Documents

For each training document, a value V is computed based on the weights of the words obtained in the previous step. We calculate the support vector, W, using the values computed above according to a function. The function is defined below:

$$W = ((\min(x_{manu}) + (\max(x_{non-manu}))/2 \tag{2}$$

Support vector machine is constructed using the support vector W.

$$svm(x) = \begin{cases} manufacturing, & \text{if} \quad x > W \\ non-manufacturing, & \text{if} \quad x < W \end{cases} \tag{3}$$

The testing data is then read and classified as manufacturing or non-manufacturing based on the support vector machine. A value for each test document is calculated by following STEP 4. Then based on the obtained value and the support vector the company in its text form is classified as manufacturing or non-manufacturing.

5 Experimentation

As a first step, we prepared the training data where each profile from the collected information is preserved into predefined classes. Next, these classes were pre-processed by removing the stop words and undesirable information. For example: stop words such as "and", "is", "for", "the", generic words such as "product", "capable", "include and removal of punctuations, numbers, and whitespaces.

In the next stage, tokenization of Pre-processed data for feature extraction is being done based on total frequency. Where frequency refers to the number of times a word is repeated in the text. As a first step in the tokenization, we sorted all the necessary words into two vector array; the first array represents the text corpus of the manufacturing sector and the second array resembles the text corpus of the nonmanufacturing sector.

Thereafter, the sorted arrays are calculated based on the frequency of words and sign function the weights have been assigned. These trained data sets have been used for further training, testing and validation using support vector machine based classification algorithm.

6   Results and Discussion

Different measures are used to evaluate the performance of the text-based classifier. They are defined as precision, recall, and F-measure. The measure precision states the relevance of a particular keyword in a group of collection i.e., the degree of relevance of terms with respect to all the retrieved terms from a given text corpus can be viewed. Precision is defined as it is the divisibility of relevant instances with retrieved instances. Whereas the measure recall gives the sensitiveness of data to wrong or negative values. The recall is used to measure how much right (true/positive) data is present in the sample we have taken. It also gave the information on the proportion of relevant instances that have been retrieved to all the retrieved terms. Another measure F-Measure is defined as the harmonic mean between precision and recalls measure, it is an averaging factor between the two binary classification methods Recall and Precision.

Hence with the help of SVM Algorithm along with the Pattern Recognition measures, we have classified the text corpus terms into two parts one consisting of Non-Manufacturing terms and the other with manufacturing terms and it is shown in Table 1. Applying the above algorithm on the partitioned data we are achieving the classified documents as 85.71% precise with a recall value of 75% shown in Table 2.

Table 1. Actual and predicted services after classification

Services	Actual	
Predicted	Manufacturing	Non-Manufacturing
Manufacturing	6	1
Non-Manufacturing	2	6

Table 2. Accuracy measures of the classified documents

Precision	0.857143
Recall	0.75
F-measure	0.8

These values clearly show that the accuracy of the algorithm in classifying the terms related to manufacturing and avoiding non-manufacturing terms in the same cluster.

7 Conclusion and Future Works

Recent manufacturing companies are increasingly using information and communication technology for their daily operations in the shop floor. This brings the manufacturers to respond quickly to the customer's request. On the other side choosing the appropriate partner to the right product is a challenge in this huge market space. More importantly, identification of explicit partner for the specialized product needs proper classification techniques, tools, and approaches. In this research work out of many such different suppliers, our main aim is to identify, classify and structured the data in particular to the manufacturing suppliers with text mining approach due to its accuracy measures. These classified data makes it easier to analyze, understand and extract useful information from it. Here, we have implemented an efficient method for processing the data and then used Support Vector Machine (SVM) based classifier for classification. Thus, we managed to classify data into the required classes and obtained high accuracy and precision results. In the future work with a case study, the proposed approach can be implemented with real time monitoring in particular to distributed manufacturing context.

Acknowledgment. This work has been supported by Department of Science and Technology, Science & Engineering Research Board (SERB), Statutory Body Established through an Act of Parliament: SERB Act 2008, Government of India with Sanction Order No ECR/2016/001808.

References

Holzinger, A., Geierhofer, R., Mödritscher, F., Tatzl, R.: Semantic information in medical information systems: utilization of text mining techniques to analyse medical diagnoses. J. UCS **14**(22), 3781–3795 (2008)

Yazdizadeh, P., Ameri, F.: A text mining technique for manufacturing supplier classification. In: Proceedings of ASME, August 2015

Allahyari, M., Pouriyeh, S., Assefi, M., Safaei, S., Trippe, E.D., Gutierrez, J.B., Kochut, K.: A Brief Survey of Text Mining: Classification, Clustering and Extraction Techniques. arXiv preprint arXiv:1707.02919 (2017)

Shotorbani, P.Y., Ameri, F., Kulvatunyou, B., Ivezic, N.: A hybrid method for manufacturing text mining based on document clustering and topic modeling techniques. In: IFIP International Conference on Advances in Production Management Systems, pp. 777–786. Springer, Cham, September 2016

Esmael, B., Arnaout, A., Fruhwirth, R.K., Thonhauser, G.: Automated operations classification using text mining techniques. In: PACIIA 2010 (2010)

Goh, Y.M., Ubeynarayana, C.U.: Construction accident narrative classification: an evaluation of text mining techniques. Accid. Anal. Prev. **108**, 122–130 (2017)

Domingues, M.A., Sundermann, C.V., Manzato, M.G., Marcacini, R.M., Rezende, S.O.: Exploiting text mining techniques for contextual recommendations. In: 2014 IEEE/WIC/ACM International Joint Conferences on Web Intelligence (WI) and Intelligent Agent Technologies (IAT), vol. 2, pp. 210–217. IEEE, August 2014

Liu, Y., Loh, H.T., Tor, S.B.: Building a document corpus for manufacturing knowledge retrieval. In: Innovation in Manufacturing Systems and Technology (IMST) (2004)

Yeganova, L., Comeau, D.C., Kim, W., Wilbur, W.J.: Text mining techniques for leveraging positively labeled data. In: Proceedings of BioNLP 2011 Workshop, pp. 155–163. Association for Computational Linguistics, June 2011

Vencovsky, F., Lucas, B., Mahr, D., Lemmink, J.: Comparison of text mining techniques for service aspect extraction. In: 4th European Conference on Social Media, Vilnius, Lithuania (2017)

Dang, S., Ahmad, P.H.: Text mining: techniques and its application. Int. J. Eng. Technol. Innovations 1, 2348–2866 (2014)

Nora, T., Mokhtar, S., Simonet, M.: The management of the knowledge evolution by using text mining techniques. In: Proceedings of I-KNOW 2009 and I-SEMANTICS 2009, 2–4 September, Graz, Austria (2009)

Salloum, S.A., Al-Emran, M., Monem, A.A., Shaalan, K.: Using text mining techniques for extracting information from research articles. In: Intelligent Natural Language Processing: Trends and Applications, pp. 373–397. Springer, Cham (2018)

Ur-Rahman, N., Harding, J.A.: Textual data mining for industrial knowledge management and text classification: a business oriented approach. Expert Syst. Appl. 39(5), 4729–4739 (2012)

Kornfein, M.M., Goldfarb, H.: A comparison of classification techniques for technical text passages. In: World Congress on Engineering, pp. 1072–1075 (2007)

Cheng, Y., Chen, K., Sun, H., Zhang, Y., Tao, F.: Data and knowledge mining with big data towards smart production. J. Ind. Inf. Integr. (2017)

Arif-Uz-Zaman, K., Cholette, M.E., Ma, L., Karim, A.: Extracting failure time data from industrial maintenance records using text mining. Adv. Eng. Inform. 33, 388–396 (2016)

Hashimi, H., Hafez, A., Mathkour, H.: Selection criteria for text mining approaches. Comput. Hum. Behav. 51, 729–733 (2015)

Te Liew, W., Adhitya, A., Srinivasan, R.: Sustainability trends in the process industries: a text mining-based analysis. Comput. Ind. 65(3), 393–400 (2014)

Park, K., Kremer, G.E.O.: Text mining-based categorization and user perspective analysis of environmental sustainability indicators for manufacturing and service systems. Ecol. Indic. 72, 803–820 (2017)

Production Scheduling with Quantitative and Qualitative Selection of Human Resources

Krzysztof Kalinowski[✉] and Barbara Balon

Faculty of Mechanical Engineering, Institute of Engineering Processes
Automation and Integrated Manufacturing Systems, Silesian University
of Technology, Konarskiego 18A, 44-100 Gliwice, Poland
{krzysztof.kalinowski,barbara.balon}@polsl.pl

Abstract. The paper presents a method of quantitative and qualitative selection of human resources in production planning based on assumed load of the production system. Human resource are treated as additional resources. Determining the required number of staff and their competences is implemented in three stages. First, the production schedule is determined with the assumption of unlimited availability of additional resources. The obtained solution is a point of reference for determining the minimum number of required resources and their competencies. In the second stage, the reference number of workers is sought. Depending on the way selected, in the last stage the process of merging or removing the competencies of individual workers is done for the optimisation of the number of workers and their individual competencies.

Keywords: Scheduling · Additional resources · Staff planning
Human resources

1 Introduction

Effective organization of the production flow, determination of achievable plans as well as efficient exploitation of available resources is one of the basic assumptions of a well prospered company. Detailed scheduling of production processes supports the coordination of activities in this area [1]. The issue of scheduling is widely undertaken in research works, and concerns very different areas of this field. Various algorithms are developed to ensure optimal solutions, heuristics and metaheuristics; apart from purely theoretical considerations, tested models take also into account a number of constraints related to the practical application for different manufacturing systems configurations [2, 3]. Despite this, translating scientific achievements into production practice is still quite a serious problem. It results mainly due to the much greater complexity of the models of production systems, limitations of available supporting tools as well as from dynamic nature of a real-world scheduling problems with significant impact of unforeseen internal and external factors, not formalised in used models.

One of the quite rarely discussed topics, but very important from the point of view of practical application, is to consider, apart from basic resources (machines, processors) other additional resources, important in the planning point of view. The solution presented in this article is a development of the scheduling methodology presented in

© Springer International Publishing AG, part of Springer Nature 2018
Á. Rocha et al. (Eds.): WorldCIST'18 2018, AISC 747, pp. 245–253, 2018.
https://doi.org/10.1007/978-3-319-77700-9_25

[4] in the field of modelling of operations with multi-resources requirements, taking into account the constraints typical for human resources.

2 Literature Review

Research achievements in the scope of additional and human resources planning described in scientific publications allow developing useful extensions to scheduling models, help to select and implement appropriate tool or techniques [5]. In [6] the use of additional scarce resources during operation execution is discussed. They present a classification scheme for resource constraints and the computational complexity for parallel machines configuration, unit-time jobs and the maximum completion time criterion. This topic was developed in subsequent works, among others [7, 8]. In [9] the simulation of human resources allocation is taken into consideration. They discuss the influence of different scheduling strategies (forward, backward, midpoint) on human resource requirements. The stages of comparison of labour demand and supply in human resources planning were also presented. The most important factors limiting the use of human resources, based on time (working calendar), labour law (regulations) and other are considered. They proposed an algorithm of simulation based HR allocation using ERP class software. [10, 11] describes stochastic methods useful in modelling and planning of human resources. [12] presents an analytical approach of staff scheduling in a multi project management environment. An interesting tutorial on staff scheduling problems is presented in [13]. They described different general models representing wide range of real world staff problems. The work includes a comprehensive description of data, hard and soft constraints, and also objectives. The handbook [14] comprehensively discusses human resources issues. A review report published in [15] refers to nearly three hundred publications on personnel scheduling problems. Classification into different categories related, among others, to personnel characteristics, constraints, performance measures and flexibility, solution methods and areas of application is presented there. They prove that this issue is constantly taken up in numerous scientific papers and many different solutions have been developed. [16] presents a novel social network analysis based method to evaluate the reconfiguration effect i.e., identification of key machines and their influence on the system performance in the context of flexible job shop scheduling problem. This research formulates a mathematical model along with the constraints by incorporating the total completion time of jobs as an objective function. In [17] the project management method derived from the theory of constraints is described – as a decision-making support tools that enable project managers to monitor and control the key factors influencing e.g. task time.

3 Problem Formulation

Production system consists of m^b basic resources R_1, ..., R_{mb} (indexed by i^b), m^a additional resources, generally denotes as W_1, ..., W_{ma} (indexed by i^a) and n jobs P_1, ..., P_n (indexed by j). Resources are renewable, discrete, with capacity = 1, and are characterized by different costs of executing of the given operation. Specialisations (skills) of additional resources are grouped in p competences groups $Comp_1$–$Comp_p$ as unrelated parallel model. Each of the additional resources belongs to at least one group of competences. Each job P_j is represented by sequence of k^j operations $O_{j,1}...O_{j,kj}$. Each operation requires exactly one basic resource and 0 or more additional resources from each competences group. In the objective function there are used time based and cost based criteria. A schedule that met all given constraints and minimizes objective function is searched.

4 Staff Selection Method

Determining the required number of staff and their competences is implemented in three main stages (Fig. 1). In the first stage, the production schedule is determined with the assumption of unlimited availability of additional resources. Classic scheduling methods can be used at this stage, adequate to the considered scheduling task model. Next, the quantitative and qualitative selection of staff is performed.

Quantitative selection of additional resources. The obtained solution is a point of reference for determining the number of required additional resources and their competences. As in the case of basic resources, depending on specialisation in the set of additional resources, the following types of specialisation can be distinguished:

- (u) workers are versatile (universal), each of them can perform any of the tasks. This situation can be found in small size systems, systems with parallel machines,
- (1) each worker has only one competence, most often in job shop, flow shop class of systems, Description of one competence worker has been extended to $W^1_{x,y}$, where x - index of competence group, y - index of worker.
- (d) workers may have more competences but differ in their number; this is the most common situation in real production systems.

Depending on the considered type of additional resources specialisation, the number of universal workers N^u, the number of one-competence workers N^1, or quantitative range $<N^u, N^1>$ of workers is sought.

Qualitative selection of additional resources. At this stage, verification and optimization of the allocation of additional resources is carried out. The following criteria may be used for optimization:

- cost based (minC), e.g. costs of man-hours minimisation,
- time based, (minT) e.g. deterioration of the selected performance measures of the obtained schedule in comparison to the reference schedule,
- quantitative, (minN^d) e.g. minimum number of workers, optimal combination of their competences.

Cases (u) and (1) already have a fixed number of workers, so any change (decrease) of this number will cause deterioration of time based performance measures of a schedule. In the case (d) the target number of resources N^d is sought, so the minimization of the number of workers can be obligatory taken into consideration.

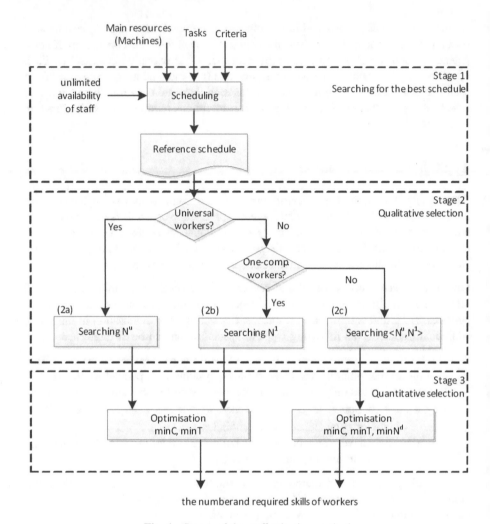

Fig. 1. Stages of the staff selection method.

The number of workers can be reduced by a combination of competence within one post. Then a worker is allocated to operations in the context of different competencies. Such operations must, however, be disjointed in time.

5 Example

Considered system consists of three basic resources R_1–R_3 and tasks P_1–P_5 with operations that require some additional resources from competence groups $Comp_1$–$Comp_4$. Operation times and resource requirements are presented in Table 1. A minimum number of workers N^z, $z \in \{u, 1, d\}$ required to execute all tasks in accordance with the adopted plan should be appointed.

Table 1. Operation times and resources requirements.

	P1			P2			P3			P4			P5		
	Op1	Op2	Op3	Op1	Op2	Op3	Op1	Op2	Op3	Op1	Op2	Op3	Op1	Op2	Op3
R1	10			8			6			5					3
R2		4		7		7			2		4			4	
R3			8					1		9			2		
Comp.1	1	1					1			1					
Comp.2			1					1		1	1	1	1		1
Comp.3	1					1	1			1					1
Comp.4		1	1		1	1	1	1			1		1		

The reference schedule, created in the first stage of the method, taking into account unlimited additional resources and optimised for makespan (C_{max}) is shown in Fig. 2. Presented solution is an optimal, according to C_{max} objective, but at this stage it is sufficient to find a selected, acceptable solution – as the basis for searching the number of additional resources.

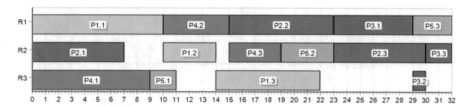

Fig. 2. Stage 1. Schedule for unlimited additional resources.

Figure 3 shows the schedule with the allocation of operations to universal additional resources (stage 2). The minimum number them that guarantee the implementation of the tasks in accordance with the reference schedule is 6 $\left(W^u_1 - W^u_6 \right)$. The load in the set of universal workers is presented in Fig. 4.

Figure 5 shows the schedule with the allocation of operations to one-competence workers (stage 2b). The minimum number of them that guarantee the implementation of the tasks in accordance with the reference schedule is 10, in 4 particular groups: $Comp_1$ – 2 workers $\left(W^1_{1,1} - W^1_{1,2} \right)$, $Comp_2$ – 2 $\left(W^1_{2,1} - W^1_{2,2} \right)$, $Comp_3$ – 2 $\left(W^1_{3,1} - W^1_{3,2} \right)$ and $Comp_4$ – 4 workers $\left(W^1_{4,1} - W^1_{4,4} \right)$. The load in the set of universal workers is presented in Fig. 4.

Analyzing the above solutions can be seen the need for optimization. In case of universal workers the last one is needed only for one task (P5.3), in the period <29–32>. On the other hand, with one-competence workers, it seems reasonable to combine some of the competencies within some single posts. Based on the load (Fig. 6) potential possible 'merges' of competencies are as follow:

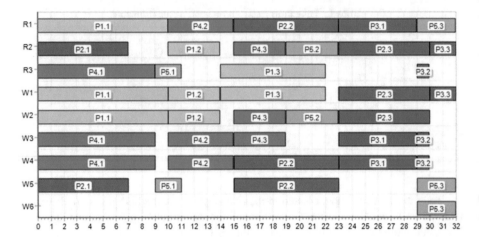

Fig. 3. Schedule for universal workers (W^u). Required: 6 workers.

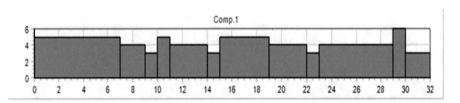

Fig. 4. The load in the set of universal workers (W^u).

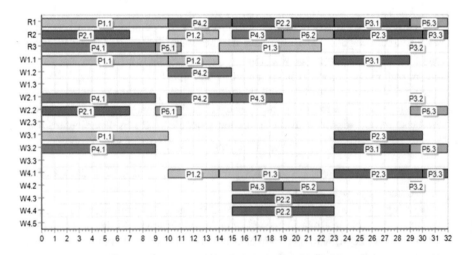

Fig. 5. Schedule for one-competence workers (W^l). Required: 10 workers.

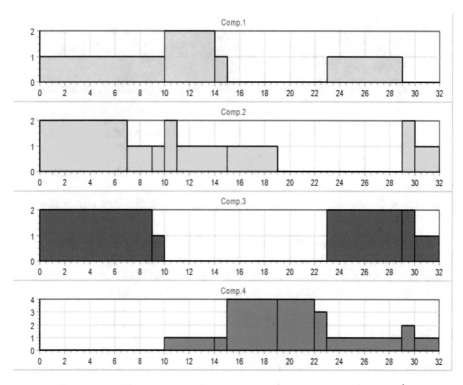

Fig. 6. The load in competences groups for one-competence workers (W^1).

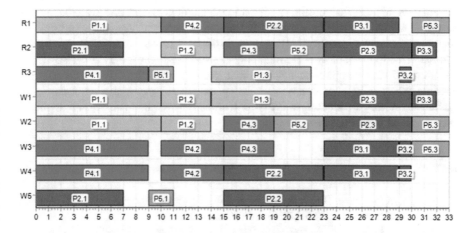

Fig. 7. "Extended" schedule - one worker (W^u) less.

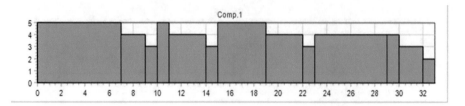

Fig. 8. The load in the set of universal workers (W^u).

- $Comp_1 + Comp_3$ or $Comp_1 + Comp_4$ - in the period <10–14> - reduction of one worker
- $Comp_1 + Comp_4$ or $Comp_3 + Comp_4$ - in the period <15–23> - reduction of 2 workers.

In stage 3, depending on the given assumptions, criteria selected from the groups presented in Sect. 3 can be used in objective function formulating. The result of the exemplary optimization is shown in Figs. 7 and 8. The cost of extending the makespan of the schedule by 1 time unit is compensated by a reduced number of workers to 5.

6 Summary

The paper presents the method of selection of additional resources to the system with a given load. The method has three stages, the construction of a schedule with unlimited additional resources, quantitative optimization - determination of the number of additional resources required and qualitative optimization. The method is applicable in the selection of the number of personnel and their competences in both newly created (in order to identify the profile of the wanted employee) and already existing production systems. In existing systems, if the competences of the crew are known, it enables simulations (verification) and analysis of various variants of production, optimal selection of orders, determination of production capacities and also desired directions of staff competence development, etc. Application of the method is also possible when only human resources are planned, without assigning tasks to the machines. Although we strongly focus on workers, the method can also be applied to other types of additional resources.

References

1. Nicholas, J.M., Steyn, H.: Project Management for Business Engineering and Technology. Taylor & Francis, New York (2012)
2. Blazewicz, J., Ecker, K.H., Pesch, E., Schmidt, G., Weglarz, J.: Scheduling Computer and Manufacturing Processes. Springer, Heidelberg (2007)
3. Pinedo, M.L.: Scheduling. Theory, Algorithms and Systems. Springer, New York (2012)
4. Kalinowski, K., Grabowik, C., Paprocka, I., Kempa, W.: Production scheduling with discrete and renewable additional resources. Materials Sci. Eng. **95**, 1–6 (2015)

5. Starzyńska, B., Hamrol, A.: Excellence toolbox: decision support system for quality tools and techniques selection and application. Total Qual. Manage. Bus. Excellence **24**(5–6), 577–595 (2013)
6. Blazewicz, J., Lenstra, J.K., RinnooyKan, A.H.G.: Scheduling subject to resource constraints: classification and complexity. Discrete Appl. Math. **5**(1), 11–24 (1983)
7. Fanjul-Peyro, L., Perea, F., Ruiz, R.: Models and matheuristics for the unrelated parallel machine scheduling problem with additional resources. Eur. J. Oper. Res. **260**, 482–493 (2017)
8. Edi, E.B., Oguz, C., Ozkarahan, I.: Parallel machine scheduling with additional resources: notation, classification, models and solution methods. Eur. J. Oper. Res. **230**, 449–463 (2013)
9. Baron-Puda, M., Mleczko, J.: Simulation of human resources allocation in scheduling processes. Appl. Comput. Sci. **4**(2), 1–16 (2008)
10. Starzynska, B., Szajkowska, K., Diering, M., Rocha, A., Reis, L.P.: A study of raters agreement in quality inspection with the participation of hearing disabled employees. In: Advances in Manufacturing. LNME, pp. 881–888. Springer, Cham (2018)
11. Donyina, A., Heckel, R.: Modelling flexible human resource allocation by stochastic graph transformation. Electron. Commun. EASST **38**, 1–17 (2010)
12. Dean, B.V., Denzler, D.R., Watkins, J.J.: Multiproject staff scheduling with variable resource constraints. IEEE Trans. Eng. Manage. **39**(1), 59–72 (1992)
13. Blochliger, I.: Modeling staff scheduling problems. A tutorial. Eur. J. Oper. Res. **158**, 533–542 (2004)
14. Armstrong, M.: A Handbook of Human Resource Management Practice, 11th edn. Kogan Page, London (2009)
15. Van den Bergh, J., Jeroen Beliën, J., De Bruecker, P., Demeulemeester, E., De Boeck, L.: Personnel scheduling: a literature review. Eur. J. Oper. Res. **226**, 367–385 (2013)
16. Reddy, M.S., Ratnam, C., Agrawal, R., Varela, M.L.R., Sharma, I., Manupati, V.K.: Investigation of reconfiguration effect on makespan with social network method for flexible job shop scheduling problem. Comput. Ind. Eng. **110**, 231–241 (2017)
17. Trojanowska, J., Dostatni, E.: Application of the theory of constraints for project management. Manage. Prod. Eng. Rev. **8**(3), 87–95 (2017)

Automatic Assist in Estimating the Production Capacity of Final Machining for Cast Iron Machine Parts

Robert Sika[1], Michał Rogalewicz[2], Justyna Trojanowska[2(✉)],
Łukasz Gmyrek[3], Przemysław Rauchut[3], Tomasz Kasprzyk[3],
Maria L. R. Varela[4], and Jose Machado[5]

[1] Institute of Materials Technology, Poznan University of Technology,
Piotrowo 3 Street, 61-138 Poznan, Poland
Robert.Sika@put.poznan.pl
[2] Chair of Management and Production Engineering,
Poznan University of Technology, Piotrowo 3 Street, 61-138 Poznan, Poland
{Michal.Rogalewicz,Justyna.Trojanowska}@put.poznan.pl
[3] Teriel Foundry, LLC Company, Lipowa 2a Street, 63-100 Gostyn, Poland
{Lukasz.Gmyrek,Przemyslaw.Rauchut,Tomasz.Kasprzyk}
@teriel.com.pl
[4] Department of Production and Systems, School of Engineering,
University of Minho, Guimarães, Portugal
leonilde@dps.uminho.pt
[5] Department of Mechanical Engineering, School of Engineering,
University of Minho, Guimarães, Portugal
jmachado@dem.uminho.pt

Abstract. The paper presents a new approach for production capacities balancing. It is dedicated to companies performing machining services on a variety of materials on a small scale. The method is to help balance the production capacity on the basis of a defined labor intensity structure taking into account the share of set-up times and the level of defectiveness. The designed numerical algorithms along with the user interface make it possible to characterize machine tools and orders in the production system and compare the production capabilities of the system with the demand. The paper presents the algorithm working pattern, mathematical formulae used and a sample source code. An example of a computer application allowing to estimate the production capacity for sample orders (cast iron castings) and process routes in a selected cast iron production plant are also presented.

Keywords: Data acquisition · Balancing production · Production capacity
Single unit production · Cast iron casting

1 Introduction

Nowadays majority of manufacturing companies from the machine building industry are operating in single unit or short-run production which is very complex in terms of decision making processes in production planning. This follows from the fact that,

© Springer International Publishing AG, part of Springer Nature 2018
Á. Rocha et al. (Eds.): WorldCIST'18 2018, AISC 747, pp. 254–263, 2018.
https://doi.org/10.1007/978-3-319-77700-9_26

at present, customers are becoming more demanding in terms of quality, cost, performance, operation and many other factors. It requires innovative, flexible, adaptive and differentiated business models. Therefore this complicated risk-reward situation requires corresponding approach which should minimize cost levels as well as maximize concentrating on customer requirements and product parameters [1]. Manufacturing companies' efficiency depends not only on production resources available, but also on how well they are used [2]. This is especially noticeable in case of complex manufacturing processes controlling [3]. The problem of using resources in the right way is strictly connected to the problem of production capacity balancing and also scheduling production flow.

The fact that there are numerous external as well as internal factors influencing production processes makes production scheduling a very complicated issue. Decisions made in the production planning area relate to manufacturing capacity balancing in terms of quantities, quality, delivery dates and production costs with customer's requirements. The necessity of analyzing numerous factors causes manufacturing companies use tools helping with decision making in the area of scheduling processes.

The aim of this paper is to present a new approach which is intended for production companies in balancing their production capacities. The method has been tested on the example of a cast iron foundry machining final products with the use of conventional machine tools, CNC tools. The results of the work were supplemented by the design of a computer tool as simple and intuitive as possible. The study was run on the example of a foundry manufacturing castings from ductile cast iron (EN-GJS-500-7) for the automotive industry (weight of a single raw casting, without the gating system and the head being about 30 kg) on an automated box line with a horizontal mold division.

2 State of the Art

Scheduling production assigned to resources that complete the work is a very important issue from a practical point of view. In literature, it is easy to find a wide variety of scheduling methods, varying from discrete programming to methods based on meta-heuristics [4] and artificial intelligence [5–9]. The objective of scheduling is to plan or sequence production tasks, in order to minimize a certain performance measure of customer satisfaction [10].

The scheduling algorithm is selected on the basis of production system characteristics, set of orders to be executed and on encountered constraints. Among scheduling algorithms available in the literature on the subject there are two types of scheduling systems: simple and complex. The simple scheduling system is described as single-machine scheduling and parallel-machine scheduling, while the complex system is described with the use of flow-shop scheduling, job-shop scheduling and open-shop scheduling. Numerous scientific publications concerning production tasks prove that job-shop scheduling problems are a current research topic. One of the key issues when scheduling production flows is proper production capacity balancing. Various production management methods and tools are implemented to solve ongoing issues at manufacturing companies. Analysis of the literature of the subject shows that companies implement solutions which support quality management [11–13], decision-making processes [14–17], and lean manufacturing [18].

The literature describes many applications supporting production processes. Recently the dominance-based rough set approach is more frequently used to gain insight into the manufacturing process [19], minimizing losses [20–25], verifying validity of measurement and inspection tools [12, 26] or even economic conditions and personal features of the employee [27]. It should also be noted the issue of efficiency is widely described in the literature not only in relation to the production process but also, for example, transport [28]. Nevertheless, efficiency of the production process causes difficulties both in terms of literature, and in economic practice. This is why it is important to create new tools which could help in decision-making processes in the area of production scheduling.

Enabling rational decisions in this area requires a parametric description of each order and each resource with sufficient detail in the planning horizon as well as their balancing [29]. This parametric description allows to use the initialized parameter to estimate many others and thus build a more efficient relationship in the form of a set of equations. In case to define set of orders, which will be considered and balanced with the available resources, it is necessary to define the planning horizon which determining its detail.

Capacity balancing is therefore a form of verification how strong is the interrelationship of loads resulting from a set of orders and the available production capacity of each machine tool. In case the first component of this comparison is less than or equals the second one, the execution plan for the order (production plan) may be accepted. Otherwise, it is necessary to iteratively eliminate further orders until the available capacity (α) is reached. The eliminated orders will be shifted to another planning horizon. Positive balancing results mean achieving acceptable level of capacity utilization rate ($\alpha \approx 1$). Negative result of balancing can mean its excess ($\alpha \gg 1$, shortage of production means), but also too low machine tools load ($\alpha \ll 1$, shortage of orders) [30].

3   Research and Results

The following section describes the results of tests performed in a cast iron foundry. They focused on the foundry's products made of gray (EN-GJL) and ductile cast iron (EN-GJS) based on modern technologies of melting, forming and production of cores. The production is performed in an automated box line with a horizontal mold division (the weight of a single raw casting, without the gating system being about 30 kg). The castings are mainly intended for the automotive industry.

3.1 Preliminary Assumptions

The following assumptions were made regarding the new approach to balancing production capacity:

- companies for which the approach is intended specialize in production of objects with a similar technological process, and any operations that are not possible to process with the use of own resources are transferred to another co-operating entity,
- the load at the beginning of the planning horizon was assumed to be zero for each machine tool (this means that machine tools at the time of receiving orders for execution are not loaded at all),
- the paper focused only on the machines' production capacity, the remaining capacity, such as energy, human resources, etc. was not taken into account,
- the new method of determining individual machines' production capacity was implemented in numerical algorithms prepared specifically for this purpose and was partially based on the work of [31, 32].

3.2 Idea of New Balancing Approach

Three different types of machined castings were selected for the study (several technological operations per product). Unit times of individual operations were measured in specifically designed conditions. Taking into account the above preliminary assumptions, the authors presented the algorithm for production capacity based on new balancing approach. This is a modification of the algorithm presented in the article [30]. Changes were caused primarily by the needs of the foundries, which most often perform machining only for selected batches of castings or in special situations where defective castings are returned from the customer and subject to special processing for visual inspection (Visual Testing, VT) of defect identification (location, dispersion). These changes then concern the following issues:

- shorter planning horizon for machining castings of small batches of products,
- machining of various materials – other machinability of materials made of iron alloys with Carbon Equivalent (CE) value below and above the CE limit = 2%,
- "difficult dimensions" – dimensions of the casting difficult to finish and thus smaller series of castings (maximum of 10 castings per series).

In the new algorithm, further referred to as ProdBal-2.0, initial steps are generally included, which are a repetition of the algorithm presented in [30], i.e.: a preview of the current state of the machine park, machine park editing (adding, removing and editing machine names and functional parameters). At the stage of machine selection, the authors modified the possibility of defining the time horizon from the minimum of 1 month to the minimum of 1 day (shorter time horizon) and added new options for different types of castings as well as difficult dimensions (additional coefficient, C_{DIM} – Difficult Dimension Coefficient).

The matter of materials machinability (steel, cast iron, light metal alloys, etc.) is also important here. It should be noted that in recent years intensive research has been carried out to effectively select machining parameters in order to improve the machinability of metal materials. Nevertheless, as many studies show, cast iron and, above all, ductile cast iron are still worse in processing and more "chimerical" material [33]. This fact has prompted the authors to include in the new algorithm a further factor prolonging the processing cycle depending on the material used (C_{MAT} – Material Coefficient). The scheme of the new approach to production capacity balancing for short series of castings made of cast iron is shown in Fig. 1.

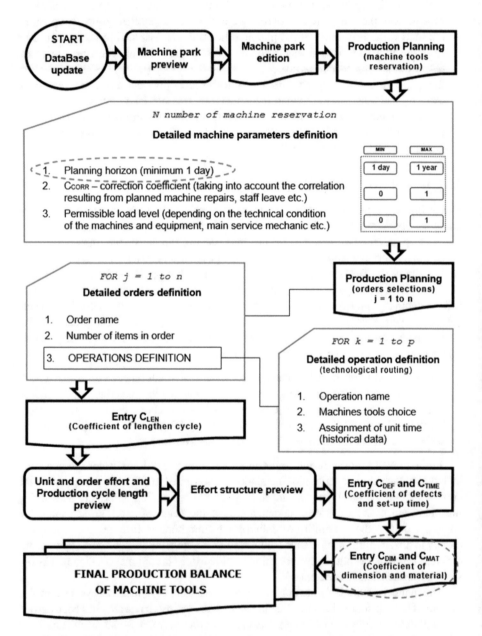

Fig. 1. Algorithm "ProdBal 2.0" for balancing production capacity (new options have been marked with the use of red dashed oval)

3.3   Mathematical Relations Used

Only selected key formulas related to separating individual operations into machine tools including time optimization are cited in this chapter. Additionally, formulas modifying the final production capacity balance due to the varied castings and difficult dimensions and type of material, are given. It should also be noted that the algorithm also takes into account the shorter planning period (1 day minimum).

   Long-term machine tools load – load on machine tools (taking into account defects and set-up times share and, additionally, C_{DIM} and C_{MAT} coefficient). See formula 1.

$$T'_i = \sum_{r=1}^{r} T_{ir}(1 + C_{DEF} + C_{TIME} + C_{DIM} + C_{MAT}) \tag{1}$$

where:

C_{DEF} – defect level (from 0 to 0,05)
C_{TIME} – factor taking into account set-up times share in the execution of the product (from 0,02 to 0,12)
C_{DIM} – the coefficient considering the different types of castings as well as difficult dimensions (0 to 0,2)
C_{MAT} – the coefficient considering the type of the material (0 to 0,3)
T_{ir} – total labor required for the i-th order and r-th machine tool.

   Fund available effective (failure rate and permissible load factor). See formula 2.

$$F_{md} = \eta \cdot \eta_d \cdot F_{mn} \tag{2}$$

where:

η – factor for machine tool failure rate, maintenance planning, etc. (from 0,95 to 1)
ηd – level of permissible load - factor taking into account technical condition of machines and equipment, efficiency of the main mechanic service
F_{mn} = number of working day * number of shifts * working hours.

   Capacity utilization rate of a given machine tool. See formula 3.

$$\alpha = \frac{T'_i}{F_{md}} \tag{3}$$

3.4   "ProdBal 2.0" Numeric Algorithms

Below an important part of the algorithm is presented, responsible for production capacity balancing using the new approach, which takes into account shorter planning horizon, various materials machined and production difficulty. Example of a Numeric Algorithm "ProdBal 2.0" (own works, VBA for Application, 2013).

```
Private Sum_Table(0 To 9, 0 To 9, 0 To 9) As Single
Sub EffortStructure()
Dim x, y                        As Byte
Dim T_unit                      As Single
Dim MachFitFund                 As Single
y = 0: x = 0
For Each oCtl In Controls
If TypeName(oCtl) = "Label" Then
If Left(oCtl.Name, Len(oCtl.Name) - 2) = _
"lbl_EffortStructure_SUM_" Then
y = y + 1
T_unit = 0: MachFitFund = 0
oCtl.Caption = 0
For x = 1 To 10
On Error Resume Next
T_unit = T_unit + Sum_Table(0, x - 1, y - 1) + _
Sum_Table(0, x - 1, y - 1) * lbl_Coeff_DEF.Caption + _
Sum_Table(0, x - 1, y - 1) * lbl_Coeff_TIME.Caption + _
Sum_Table(0, x - 1, y - 1) * lbl_Coeff_DIM.Caption + _
Sum_Table(0, x - 1, y - 1) * lbl_Coeff_MAT.Caption
Next x
MachFitFund = MachFitFund + txtMachRealFund.Text * _
Sheets("DB_MACHTOOLS").Cells(y + 1, 4).Value * _
Sheets("DB_MACHTOOLS").Cells(y + 1, 6).Value
T_unit = Round(T_unit, 4)
MachFitFund = Round(MachFitFund, 4)
oCtl.Caption = Round(T_unit / MachFitFund, 4)
End If
End If
Next oCtl
End Sub
```

4   Computer Application "ProdBal 2.0" Exemplary Praxis

Below selected screenshots of the computer application for production capacity balancing according to the new approach are presented. They take into account previously described assumptions. The application's was validated for data obtained in specially supervised tests for 3 CNC machines and 3 production orders with the assumption that one order is a complete processing of one cast product.

During the use of application the user is able to review the current machine park inventory, edit and if necessary introduce new machine tools. The next step is defining orders entered into the production system – the number of items and the name – as well as selecting parameters for the order chosen, i.e. technological operations and individual operations' unit times.

After defining orders, the program user receives the workload of unit and of the entire order for each order and the actual length of the production cycle, as well as the workload structure. The final step in calculating the production system capacity is to introduce significant correction factors: C_{DEF}, C_{TIME}, C_{DIM}, C_{MAT}, described in earlier sections of this article. As a result, the user receives the labor intensity structure for these factors as well as the final production capacity balance of machine tools (Fig. 2).

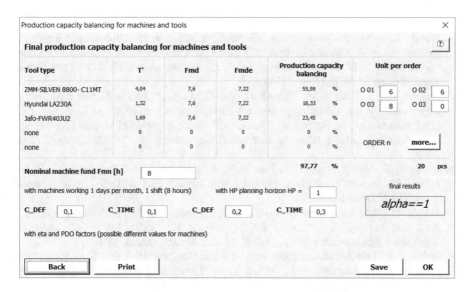

Fig. 2. Final production capacity balancing for machines and tools

5 Summary

The new approach to production capacity balancing is intended to help micro and small enterprises in balancing their production capacity and was presented on the example of cast iron processing. The method uses a determined labor intensity structure taking into account the share of set-up times and the level of defectiveness.

As mentioned above, ever more companies produce products in short series, and the organizational cycle depends not only on available resources but also on the efficiency of their use. The authors proposed a computer application based on a new approach to production capacity balancing. Additionally, it takes into account a shorter planning period due to the specific nature of the machining processes of some production plants, the casting dimension difficulty factor as well as the material factor. The presented mathematical formulae as well as the algorithm were implemented in the working version of ProdBal 2.0. application. It is currently being tested in two selected production plants.

The idea of further development of the computer application involves, among others, including initial load of machine tools and assistance in case of excessive load or underload.

Acknowledgments. The research work had the financial support of Ministry of Science and Higher Education, Republic of Poland, under the project 02/23/DSPB/7695.

References

1. Selech, J., Joachimiak-Lechman, K., Klos, Z., et al.: Life cycle thinking in small and medium enterprises: the results of research on the implementation of life cycle tools in polish SMEs – Part 3: LCC-related aspects. Int. J. Life Cycle Assess. **19**, 1119–1128 (2014). https://doi.org/10.1007/s11367-013-0695-9
2. Trojanowska, J., Kolinski, A., Galusik, D., Varela, M.L.R., Machado, J.: A methodology of improvement of manufacturing productivity through increasing operational efficiency of the production process. In: Hamrol, A., Ciszak, O., Legutko, S., Jurczyk, M. (eds.) Advances in Manufacturing. Lecture Notes in Mechanical Engineering, pp. 23–32. Springer, Cham (2018)
3. Ignaszak, Z., Sika, R., Perzyk, P., Kochański, A., Kozłowski, J.: Effectiveness of SCADA systems in control of green sands properties. Arch. Foundry Eng. **16**(1), 5–12 (2016). https://doi.org/10.1515/afe-2015-0094
4. Pinedo, M.: Scheduling: Theory, Algorithms, and Systems. Springer, Heidelberg (2012)
5. Andresen, M., Bräsel, H., Engelhardt, F., Werner, F.: LiSA-a Library of Scheduling Algorithms: Handbook for Version 3.0. Univ. Fak. für Mathematik (2010)
6. Baker, K.R.: Introduction to Sequencing and Scheduling, vol. 15. Wiley, New York (1974)
7. Conway, R.W., Maxwell, W.L., Miller, L.W.: Theory of Scheduling. Addison-Wesley Publishing Company, Inc., England (1967)
8. Lenstra, J.K., Shmoys, D.B., Tardos, E.: Approximation algorithms for scheduling unrelated parallel machines. Math. Program. **46**, 256–271 (1990)
9. Brucker, P.: Scheduling Algorithms. Springer-Verlag, Heidelberg (1995)
10. Hadidi, L.A., Al-Turki, U.M., Rahim, A.: Integrated models in production planning and scheduling, maintenance and quality: a review. Int. J. Industr. Syst. Eng. **10**(1), 21–50 (2012)
11. Hamrol, A.: Strategies and practices of efficient operation. Lean six sigma and other [in Polish]. PWN, Warszawa (2015)
12. Starzyńska, B., Szajkowska, K., Diering, M., Rocha, A., Reis, LP.: A study of raters agreement in quality inspection with the participation of hearing disabled employees. In: Advances in Manufacturing, pp. 881–888. Springer (2018)
13. Jasarevic, S., Brdarevic, S., Imamovic, M., Diering, M.: Standpoint of the top management about the effects of introduced quality system and continuation of activities of its improvement. Int. J. Qual. Res. **9**(2), 209–230 (2015)
14. Kłos, S., Patalas-Maliszewska, J.: Throughput Analysis of Automatic Production Lines Based on Simulation Methods. In: Jackowski, K., Burduk, R., Walkowiak, K., Woźniak, M., Yin, H. (eds.) IDEAL 2015. Lecture Notes in Computer Science, vol. 9375, pp. 181–190. Springer, Cham (2015)
15. Kunz, G., Machado, J., Perondi, E.: Using timed automata for modeling, simulating and verifying networked systems controller's specifications. Neural Comput. Appl. **28**, 1–11 (2015)

16. Sika, R., Hajkowski, J.: Synergy of modeling processes in the area of soft and hard modeling. In: Proceedings of 8th International Conference on Manufacturing Science and Education (MSE 2017), Trends in new industrial revolution, Sibiu (2017). https://doi.org/10.1051/matecconf/201712104009

17. Sika, R., Rogalewicz, M.: Demerit control chart as a decision support tool in quality control of ductile cast-iron casting process. In: Proceedings of 8th International Conference on Manufacturing Science and Education (MSE 2017), Trends in new industrial revolution, Sibiu (2017). https://doi.org/10.1051/matecconf/2011721104009

18. Krolczyk, J.B., Krolczyk, G.M., Legutko, S., Napiorkowski, J., Hloch, S., Foltys, J., Tama, E.: Material flow optimization – a case study in automotive industry. Tehnički vjesnik **22**(6), 1447–1456 (2015)

19. Żywicki, K., Rewers, P., Sobkow, L., Antas, D.: The impact of task balancing and sequencing on production efficiency. In: Hamrol, A., Ciszak, O., Legutko, S., Jurczyk, M. (eds.) Advances in Manufacturing, pp. 67–77. Springer, Cham (2018)

20. Kłos, S., Trebuňa, P.: The impact of the availability of resources, the allocation of buffers and number of workers on the effectiveness of an assembly manufacturing system. Manag. Produc. Eng. Rev. **8**(3), 40–49 (2017)

21. Żywicki, K., Zawadzki, P., Hamrol, A.: Preparation and production control in smart factory model. In: Rocha, Á., Correia, A.M., Adeli, H., Reis, L.P., Costanzo, S. (eds.) Recent Advances in Information Systems and Technologies, vol. 571, pp. 519–527. Springer, Cham (2017)

22. Starzyńska, B., Grabowska, M., Hamrol, A.: Effective management of practitioners' knowledge – development of a system for quality tool selection. In: 3rd International Conference on Social Science, Shanghai, China, pp. 859–865 (2014)

23. Golinska, P., Kuebler, F.: The method for assessment of the sustainability maturity in remanufacturing companies. Procedia CIRP **15**, 201–206 (2014)

24. Kalinowski, K., Zemczak, M.: Preparatory stages of the production scheduling of complex and multivariant products structures. Adv. Intell. Syst. Comput. **368**, 475–483 (2015)

25. Kalinowski, K., Skolud, B.: The concept of ant colony algorithm for scheduling of flexible manufacturing systems. In: Advances in Intelligent Systems and Computing, vol. 527, pp. 408–415 (2017)

26. Adamczak, M., Domanski, R., Hadas, L., Cyplik, P.: The integration between production-logistics system and its task environment-chosen aspects. IFAC-PapersOnLine **49**(12), 656–661 (2016)

27. Hamrol, A., Kowalik, D., Kujawinska, A.: Impact of selected work condition factors on quality of manual assembly process. Hum. Factors Ergon. Manufac. Serv. Industr. **21**(2), 156–163 (2011)

28. Jacyna, M., Semenov, I.N., Trojanowski, P.: The research directions of increase effectiveness of the functioning of the RSA with regard to specialized transport. Archiv. Transp. **35**(3), 27–39 (2015)

29. Rewers, P., Hamrol, A., Żywicki, K., Bożek, M., Kulus, W.: Production leveling as an effective method for production flow control – experience of polish enterprises. Procedia Eng. **182**, 619–626 (2017)

30. Senger, Z.: Production Flow Control. PUT Publishers, Poznan (1998)

31. Pająk, E.: Production Management. PUT Publishers, Warsaw (2006)

32. Rogalewicz, M., Sika, R.: ProdBalance – A tool for balancing the production capacity of a production system, vol. 7, pp. 95–108. PUT Publishers, Warsaw (2007)

33. Wojciechowski, S., Maruda, R.W., Krolczyk, G.M., Nieslony, P.: Application of signal to noise ratio and grey relational analysis to minimize forces and vibrations during precise ball end milling. Precis. Eng. **51**, 582–596 (2017)

Virtual Reality System for Learning and Selection of Quality Management Tools

Beata Starzyńska, Filip Górski[(✉)], and Paweł Buń[(✉)]

Chair of Management and Production Engineering,
Faculty of Mechanical Engineering and Management, Poznan University of Technology,
Piotrowo 3 STR, 60-965 Poznan, Poland
{beata.starzynska,filip.gorski,pawel.bun}@put.poznan.pl

Abstract. The paper presents an innovative system for teaching of quality tools, facilitating decision-making in scope of solving quality problems in production processes. The solution was created with use of immersive Virtual Reality technology. The system allows two groups of users – production workers and quality staff – to interactively gain knowledge about quality tools or obtain advice on which quality tool should be selected for a given problem. The paper presents the idea behind the system, its functions and basic stages of its creation, as well as course and results of testing of its first prototype. The test results confirm validity of taken assumptions and solutions and allow to perform guidelines for development of the solution.

Keywords: Quality tools · Virtual Reality · Production processes · DSS
Lifelong learning

1 Introduction

Modern production companies have found themselves in a situation, where capability of dynamic adaptation for changing market conditions (which can be achieved by different means, such as simulation of complex systems and their possible failures [1], production levelling plans [2], quality management strategies etc.), is a key factor of competitive advantage. It means, that all the employees of companies, engaged in processes, must be equipped in appropriate tools, allowing easy and quick, effective decision making in all the areas of creating added value for a client.

While delegating privileges and inviting employees to get involved in "quality issues", management staff often forgets that the employees may be helpless in a situation of necessity of making a decision. Often they are not sure of rightness of a decision they made. On the other hand, they more and more frequently see necessity of making decisions on the basis of data, facts, not only on the basis of personal experience and intuition.

In the quality management, obtaining and processing data and information from various processes is done by the so-called quality tools [3–6]. Their application allows supporting the decision-makers with information, necessary for the decision-making in field of solving the quality problems, as well as introducing corrective and preventive

© Springer International Publishing AG, part of Springer Nature 2018
Á. Rocha et al. (Eds.): WorldCIST'18 2018, AISC 747, pp. 264–274, 2018.
https://doi.org/10.1007/978-3-319-77700-9_27

and improving measures in the production processes. Examples of these tools are Pareto diagram, Shewhart control chart or Ishikawa diagram [7, 8].

The quality tools are effective, widely accepted and used methods of problem solving. However, their effective use outside quality management departments, especially by shop floor workers in production companies, is heavily limited. That is caused by difficulty in learning them by regular production workers – current teaching methods are not effective. The typical training courses or books contain complex descriptions, difficult language and they lack visual, practical examples of use. Besides, there are many quality tools and knowledge about them is scattered among different elaborations. It is difficult to motivate a production worker to undertake a learning process and then make use of the quality tools in daily work. The authors decided to take a closer look on this problem and develop a new, more effective training method, in order to expand range of use of quality tools in production companies.

The authors propose to use the Virtual Reality (VR) technology. Its proper implementation reduces disadvantages of traditional trainings (boredom, ineffective form, time consumption, lack of practical approach) and allows to focus on a visual form of presenting knowledge, as well as engaging trainees in a set of practical activities. This helps gain knowledge in a more friendly, intuitive and effective way. That is why VR is frequently used in technical education [9, 10].

In general, the VR systems allow their users to explore and interact with elements of artificially created, three-dimensional worlds in real time [11]. During the last decade, the VR applications improved their level noticeably, from simple applications with non-complex graphics to graphically advanced and logically complex environments, allowing trainings, design aid or decision making in scope of engineering work [12]. It is now possible to obtain a good level of immersion, which is a feeling of being a part of a virtual world [13]. This phenomenon is induced both by use of stereoscopic display systems (e.g. Head-Mounted Display devices) and by use of natural methods of interaction (e.g. gesture recognition). In the recent few years, low-cost VR solutions have emerged, which makes availability of hardware such as gesture recognition systems or HMD goggles much higher than several years ago. The software for preparation of Virtual Environments has also undergone rapid development in terms of ease of use, price and capabilities. Virtual Reality is widely used e.g. in prototyping of ergonomic workplaces [14], designing new complex mechanical systems [15] or in medical training applications [16, 17]. It is also frequently used for aiding decision-making processes in industrial companies [18].

The paper presents a concept of a VR system for learning and selection of quality tools, as well as practical implementation of its first prototype, along with results of tests performed on groups of target users.

2 Materials and Methods

2.1 Main Research Problem

During a realization of production processes, when something goes wrong, the employees do not always know how to react and what to do. That is why there are cases of hiding the faulty products and results of mistakes and corrective measures taken by the workers are, at most, the so-called "firefighting". The quality management staff, has knowledge on tools allowing to avoid defects in production, better organization and improvement of processes and products. This knowledge is not easy to share with the shop floor workers, so its potential sometimes remains undeveloped. The quality staff also needs support in effective and adequate selection of proper tools for a given problem. Moreover, the quality staff constantly looks for methods of stimulating creativity of operational workers, as well as methods for better teamwork. In general, it must be stated, that the human factor is very important in the quality aspect, as some authors note [19].

On the basis of results of studies, carried out in Polish production companies [7], it can be stated that level of knowledge and application of quality tools is insufficient. If there are problems in production processes, diagnosis and solution by the employees is troublesome and time consuming. The solution often requires engagement by management or quality staff.

One of the barriers in learning the quality tools is dispersion of knowledge in this field [8]. It makes it difficult both to learn the tools and decide which one to use in a given situation. Another problem in gaining knowledge about the quality tools is inapproachable form of this knowledge (difficult language, lack of practical examples). Traditional methods of teaching the employees are not always effective.

An answer to these problems may be application of modern methods of training, based on digital technologies. One of them is the aforementioned Virtual Reality – the authors propose an innovative approach to learning the quality tools, with use of immersion effect and three-dimensional, interactive visualizations.

2.2   Concept of the System

The developed system will allow employees of small and micro production companies to easily gain knowledge about application of quality tools in practice. It will be achieved by application of immersive VR, in connection with three-dimensional, interactive visualization of implementation of selected quality tools. This form of presentation – education using solely an interactive visualization in immersive googles – will eliminate problems of knowledge forms and will make the learning process faster and more effective, especially for older employees, not always eager to expand their knowledge in traditional ways. There are numerous examples of general education [9] or engineering education [10] are a proof that a visual form of knowledge presentation aided with VR immersion accelerates the learning process and improves permanence of its results.

The presented innovative system will deliver knowledge transformed from its initial, "book" form to a form of interactive 3D visualization, presenting effective ways of implementation of quality tools in a production company. The created system consists of the following elements:

- VR platform (Windows and Android) – a software application compatible with popular VR goggles, with 3 modes of operation: learning (for novice users – production workers and inexperienced quality staff), selection (for experienced users) and exam (knowledge test for users after they finish the lessons),
- a base of quality tools, presented in a visual form – an integral part of the platform, but with possibility of independent development in further work,
- hardware components: a PC computer + goggles of Oculus Rift class or an Android phone + Gear VR/Google Cardboard class device.

Application of VR technology will make the quality tools learning process less arduous, especially for those, who start learning after a longer break. This will be fulfilled by placing all the knowledge in one place, presenting the methods in an attractive, visual form, with examples of use.

There are two groups of potential users of the developed system. The first one are production workers, employed in micro, small and average production companies, directly engaged in executive processes. The second group of recipients of the system is quality staff of these companies. They are involved in supervising course of processes and products, which are their results. The two groups will utilize the system differently – the production workers will focus mostly on the learning and exam modes, while the quality staff will mostly use the selection mode and a learning mode to a lesser extent. The quality staff will also motivate the production workers to gain knowledge of the quality tools by themselves. It is assumed, that the system can be used without instructor after a short training.

The authors assume, that learning of the quality tools using the VR technology will be quicker and easier for the production workers and less experienced quality staff. As a consequence, due to attractive form of learning, the employees will be more motivated to gain knowledge. As they will gain the knowledge, they will be able to get themselves involved in pro-quality actions and decision-making when a problem situation arises during the production. The quality staff, on the other hand, should benefit from use of the selection mode, as well as from the learning mode and better communication with production workers educated by the system.

2.3   Methodology of Work

Creation of the prototype system for learning and selection of quality tools with use of VR consist of the following tasks:

- preparation of data – dynamical visual representation of use of selected quality tools in an exemplary small production company, scenarios of operation in a VR environment, basics of spatial user interface in VR,
- VR programming – creation of application of the first, PC-based prototype of the system, with a limited range of functionalities,
- testing the first prototype with representatives of target groups of users,
- preparation of updated data and full user interface for the second prototype,

- VR programming of the second prototype – PC-based application and a mobile application for Android-based VR solutions, full scope of functionalities (extended library of quality management tools, selection mode, exam mode etc.),
- development of mechanisms of further expansion of quality tools library,
- testing the second prototype,
- development of a final, revised prototype of the VR system, pilot implementation in selected, interested companies.

This paper describes the first prototype of the system and its tests. The first prototype contains three, selected quality management tools in a form of interactive lessons in Virtual Reality.

The tests of the first prototype consisted of the following stages:

- interview with a participant, introduction of idea of the system,
- short training of use of VR application,
- self-learning of the three quality management tools in VR,
- knowledge test (exam) performed in a traditional way, 10 questions,
- survey study about quality of particular elements of the VR system (comfort, graphics, fluency, intuitiveness etc.),
- evaluation survey study of a training solution (not connected to the technical aspects of VR), containing general questions about importance, attractiveness and clarity of the training, as well as willingness of using it again.

The test group was 20 persons, divided equally into representatives of the two target user groups (production workers and quality management staff). One person was engaged for approximately 45 min – the time of learning was neither limited nor thoroughly recorded by the authors, but no person spent more than 45 min immersed in Virtual Reality. The users had to go through all the lessons and perform all the activities in a given lesson. The production workers were then asked to perform a knowledge test. After that, all participants were interviewed about the whole idea and technical aspects of the solution. The results of these tests are presented in the Sect. 3.2 of this paper.

2.4   Hardware and Software

The prototype system was created using the Unity 3D software. It is a software platform for building interactive 3D and 2D applications, including interactive visualizations, video games and VR applications. It is a very popular platform and it allows building applications for most of the currently available commercial VR hardware, both PC-based and mobile, that is why it was selected for the work.

In terms of the hardware used, the authors used a test stand equipped in a VR system, consisting of the following components:

- a computer with Intel Core i7 processor, nVidia GTX 980 graphics card and 12 GB of RAM,

- Oculus Rift CV1 Head-Mounted Display (HMD), equipped with two tracking sensors, the remote controller (one-handed joystick) and Touch controllers (the controllers were not used for interaction with the system),
- a 50″ TV for preview of actions currently performed by the user.

3 Results

3.1 Description of the System

The application contains different types of contents, including audio, 2D graphics, text and 3D models. In the prototype 1, the Unity application contains two basic types of scenes:

- the main menu scene,
- the lesson scene for a selected tool.

The selection mode will be implemented in the prototype 2 (which is currently under development by the authors). Similarly, the exam mode – in the prototype 1, the knowledge of a trainee is tested in a traditional way (a test).

Each lesson scene contains a number of basic constant 2D and 3D objects, which remain relatively unchanged between lessons of different quality tools. They are, among other things, the flipchart with 2D illustration and instruction, the production master (an animated character), as well as the door allowing getting out to the main menu. It also contains a predefined production setting – a hall with machines, storage shelves, workplaces, also containing the quality corner (QC) – a desk, whiteboard, electronic display and a meeting room, with a round table, whiteboard etc. Figure 1 presents the main starting location and the most important scene components.

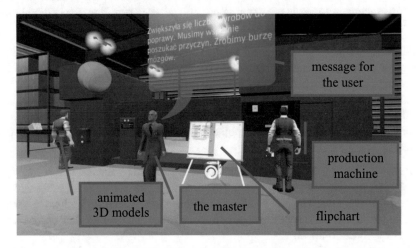

Fig. 1. The VR system for quality tool learning – main interactive elements

Each quality tool is a separate scene (level), launched from the main menu, which is also 3D and immersive (Fig. 2). The tools are based on the same set of 3D models, modified and expanded separately. The whole programming is done in C# language, but the authors have created a set of ready, flexible scripts (with predefined object behaviors), which make the interaction programming easier and quicker.

Fig. 2. Main menu – lesson selection

Learning of a selected tool requires user to perform appropriate activities in the production hall. The whole process is divided into a number of steps, different for each tool. Advancement towards next step is possible after performing a specified activity (or a number of activities). The most frequently performed activities are:

- listening and reading messages from the master (usually theoretical introduction to a given problem or detailed instruction for a given step),
- watching a predefined animation (usually behavior of other workers),
- gathering an object and bringing it to a specified place (e.g. tool change in a machine),
- interaction with a stationary object (e.g. adjustment of machine parameters),
- interaction between two movable objects (e.g. taking a caliper and measuring products).

Each step has a specific informative contents (text, 2D graphics, audio – text read by a lector), as well as an appropriate set of interactive 3D models, presented after activating a given step. The user is informed by the master what must be done to advance. After going through all the steps, the user is awarded an achievement and can replay the lesson or go back to menu.

The user can explore the environment by moving his head and body, walking is realized partially by body movement (over a small distance – less than 2 m) and partially by a single-handed joystick (Oculus Remote in the particular, presented setup).

Interaction with objects is realized first by gazing on them, what should activate a visual clue (such as highlighting an object). Then the user can tap on the selected object, activating an animation, sound, physics simulation, picking up the object or activating a more advanced sequence of events (Fig. 3).

Fig. 3. Interaction with objects – measuring

The first prototype of the system contains the following tools:

– Pareto diagram,
– control chart,
– affinity diagram.

The first two tools were implemented according to the "learning by doing" approach – the lesson requires a user to perform a number of practical tasks (such as making measurements – see Fig. 3, sorting products by defect types etc.). The affinity diagram was implemented as a lesson of observational character – a user mostly observes activities by the master and other employees, being directly engaged only insignificantly. Such a lesson is more like a traditional training (although presented in an immersive environment, with no live instructor), what has certain advantages (easier interaction), but also disadvantages (less attractive form of presentation). Intention of the authors was to test both approaches in teaching of quality tools using the VR technology.

3.2   Test Results

The tests, as mentioned before, were performed on a group of exemplary users from the target groups. The tests were carried out in the Laboratory of Virtual Reality at authors' university. The test participants were, as mentioned before, production operators and quality staff. Total 20 persons were tested (10 from each group, total

7 women and 13 men). Because the realized project stays in scope of learning of adults, with special focus on certain groups, the sample contained persons of low qualifications, persons of age 40 and higher, persons living in rural areas or small cities and, most of all, employees of micro, small and medium-sized companies. Synthetic summary of the evaluation study is presented in Fig. 4.

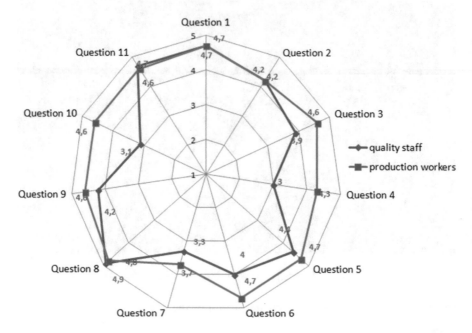

Fig. 4. Synthetic results of evaluation survey study of prototype 1, red – quality staff, blue – production workers

Attractiveness of the tested idea (question 1) was evaluated as high (average 4,7 in both groups), it can be justified by a form attractive for all participants. The second question – importance of tested concept for the test participants – also noted high value score (4,2) in both groups.

The opinions were different in both groups regarding question 3 – is the presented concept important for a large group of people? The quality staff was more skeptical about it (3,9) than the production workers (4,6). Regarding the question 4 (is the presented knowledge new to you?) the discrepancy of results between groups is natural, as quality staff is familiar with basics of quality tools.

The tested idea was evaluated as highly encouraging for learning and gaining new skills (question 5), more for production workers (4,7), as the quality staff is familiar with the knowledge. The production workers were also very positive about easiness of learning (question 6, 4,7), more than the quality staff (4,0).

Question 7 was about the idea being well-matched to the test participants. Here the average score was the lowest, although still positive (3,7 and 3,3). The low scores sometimes resulted from problems with hardware ("one size fits all" VR goggles are not suitable for everyone).

Uniqueness of the presented solution (question 8) was evaluated very high, what is not surprising. Clarity and understandability of the application (question 9) was evaluated lower, but still positively (4,6 and 4,2). The question 10 (does the presented concept convince you that it is worth to learn for your whole life?) noted another discrepancy (4,6 to 3,1 in favor of production workers).

The summarizing question 11 (how much you would be willing to use the solution if all your critical comments were considered in the final version?), noted score of 4,7 for production workers and 4,6 for the quality staff, proving that the solution is very interesting for all the participants.

Results of the knowledge test – allowing to draw some conclusions about educational effects – were also positive, as 9 out of 10 production workers passed the test (more than 50% positive answers). Regarding the short time of work, a completely new technology (VR) and stress of participation in a scientific study, the results are very promising.

4 Summary

The presented solution is a unique computer system for learning of quality tools using the Virtual Reality technology. So far, VR has not been extensively used in teaching of abstract concepts or decision-making methodologies – it has been associated more with practical, technical skills. However, the presented study results show, that VR can be successfully used to teach such methods, even to non-qualified personnel. In general, potential of VR in learning can be evaluated as very high.

The quality tools learning system is still in the development phase – the authors are developing the prototype 2, which will contain much more knowledge and an expanded range of functionalities. The final tests will show what should be improved and then the authors will attempt to deploy the solution in industrial companies, outside laboratory conditions. In the future, the solution should contain at least several dozen tools, in order to be helpful both for novice and advanced users of quality tools.

The main disadvantage of the VR systems, as resulting out of interviews with the participants, is the user interface and ways of interaction. It was programmed to be as easy as possible, yet still some participants found even a simple task of moving around and looking at objects difficult, let alone more complex activities. The main challenge in creating the VR systems for education is to balance out complexity of tasks, while ensuring a proper level of presented knowledge.

Acknowledgments. The presented development work was financed indirectly by Operational Program Knowledge Education Development, realized by Polish Ministry of Development, in scope of the "Po-po-jutrze" program.

References

1. Machado, J., Seabra, E., Campos, J., Soares, F., Leao, C.: Safe controllers design for industrial automation systems. Comput. Ind. Eng. **60**, 635–653 (2011)
2. Rewers, P., Trojanowska, J., Diakun J., Rocha A., Reis L.P.: A study of priority rules for a levelled production plan. In: Hamrol, A., Ciszak, O., Legutko, S., Jurczyk, M. (eds.) Advances in Manufacturing, pp. 111–120. Springer, Cham (2018)
3. Asaka, T., Ozeki, K.: Handbook of Quality Tools. The Japanese approach. Productivity Press, Portland (1998)
4. Breyfogle, F.W.: Implementing Six Sigma - Smarter Solutions® Using Statistical Methods, 2nd edn. Wiley, Hoboken (2003)
5. De Feo, J.A. (ed.): The Juran's Quality Handbook. McGraw - Hill Education, New York (2017)
6. Tague, N.: The Quality Toolbox. ASQ Quality Press, Milwaukee (2005)
7. Starzyńska, B.: Practical applications of quality tools in Polish manufacturing companies. Organizacija **47**(3), 153–164 (2014)
8. Starzyńska, B., Hamrol, A.: Excellence toolbox - decision support system for quality tools and techniques selection and application. Total Qual. Manage. Bus. Excellence **24**(5–6), 577–595 (2013)
9. Martín-Gutiérrez, J., Mora, C.E., Añorbe-Díaz, B., González-Marrero, A.: Virtual technologies trends in education. EURASIA J. Math. Sci. Technol. Educ. **13**(2), 469–486 (2017)
10. Abulrub, A.H., Attridge, A., Williams, M.A.: Virtual reality in engineering education: the future of creative learning. Int. J. Emerg. Technol. Learn. **6**(4), 4–11 (2011)
11. Scherman, W.R., Craig, A.B.: Understanding Virtual Reality: Interface, Application, and Design. Morgan Kaufmann, Burlington (2003)
12. Burdea, G.C., Coiffet, P.: Virtual Reality Technology. Wiley, Hoboken (2003)
13. Bowman, D.A., McMahan, R.P.: Virtual reality: how much immersion is enough? Computer **40**(7), 36–43 (2007)
14. Grajewski, D., Górski, F., Zawadzki, P., Hamrol, A.: Immersive and haptic educational simulations of assembly workplace conditions. Procedia Comput. Sci. **75**, 359–368 (2015)
15. Pandilov, Z., et al.: Virtual modelling and simulation of a CNC machine feed drive system. Trans. FAMENA **39**(4), 37–54 (2015)
16. Górski, F., Bun, P., Wichniarek, R., Zawadzki, P., Hamrol, A.: Effective design of educational virtual reality applications for medicine using knowledge-engineering techniques. EURASIA J. Math. Sci. Technol. Educ. **13**(2), 395–416 (2017)
17. Escobar-Castillejos, D., Noguez, J., Neri, L., Magana, A., Benes, B.: A review of simulators with haptic devices for medical training. J. Med. Syst. **40**(4), 104 (2016)
18. Colombo, S., Nazir, S., Manca, D.: Immersive virtual reality for training and decision making: preliminary results of experiments performed with a plant simulator. SPE Econ. Manage. **6**(4), 165–172 (2014)
19. Kujawińska, A., Vogt, K.: Human factors in visual quality control. Manage. Prod. Eng. Rev. **6**(2), 25–31 (2015)

Decision Support on Product Using Environmental Attributes Implemented in PLM System

Jacek Diakun[✉], Ewa Dostatni[✉], and Maciej Talarczyk

Poznań University of Technology, M. Skłodowska-Curie Square 5, 60965 Poznan, Poland
{jacek.diakun,ewa.dostatni}@put.poznan.pl, talarczyk6@wp.pl

Abstract. In the paper the problem of environmental product characteristics in Product Lifecycle Management (PLM) systems is undertaken. The general characteristics of PLM systems is presented. The issue of existence in PLM systems data concerning environmental product description is discussed. Because of lack of such data in typical PLM system, the set of environmental product attributes and their implementation in PLM system is presented. Potential ways of usage of these attributes in decision support concerning product are described.

Keywords: PLM · Eco-design · Data integration · Decision support

1 Introduction

Production companies have transformed over the years, in the process affecting the evolution of the global economy and the consumer market. Modern production is not only very diverse in terms of products (the customer requires multivariance, a broad offer to choose from), but it also needs to constantly reduce costs, shorten the lead time, and improve product quality. Enterprises that produce modern products must use tools that enable quick decision-making regarding e.g. new product concepts. Product Lifecycle Management (PLM) systems are among such tools. The benefits of implementing PLM in a production company include the following:

- support for market analysis and development of new ideas,
- ensuring compliance with legal and safety requirements,
- reducing the risk associated with launching a new product on the market,
- improving R&D employee productivity in the company,
- support for product design management and manufacturing management,
- reduction of time and costs of product design and launch, while maintaining or even improving quality,
- building a product information database,
- close cooperation of product designers, process engineers, suppliers, partners and clients, which results in improved business and marketing strategies.

During the implementation of a PLM system in a company it is necessary to properly configure the system and perform a number of tasks related to the implementation. PLM systems must be individually configured for each end user, which is not

© Springer International Publishing AG, part of Springer Nature 2018
Á. Rocha et al. (Eds.): WorldCIST'18 2018, AISC 747, pp. 275–284, 2018.
https://doi.org/10.1007/978-3-319-77700-9_28

simple. Often, companies design systems tailored to their specific needs. One of the most important tasks to be completed before PLM implementation is the appropriate description of data flow processes, and the identification of the data itself. PLM systems cover all areas of the company. As a rule, however, each implementation includes only selected functional areas. Some of the functional areas of the company are more often supported by PLM systems, some less frequently [1]. One of the areas that is hardly considered at all in the PLM support area is the last stage of the product life cycle (recycling, utilization or re-use). From the point of view of managing the data necessary to make decisions in the company, this is a very important area. Properly defined and ordered data regarding, for example, the possibility of recycling or utilization of a given product, may already be used at the product design stage. This will later facilitate product recycling, and hence improve its environmental properties.

2   PLM System as a Tool Supporting the Decision-Making Process

The life cycle of a product is the time from the appearance of an idea for a given product, through the development of the concept, design, manufacturing, distribution and sale as well as its operation until end-of-use and withdrawal from use [2–4]. To put it simply, the life cycle of a product includes the stages of product design, manufacturing, use, and withdrawal from use [5]. Each of the stages has shortened over the years, and as a result the entire life cycle of the product has been reduced. Markets have been more and more saturated with new, more advanced products with improved features, which reduced the product use phase. This trend continues to this day and the role of enterprises wishing to survive on the market is to keep up with the dynamically changing consumer needs. Manufacturing companies must therefore focus their efforts on designing and producing products in the shortest lead time possible. This forces companies to adopt appropriate engineering tools which, thanks to the development of information and digital technologies, can accelerate the work on new products [2, 6–9]. These are CAx systems and other solutions that support them, such as: Computer Aided Design (CAD), Computer Aided Manufacturing (CAM), Computer Aided Engineering (CAE), Engineering Data Management (EDM), Product Data Management (PDM), Product Lifecycle Management (PLM), Product Information Management (PIM), Enterprise Resources Planning (ERP), and many other, more or less complex systems, the use of which depends e.g. on the industry which the given company represents. PLM systems were created as the second generation of PDM systems, based on such solutions as: CAD i EDM/PDM [10, 11]. Their functionality includes product design, manufacturing, quality management, and market evaluation of the product, as they facilitate the exchange of engineer data by cooperating with CAD/CAM systems and connecting them with ERP systems. In addition, a PLM system should exchange data for example with CRM (Customer Relationship Management) systems in order to include customer needs in the product life cycle, as well as with SCM (Supply Chain Management) or MES (Manufacturing Execution System). Another important feature of PLM systems consists in the integration with other systems used in the company by marketing and service departments, and even in external entities (systems of cooperating companies) in order to accelerate the circulation of information about the product. In this case, integration with PIM (Product Information Management) systems will be helpful. PIM systems allow the users to store and share product data with interested and authorized business partners regardless of their location [2]. Therefore, PLM is not perceived as a single application which operates completely independently; it is rather a system that integrates the software responsible for the specific stages of product life cycle into a well-functioning whole [12]. PLM, in addition to being able to integrate a number of the abovementioned IT tools, also connects people, processes, practices and technologies related to all aspects of product life, from designing through manufacturing, development and service, to product recall and final disposal or recycling [10]. PLM is also said to be a strategy, philosophy, approach in business, because it can help standardize projects, control and share product data related to product design and development (regarding e.g. time, energy, materials), automate processes related to tasks that use the data, automate project management itself and support certain logistic operations [12, 13]. PLM refers to engineering, production and service functions; it allows you to manage the documentation, use data describing the product, control product configuration, its development and modification, which increases the productivity and profitability of the company [2, 12]. Some PLM systems enable the integration of rapid prototyping processes and the automation of activities related to the input of data regarding product construction and engineering [14]. PLM strategies are used by the engineering, procurement, marketing, and production department, but also by the research and development departments, because PLM systems enable better management of the process of developing and introducing innovations, achieving higher profitability and better results [2]. In the Internet are online business is growing, which calls for the development of electronic Product Lifecycle Management (e-PLM) systems. E-PLM, also called PLM 2.0, is a global, mass and interactive environment intended for all users creating and using the IP protocol. It is a significant redefinition of the existing PLM solutions. It is a set of groupware applications in which a distributed project team can be built. They are based on solutions known to all Internet users, and they focus on online cooperation and specific sharing of ideas. This solution is not completely disconnected from ordinary PLMs - it is an extension of these systems by options offered by the web. The use of websites allows to obtain additional functional possibilities, such as joint development of projects, management of cooperation with component suppliers, configuration management and dissemination of information about the product. In PLM 2.0, all business processes can be activated via network access [2, 14–16]. Another aim is to automate work carried out in projects. This direction of PLM systems development will continue due to the need to implement the concept of Industry 4.0 in companies [1, 17].

It often happens that PLM systems are equated with PDM systems, but these are different solutions. PLM is a business strategy, controlled by information and integrating people, processes, practices, data and technologies in all aspects of product life [2]. PDM is an IT system that focuses on the collection and processing of product data, created in engineering systems (CAx), in the process of product design and implementation for production. The data include: plans, geometric models, drawings from CAD programs, visualizations and all related technical information [12]. In contrast, Product Lifecycle Management is a comprehensive IT system that controls the whole life cycle of a product in an organization, from the moment it is invented to the time of withdrawal from production and discontinuation of service [2]. The PDM system gathers data about the CAD model, while the PLM system has information about the product definition. Such a specialized design environment as CAD + PDM supplies the PLM system [18]. The second main area which feeds data to the PLM system are ERP (Enterprise Resource Planning) systems that support the management of human resources, materials, production means, etc. Cooperating with PLM, ERP systems support the management of enterprise resources, mainly through finance as well as planning and supervision of business processes. PLM system, due to its features, can serve as a tool supporting the decision-making processes throughout the product life cycle. The majority of PLM systems support the decision-making process in the area of document flow management in the company. They also serve as a source of information for decision-making in other areas, such as product design, production planning [19], manufacturing [20], etc. The use of the PLM system facilitates product development decision-making in dispersed structure teams.

It is clear however, that according to the above description PLM systems (despite the wording of the term) only to a very small extent support the area of managing data concerning the last phases of product life cycle (recycling, utilization or reuse of the product). The data is necessary for making decisions on the one hand already at the design stage, and on the other hand, at the stage of product withdrawal from use. This is particularly important due the increasingly popular concept of ecodesign.

The goal of this paper is to propose extension of the data managed by the PLM system with environmental attributes in order to make decisions concerning environmental issues of the product.

3   Taking into Account the Ecodesign Aspects in the PLM System

3.1   Ecodesign Characteristics

Ecodesign, also known as Ecological Design, Green Design, Design for Environment etc., is a term known since the 1990s [21]. Ecodesign consists in the identification of the environmental impact associated with the product and considering it in the design process at an early stage of product development [22–24]. Therefore, the term describes the basic approach to reducing the negative impact of the product on the environment during its whole life cycle. The product of ecodesign should be durable, socially acceptable and meet environmental requirements [21]. Ecodesign results in the reduction of production and operating costs, improvement of products and technological processes, reduction of material consumption and energy consumption by products and their packaging, rectifying customer requirements and creating new needs and requirements, reducing costs thanks to early concept verification and modification [25]. With environmental product design, you can find many purposes for which it is used. It may be the intention to protect the environment resulting from high awareness, as well as a completely trivial reason: the legal requirement. Regardless of the reason, always the direct or indirect primary or secondary objective of ecodesign will be to minimize the negative environmental impact of the product in every stage of its existence. The sooner it is implemented, i.e. the earlier in the product life cycle in which environmental aspects are taken into account, the faster and better this goal will be achieved. Designers have a variety of tools, methods and indicators (e.g. LCA, LCM, LLC, control chart, MIPS, MIT, etc.) which are based on the guidelines of ecodesign [26]. If the above guidelines are taken into account already when initial ideas, sketches and consideration about the new product are made, we can affect natural environment in 70% or even 90% [27]. What is more, both economic and environmental cost of product modifications introduced at early product design stage are very low. Therefore, with easily accessible modern technological solutions (for simulations and prototyping), one can redesign a product and its production process many times, to arrive at the best possible outcome in the end. We should remember that the first 20% of decisions (regarding cost and other), determine up to 80% of later effects [28]. Ecodesign takes into account the environmental impact of materials, resources and product lifetime. At the beginning of the work, the "ecodesigner" must make choices and decisions that minimize the negative environmental impact of the product. In other words, he has to design with the environment in mind, so environmental aspects should frequently cross his mind. It is therefore extremely important to take these issues into account early on in the design stage, because all the decisions taken at this stage will have an impact on the properties and functioning of the product and on the costs arising later in its life. In order to reach all the intended objectives (such as reduction of environmental and financial cost, shortening of production processes, minimization of adverse environmental impact after the end of product life cycle) it is crucial to properly identify all the important aspects already at the initial stages of product development. Of course, this is practically impossible when new products are designed, but such an approach will help when existing products are improved, because ecodesign also applies to such situations as well [18].

3.2 Ecodesign in the Product Life Cycle

As mentioned above, the life cycle of products is getting shorter, and it needs to be properly managed. As a rule, PLM systems support operations in four basic areas [11]:

- idea, specification, technical requirements and concept (area I),
- designing, testing and analyzing the product (area II),
- implementation (area III),
- after-sales service and maintenance (area IV).

The life cycle of products can also be considered from the point of view of ecology and environmental protection [2]. Environmental management uses the product life cycle analysis defined by the US Environmental Protection Agency (EPA) as a concept applied to obtain a full picture of the environmental impact associated with a specific product or activity, by analyzing the entire life cycle of a given product, process or activity [15]. Ecodesign is one of the tools used nowadays by production companies. In the areas mentioned above, the ecodesign aspect can also be taken into account. In the area I and II, the environmental aspect consists in e.g. appropriate selection of material (for example, secondary raw materials, regenerated) or technology; a manufacturing process should be designed in such a way to use as little energy and water as possible, reduce waste, exhaust emissions or other by-products of the process which have a negative impact on the environment. Design processes which take into account recycling are also an important element. Recycling-oriented design will later result in quick, environmentally friendly disassembly. The theme of ecodesign also applies to area IV, which can be extended with utilization, which in practice ends the life of the product. In this aspect, the aim of ecodesign is to facilitate the utilization of the product at the end of its life cycle: dismantling to a level where it is possible to re-use some of its elements, recycle used parts, and safely dispose of hazardous waste. This clearly shows a close connection between area I and II. A PLM system does not provide environmental analyses, and it will not offer any proposals of green solutions for the project; it does not support any "eco" aspects directly. However, its functions may be used to store and organize all ecodesign documents [29]. The user can extend these functions on his own. In all the abovementioned areas, the use of a PLM system can be beneficial. When managing the life cycle of a product, communication and cooperation between the parties involved in the creation of a new product is very important. In particular, it is crucial to ensure appropriate management (control) over the course of the entire process. Regardless of rank or type, any information in the form of a document should have its place in the system. Each file must be accessible to anyone who needs it. It should be available within a single, structured database, accessible to authorized persons regardless of their place of work, while maintaining the ability to design a workflow, for fuller control and security. This is especially necessary when the data generated at one design stage can be used on another stage for decision-making. This is the case with ecodesign aspects. If all the appropriate attributes are defined from the recycling point of view already at the design stage, they will be used in decision-making processes at the end of the product's life cycle. Therefore, taking all the above into account, the authors decided to implement the necessary attributes concerning the environmental properties of the product into the PLM system.

4   Implementation of Selected Attributes Concerning the Environmental Properties of the Product into a PLM System

The following environmental aspects of the product have been selected for implementation into the PLM system:

- the Equipment Group to which the product belongs, under the Act of 11 September 2015 on Waste Electrical and Electronic Equipment [29] and:
  - the required level of recovery,
  - the required recycling rate,
  - recovery rate achieved,
  - recycling rate achieved,
- materials and their parameters important from the point of view of recycling [30]:
  - material density expressed in [g/cm^3],
  - tensile strength [MPa],
  - elongation at the yield stress Re [%],

- material processing temperature [°C],
- dielectric constant,
- dielectric strength [kW/mm],
- Young's modulus E [GP],
- water absorption through the material [%],
- adverse environmental effects of the material (yes/no),
- recycling cost (PLN/kg).

The abovementioned attributes have been implemented in PLM Enovia Smarteam V5R20 in the standard (supplied by the manufacturer) database PLMDB_R20. This required that the attributes of the two existing classes in this database be extended: *Projects* (containing general information about the product) and *Materials* (containing data about materials), and that relevant user interface screens be modified.

After the implementation, data about the environmental properties of the product - electric mixer (data from paper [30]) were entered into the system. Figures 1 and 2 present the following data for this product: product definition together with its assignment to the appropriate group under the Act [29] (Fig. 1) and the definition of materials for this product along with the values of their recycling parameters (Fig. 2).

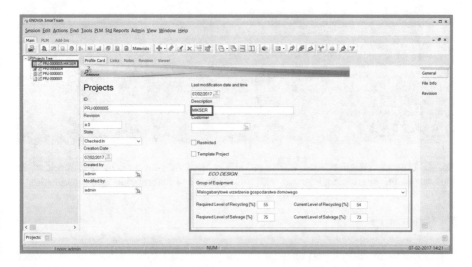

Fig. 1. Data about the new product (marked with a blue rectangle) entered into the system (environmental attributes resulting from the implementation - assignment to the Equipment group under the Act [29] and parameter values - marked with a green rectangle).

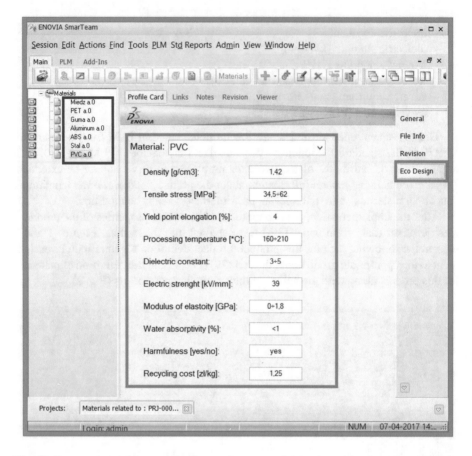

Fig. 2. Data on materials used in the product entered into the system as a result of the implementation (blue rectangle) along with values for one of the materials (green rectangle).

5 Summary

PLM systems should support all stages of the product life cycle. However, in practice, there is a discrepancy between the support of individual stages. Some of them are supported in a very good way (e.g. support of geometrical design by CAD 3D systems), while others are supported to a small extent or are not supported at all. The last phase of the product's life cycle, i.e. the withdrawal of the product from use, can be considered among the latter. The described work was aimed at closing the gap at least to some extent. The implemented attributes are not an exhaustive set of all the environmental data of the product. These include certificates confirming the product's compliance with the requirements of specific legal acts or standardization organizations, and environmental attributes of the product which are not present in typical CAD 3D models. The implemented attributes will help you make decisions regarding the last stage of the product's life cycle (recycling) already at the design stage. The results described in this article constitute a starting point for further work related to the implementation of environmental product attributes.

References

1. Fischer, J.W., Lammel, B., Hosenfeld, D., Bawachkar, D., Brinkmeier, B.: Do(PLM)Con: an instrument for systematic design of integrated PLM-architectures. In: Abramovici, M., Stark, R. (eds.) Smart Product Engineering. Lecture Notes in Production Engineering, pp. 211–220. Springer, Heidelberg (2013). https://doi.org/10.1007/978-3-642-30817-8_2
2. Lenart, A.: Product life cycle management and ERP systems. In: Knosala, R. (ed.) Computer Integrated Management, vol. 3, pp. 115–123. Oficyna wydawnicza Polskiego Towarzystwa Zarządzania Produkcją, Opole (2009)
3. Łunarski, J.: Quality Management - Standards and Principles. WNT, Warszawa (2008)
4. Legutko, S.: Development trends in machines operation maintenance. Maint. Reliab. 2, 8–16 (2009)
5. Cheballah, K., Bissay, A.: Change management and PLM implementation. In: Bernard, A., Rivest, L., Dutta, D. (eds.) Product Lifecycle Management for Society, PLM 2013. IFIP Advances in Information and Communication Technology, vol. 409, pp. 695–701. Springer, Heidelberg (2013). https://doi.org/10.1007/978-3-642-41501-2_68
6. Madenas, N., Tiwari, A., Turner, C.J., et al.: Improving root cause analysis through the integration of PLM systems with cross supply chain maintenance data. Int. Adv. Manuf. Technol. 84, 1679 (2016). https://doi.org/10.1007/s00170-015-7747-1
7. Rojek, I.: Technological process planning by the use of neural networks. Artif. Intell. Eng. Des. Anal. Manuf. 31(1), 1–15 (2017)
8. Kuczko, W., Wichniarek, W., Buń, P., Zawadzki, P.: Application of additively manufactured polymer composite prototypes in foundry. Adv. Sci. Technol. Res. J. 9(26), 20–27 (2015)
9. Górski, F., Buń, P., Wichniarek, R., Zawadzki, P., Hamrol, A.: Immersive city bus configuration system for marketing and sales education. Procedia Comput. Sci. 75, 137–146 (2015). Gonzalez Mndivil, E., Flores, P.G.R., Gutierrez, J.M., Ginters, E. (ed.): 2015 International Conference Virtual and Augmented Reality in Education. https://doi.org/10.1016/j.procs.2015.12.230
10. Grieves, M.: Product Lifecycle Management – Driving the Next Generation of Lean Thinking. MC Graw-Hill, New York (2006)
11. Stanisławski, M.: Do not be afraid of PLM, nr. 4(05), pp. 17–22, Warszawa (2009). Cadblog.pl
12. Turban, E., Leidner, D., Mclean, E., Wetherbe, J.: Information Technology for Management. Transforming Organizations in the Digital Economy. Wiley, New York (2006)
13. Banaszak, Z., Kłos, S., Mleczko, J.: Integrated Management Systems. PWE, Warsaw (2011)
14. Kobyłko, G. (ed.): Environmental Company Management. Wydawnictwo Akademii Ekonomicznej we Wrocławiu, Wrocław (2007)
15. Madu, C.N., Kuei, C.: ERP and Supply Chain Management. Chi Publisher, Fairfield (2005)
16. Bitzer, M., Vielhaber, M., Kaspar, J.: Product lifecycle management - how to adapt PLM to support changing product development processes in industry? In: Proceedings of NordDesign, vol. 1, pp. 360–369 (2016)
17. Talarczyk, M.: Implementation of selected attributes concerning the environmental properties of the product into a PLM system. MA Thesis, Poznań University of Technology, Faculty of Mechanical Engineering and Management, Poznań (2017)

18. Czaplicka, K.: Ecodesign as an important element of Ecologistics. Zeszyty naukowe Politechniki Śląskiej **1681**, 51–63 (2005)
19. Trojanowska, J., Varela, M.L.R., Machado, J.: The tool supporting decision making process in area of job-shop scheduling. In: Rocha, Á., Correia, A., Adeli, H., Reis, L., Costanzo, S. (eds.) Recent Advances in Information Systems and Technologies. Advances in Intelligent Systems and Computing, vol. 571, pp. 490–498. Springer (2017)
20. Bożek, M., Kujawińska, A., Rogalewicz, M., Diering, M., Gościniak, P., Hamrol, A.: Improvement of catheter quality inspection process. In: 8th International Conference on Manufacturing Science and Education – MSE 2017, Trends in New Industrial Revolution, MATEC Web of Conferences, vol. 121, p. 05002 (2017)
21. Libera, J., Labacka, J.: Ecodesign - A New Approach to Design, SKN Ekobiznesu przy Katedrze Ekologii Produktów. WAE, Poznań (2008)
22. Dostatni, E., Diakun, J., Grajewski, D., et al.: Multi-agent system to support decision-making process in design for recycling. Soft Comput. **20**, 4347–4361 (2016). https://doi.org/10.1007/s00500-016-2302-z
23. Lewandowska, A., Foltynowicz, Z.: Environmental activities – a source of innovation in the enterprise. In: Conference Materials, Boszkowo (2006)
24. Karwasz, A., Trojanowska, J.: Using CAD 3D system in ecodesign—case study. In: Golinska-Dawson, P., Kolinski, A. (eds.) Efficiency in Sustainable Supply Chain, Part II, pp. 137–160. Springer International Publishing, Cham (2017)
25. Dostatni, E., Diakun, J., Grajewski, D., Wichniarek, R., Karwasz, A.: Functionality assessment of ecodesign support system. Manag. Prod. Eng. Rev. **6**(1), 10–15 (2015)
26. Ferrendier, S., Mathieux, F., Rebitzer, G., Simon, M., Froelich, D.: Environmentally improved product design-case studies of the European electric and electronic industry (2002)
27. Pająk, E.: Production Management: Product, Technology, Organization. PWN, Warsaw (2006)
28. Stachura, M., Karwasz, A.: Ecodesign in practice. Zeszyty naukowe Politechniki Poznańskiej **5**, 53–63 (2007)
29. Act of September 11, 2015 on waste electrical and electronic equipment. Journal of Laws of the Republic of Poland, Warsaw, 23 October 2015
30. Dostatni, E. (ed.): Ecodesign of Products in CAD 3D Environment with the Use of Agent Technology. Wydawnictwo Politechniki Poznańskiej, Poznań (2014)

New Pedagogical Approaches with Technologies

Bipolar Laddering Assessments
Applied to Urban Acoustics Education

Josep Llorca[1](✉) ⓘ, Héctor Zapata[1], Ernesto Redondo[1], Jesús Alba[2], and David Fonseca[3]

[1] AR&M, Barcelona School of Architecture, Barcelona Tech, Catalonia Polithecnic University,
Av/Diagonal 649, 08028 Barcelona, Spain
{josep.llorca,hector.zapata,ernesto.redondo}@upc.edu

[2] Applied Physics Department, Physics Technologies Centre, Valencia Polithecnic University,
C/Paranimf 1, 46730 Grao de Gandia, Spain
jesalba@fis.upv.es

[3] GRETEL – Grup de Recerca en Technology Enhanced Learning,
La Salle – Ramon Llull University, C/Sant Joan de la Salle 42, 08022 Barcelona, Spain
fonsi@salle.url.edu

Abstract. The role of acoustics in architectural planning processes is often neglected if the designer lacks necessary experience in acoustics. Moreover, few experiments till the date have brought audio design tools to the classrooms. A basic BLA (Bipolar Laddering Assessment) experiment is presented with architectural degree students showing that sound aspects of urban design are as important as visual ones. Additionally, this experiment indicates the basic guidelines for the improvement of a urban acoustic education tool which pretends to broaden architecture students insight into urban acoustic problems.

Keywords: Urban acoustics · Architectural education · BLA · Virtual reality

1 Introduction

The current paper follows three main research lines. The first line focuses on teaching innovations, such as the so-called information technologies (IT), within the university framework that cultivates higher motivation and satisfaction in students [1]. The second research line concerns how to implement new technologies in the architectural representation subject, specifically in urban acoustics representation, in which some investigations have been made [2]. Finally, the study employs a qualitative analysis method to concretely obtain the most relevant aspects of the experience that should be improved both in future interactions and in any new technological implementation within a teaching framework [3].

Despite the fact that the three research lines mentioned above are not innovations themselves, their integration into an experiment gives them a clearly innovative nature, given that very few similar examples exist to date [4, 5]. In addition, the design of the study focuses on the university level, specifically Architecture studies, where spatial comprehension is essential and IT elements are very helpful. Thus, this work is both novel and justified.

© Springer International Publishing AG, part of Springer Nature 2018
Á. Rocha et al. (Eds.): WorldCIST'18 2018, AISC 747, pp. 287–297, 2018.
https://doi.org/10.1007/978-3-319-77700-9_29

The need of and justification for incorporating IT into the educational process are particularly relevant, and they are described in the main roles of the European Higher Education Area (EHEA), which runs the university studies of member countries, including Spain, where this project was undertaken. Moreover, nowadays the incorporation of technology into classrooms is a fact, despite the fact that one cannot affirm that using technology will lead to an improvement on the motivation, satisfaction or academic achievement of students.

Similarly, we find a large number of studies concerning the implementation and evaluation of technology into classrooms [6–8], and if some of them point out the teachers' lack of means or knowledge to use IT in promoting learning [9], the fact is that they conclude that special caution should be observed on the training of the professor in order to fulfill the support to students [10], also on the school facilities to introduce the IT into the classrooms [11], and finally on the policies of software they use in their lessons [11].

It is common to find studies focused on how digital content is included into teaching ranging from computer use in the classroom or the use of digital resources for online training [12, 13]. However, it must be emphasized that these quantitative studies have small sample sizes (quantitative studies are focused on defined variables, which are better described with a large sample and a large number of respondents), and they lack clear questions to identity the degree of information that two or more variables could provide (descriptive, predictive, or casual questions that differentiates research problems). According to Fonseca, Redondo and Villagrasa [3], this lack of accuracy is due to the teacher's inadequate preselection of questions. Using complementary qualitative research, it is possible to obtain new variables to study in future iterations and more detail for the quantitative data.

Meanwhile, the second research line that frameworks this study (the urban acoustics representation subject) demands a different approach to those studies that have preceded us. Sound conceived as a waste product in the environmental noise field is a conception that has increasingly spread into environmental ecology [14]. Instead, Brown [15] suggests that, in contrast, the soundscape field regards sound largely as a resource – with the same management intent as in other scarce resources such as water, air and soil: rational utilization and protection and enhancement where appropriate. In his own words: "resource management has a particular focus on the usefulness of a resource to humans and its contribution to the quality of life for both present and future generations." This means that a quantitative approach to the soundscape is no longer be useful. Some studies already include qualitative questions to evaluate the effect of the soundscape in the urban environment [16, 17]. Thus, a qualitative approach seems more appropriate for the evaluation and proposals of soundscape approach as this change of sensibility demands.

For all these reasons, in this paper a qualitative method (using the Bipolar Laddering Assessment (BLA)) is used to test the use of virtual reality (VR) to present, visualize, auralize and discuss an architectural scenario realized using CAD tools (computer-aided design). Moreover, BLA method is very useful with qualitative little samples. It can be applied to a minimum of 3 people with useful and reliable results [18].

2   The Real Environments

The old city of Barcelona, also known as Ciutat Vella, presents a particular scenario for street music practice, as it can be proved when one walks through its streets. The consideration of this particular city soundscape as an identifying feature of this part of the urban environments is a topic that has risen during the last decades. In fact, after a long history of treating "street music performance" or "busking" by local governments as an annoying or even dangerous practice, legislation has hanged in the last decades into a more permissive view. Some conditions are imposed to buskers when active regarding their contribution to the "vitality of everyday life of the city" [19]. What is more, recent studies conclude that, in some cities, contemporary busking laws are not only tolerated by buskers, but widely embraced by them [20]. In the case of central Barcelona, although most of the official buskers feel it difficult to practice their job when they do not follow the council's laws [21], the fact is that the sequencing of sounds and plays, of musicians and wizards, of painters and artists has been shaping definitely the way citizens and tourists experience the central areas of the city. Street performers and, particularly, street musicians, far from being an undesirable practice, they are really providing with more incentives to the public space life, to the interactions between people and the built environment, to the benefit of the retail activity and, in short, to its urbanity [22, 23].

Taking into account the contribution of the street musicians to the standard of living of this part of the city, it is reasonable to study the soundscape these musicians produce in those spaces. For architects, not only the quality of life in terms of pleasure or displeasure is important, but also how this pleasure can be designed and modified in the spaces. Thus, the study of tools that can manage the soundscape in our cities is of great value for all urban and architecture designers.

0   10   20   30   40 m

Fig. 1. Plan of Sant Felip Neri Square with emitting and recording points.

In this context, the analysis of some typical street music points in Ciutat Vella of Barcelona takes part into this soundscape evaluation. An environment is studied here: Plaça Sant Felip Neri. Different recordings were made in the environment regarding different positions of the listeners. For the current study, a recording was selected: *The*

Fountain by Marcel Lucien was reproduced in Plaça de Sant Felip Neri and then recorded from five different positions. Figure 1 shows the emitting point with the enumeration of the different recording points in the square.

3   The Virtual Environments

In order to create the test conditions, it was necessary to virtually recreate Plaça Sant Felip Neri as faithfully as possible, both on the visual and the auditory aspects.

First of all, a 3D model was created using photogrammetric processing of digital images to generate 3D spatial data. This method is relatively fast to implement, doesn't require specialized hardware like laser scanning and, if performed correctly, produces high quality results, not as precise as other techniques, but more than enough to transport the user to a faithful 3d recreation of the square.

The next goal was to recreate the soundscape. Two options were developed and presented to the test subjects. The first option was to use the original concert hall recording and present it to the users as is, without any distortion, reverberation or additional ambient sounds from the square.

The second option was more complex to create. As stated, five recordings of the song were made from different locations on the square. These recordings captured the subtleties a user would experience listening to it in the real square. The challenge was to allow the test subjects to move freely around the square and still be able to listen the song as close as possible to the real conditions at any point on the square, not only on the five recording points.

To extend the experience to any given point, a mixing algorithm was used to perform a logarithmic interpolation between the nearest recordings, in order to provide an experience that was identical to the original when the user was exactly at the recording point, and faded seamlessly as they moved closer to the next one.

Finally, it was necessary to use a Head Mounted Display to show the virtual reality to the test subjects. The Oculus Rift was selected for this task due to its compact size, high quality display and integrated headphones. This provided a fully immersive environment with a great sense of presence to the users.

The Oculus Rift allows for Room Scale tracking. This means that the user can move around the real space, and that movement translates to the virtual world. This amplifies greatly the sense of presence and makes the experience a lot more realistic and comfortable. However, it has limitations: the length of the cable and the resolution of the tracking cameras only allow the user to move around a space 3 m long and 2 m wide, approximately.

To improve this aspect, a teleport system was created. Using the Oculus Rift touch controllers, the user could point at any location in the square, and instantly teleport to that location. This provided the necessary freedom to move around the square, while keeping all the benefits of the room tracking.

The result of this process was an experience that allowed to move freely around a realistic 3d recreation of the original square, and perceive two distinctly different audio environments: one that recreated concert hall conditions, and a second one that allowed to experience as close as possible the square sound conditions.

4   The BLA Test

Using complementary qualitative research, it is possible to obtain new variables to study in future iterations and more detail for the quantitative data [24].

Quantitative and qualitative approaches have historically been the main methods of scientific research. On the one hand, quantitative research focuses on analyzing the degree of association between quantified variables, as promulgated by logical positivism; therefore, this method requires induction to understand the results of the investigation. Because this paradigm considers that phenomena can be reduced to empirical indicators that represent reality, quantitative methods are considered objective [25].

On the other hand, qualitative research focuses on detecting and processing intentions. Unlike quantitative methods, qualitative methods require deduction to interpret results. The qualitative approach is subjective, as it is assumed that reality is multifaceted and not reducible to a universal indicator [26].

Qualitative methods are commonly employed in usability studies and, inspired by experimental psychology and the hypothetical-deductive paradigm, employ samples of users who are relatively limited. Nevertheless, the Socratic paradigm from postmodern psychology is also applicable and useful in these usability studies because it targets details related to the UX with high reliability and uncovers subtle information about the product or technology studied.

Socrates was one of the greatest teachers who used questions for his insightful teaching. Many researchers have pointed out that one of the most powerful methods to promote critical thinking is verbal interaction, especially Socratic interaction including rational dialogue and questioning between students and the instructor [27]. Socratic questioning is defined as the use of thoughtful questions to stimulate students to continually probe the subject and as inductive process of guided discovery [28]. Instead of providing direct answers, the Socratic questioning approach encourages students' curiosity by continually probing the subject matter with thought-provoking questions.

Through qualitative methods, the goal was to explore users' desires, needs, and goals when learning about informatics tools that they would use to design 3D projects in their future work.

Starting from the Socratic paradigm basis, the BLA system has been designed. BLA method could be defined as psychological exploration technique, which points out the key factors of use experience. The main goal of this system was to ascertain which concrete characteristic of the product entails user's frustration, confidence, or gratitude (among many others) [24].

For the qualitative study, a sample of 18 students (11 men and 7 women) who agreed to participate was selected.

The BLA method works on positive and negative poles to define the strengths and weaknesses of the product. Once the element is obtained, the laddering technique is going to be applied to define the relevant details of the product. The object of a laddering interview was to uncover how product attributes, usage consequences, and personal values are linked in a person's mind. The characteristics obtained through laddering application will define what specific factor make consider an element as strength or as a weakness. BLA performing consists in three steps, following the a similar methodology as Fonseca, Redondo and Villagrasa [3]:

1. Elicitation of elements. The implementation of the test starts from a blank template for the positive (most favorable) and negative (less favorable) elements. The interviewer (in this case the professor) will ask the users (the student) to mention a positive and a negative aspect of the two types of music than can be heard (Option A and Option B). Thus, we are going to obtain two positive aspects and two negative aspects.
2. Marking of elements. Once the list of positive and negative elements is completed, the interviewer will ask the user to mark each one from 0 (lowest possible level of satisfaction) to 10 (maximum level of satisfaction);
3. Elements definition. Once the elements have been assessed, the qualitative phase starts. The interviewer asks for a justification of each one of the elements performing laddering technique. Why is it a positive element? Why this mark? The answer must be a specific explanation of the exact characteristics that make the mentioned element a strength or weakness of the product.

From the results obtained, the next step was to polarize the elements based on two criteria:

1. Positive (Px)/Negative (Nx): The student must differentiate the elements perceived as strong points of the experience that helped them to understand the music as satisfactory, in front of the negative aspects that didn't help them to understand the music or simply need to be modified to be satisfactory;
2. Common Elements (xC)/Particular (xP): Finally, the positive and negative elements that were repeated in the students' answers (common points) and the responses that were only given by one of the students (particular points) were separated according to the coding scheme shown in Tables 1, 2 and 3.

The common elements that were mentioned at a higher rate are the most important aspects to use, improve, or modify (according to their positive or negative sign). The particular elements, due to their citation by only a single user, may be ruled out or treated in later stages for development.

Table 1. Positive common (PC), particular (PP), negative common (NC) and negative particular (NP) elements for A option (concert hall) and B option (public square).

E. code	Description	Av. score (Av)	Mention Index (MI) (%)
1PC (A)	Clarity of music	7,7	44,4
2PC (A)	Guiding thread for music	8,3	16,6
3PC (A)	Quality of sound	8,3	16,6
4PC (A)	Focused on the music	9	11,1
1PP (A)	Peaceful Music	9	5,6
1NC (A)	No realistic	3	33,3
2NC (A)	No sense of space	4	22,2
3NC (A)	Without background	4,5	11,1
4NC (A)	Movement too fast	4	11,1
1NP (A)	No variance of echo	4	5,6
2NP (A)	Like a television	4	5,6
3NP (A)	Too loud	4	5,6
1PC (B)	Realistic	8,4	50
2PC (B)	Sense of the place	8,7	33,3
1PP (B)	Alive	9	5,6
2PP (B)	Softer and modulated	7	5,6
3PP (B)	More natural	7	5,6
1NC (B)	No clarity of music	3,8	22,2
2NC (B)	Relation between background and vision	4,7	16,7
3NC (B)	Disturbing background	3,7	16,7
4NC (B)	Problems with volume	3,7	16,7
5NC (B)	It is not real enough	3,5	11,1
1NP (B)	Quality of hardware	5	5,56
2NP (B)	Sudden changes on sound	3	5,56

The individual values obtained for both indicators, positive and negative, are shown in the following Table 2. Once the features mentioned by the students were identified and given values, the third step defined by the BLA initiated the qualitative stage in which the students described and provided solutions or improvements to each of their contributions in the format of an open interview.

Table 3 shows the main improvements or changes that the students proposed for both positive and negative elements.

Table 2. Individual scores for PC, PP, NC and PC elements for A option (concert hall) and B option (public square)

E. code	Male											Female						
Users	U 1	U 2	U 3	U 4	U 5	U 6	U 7	U 8	U 9	U 10	U 11	U 12	U 13	U 14	U 15	U 16	U 17	U 18
1PC (A)	-	-	-	9	8	-	8	-	7	7	8	-	8	-	-	-	7	-
2PC (A)	9	-	-	-	-	-	-	8	-	-	-	-	-	-	-	8	-	-
3PC (A)	-	8	-	-	-	7	-	-	-	-	-	-	-	9	-	-	-	8
4PC (A)	-	-	8	-	-	-	-	-	-	-	-	10	-	-	-	-	-	-
1PP (A)	-	-	-	-	-	-	-	-	-	-	-	-	-	-	9	-	-	-
1NC (A)	2	-	-	-	3	4	-	-	-	-	-	-	-	3	-	5	-	1
2NC (A)	-	2	4	-	-	-	-	5	-	-	-	-	-	-	-	-	5	-
3NC (A)	-	-	-	-	-	-	5	-	-	-	-	-	4	-	-	-	-	-
4NC (A)	-	-	-	5	-	-	-	-	3	-	-	-	-	-	-	-	-	-
1NP (A)	-	-	-	-	-	-	-	-	-	-	4	-	-	-	-	-	-	-
2NP (A)	-	-	-	-	-	-	-	-	-	-	-	4	-	-	-	-	-	-
3NP (A)	-	-	-	-	-	-	-	-	-	3	-	-	-	-	-	-	-	-
1PC (B)	-	9	-	-	10	9	7	-	5	9	-	10	-	-	-	8	-	9
2PC (B)	10	-	-	9	-	-	-	9	-	-	4	-	-	10	-	-	10	-
1PP (B)	-	-	-	-	-	-	-	-	-	-	-	-	-	-	9	-	-	-
2PP (B)	-	-	7	-	-	-	-	-	-	-	-	-	-	-	-	-	-	-
3PP (B)	-	-	-	-	-	-	-	-	-	-	-	-	7	-	-	-	-	-
1NC (B)	-	-	-	-	-	-	4	3	-	4	4	-	-	-	-	-	-	-
2NC (B)	-	5	4	5	-	-	-	-	-	-	-	-	-	-	-	-	-	-
3NC (B)	-	-	-	-	-	4	-	-	-	-	-	3	-	-	4	-	-	-
4NC (B)	3	-	-	-	-	-	-	-	-	-	-	-	-	-	-	5	3	-
5NC (B)	-	-	-	-	3	-	-	-	2	-	-	-	-	-	-	-	-	-
1NP (B)	-	-	-	-	-	-	-	-	-	-	-	-	5	-	-	-	-	-
2NP (B)	-	-	-	-	-	-	-	-	-	-	-	-	-	-	-	-	-	3

Table 3. Proposed common improvements (CI) and particular improvements (PI) for both positive and negative elements for common and particular items in A recording (concert hall) and B recording (public square).

E. code	Description	Mention Index (MI) (%)
1CI (A)	Improve the relation with the environment	66,7
2CI (A)	Improve the background sound	22,2
3CI (A)	Change the position of the sounds	11,1
2PI (A)	Decrease the volume of the sound	5,7
3PI (A)	Improve the relation with the musician	5,7
4PI (A)	Improve the quality of the sound	5,7
1CI (B)	Improve sound quality	27,8
2CI (B)	The changes between position could be more soft	22,2
3CI (B)	Balance the volume levels between different points	11,1
4CI (B)	Improve the relation between vision and sound	11,1
5CI (B)	Decrease the background noise	11,1
1PI (B)	Improve the clarity of sound	5,7
1PI (B)	Improve the relation with the place	5,7

4.1 Results

At this point, it is interesting to identify the most relevant items obtained from the BLA, by high rates of citation, high scores, or a combination of both. It is important to separate the types of results obtained. The first group belongs to A option (concert hall recording), and the second group belongs to a B option (public square recording). After the elicitation of the most relevant features of each of them, we are going to close with their comparison.

A option (concert hall recording). We can highlight that this kind of recording has a good clarity of music (MI: 44,4%, Av: 7,8), it favors the guidance of the thread for music (MI: 16,6%, Av: 8,3), and the quality of the sound is valued (MI: 16,6%, Av: 8,3). In terms of the main negative comments, students clearly identified a lack of realism in this kind of experience (MI: 33,3%, Av: 3), that was related to the lack of sense of space (MI: 22,2%, Av: 4) and they missed a background noise (MI: 11,1, Av: 4,5), aspects that were directly related to the design of the application.

B option (public square recording). Two main positive aspects were highlighted by students: the high degree of realism of the application both in visual and acoustic terms (MI: 50%, Av: 8,4), and the good relation between sound and place (MI: 33,3%, Av: 8,7). Conversely, some negative comments were pointed out: a lack of clarity in the music (MI: 22,2%, Av: 3,8), a bad relation between background and vision (MI: 16,7%, Av: 4,7) which should be solved with the position of different visual avatars, and the presence of some disturbing background (MI: 16,7%, Av: 3,7) due to the different moments of the original recordings. Technically, these would be the main aspects to modify in future iterations of the proposed method.

In summary, two clear opinions about the experiment were shown, which confirm the first question of the survey: Which recording do you prefer, A or B? Most of people (61,1%) agreed that B option was better than A option (38,9%). The reasons for this answer were clearly explained in the rest of the survey. Although there was a high valuation of the realism of the application both in visual and acoustic terms in B option (MI: 50%, Av: 8,4), it was also certain that clarity of music in B option was not as good as A option, as we can note if we compare 1PC (A) with 1NC (B). This fact, confirms that street music recording implies a decrease of the quality in the music played. This loss could be a drawback for musicians who want to perform their art in the middle of the city. But the survey reveals us another feature that must be taken into account: a third of the students (MI: 33,3%) evaluated option B with an almost excellent value (Av: 8,7) on the sense of space quality (2PC (B)). This fact shows a hidden potential of spatial sound, that is, the sound spatialization. One can find several attempts on the history of music were composers wrote their music having in mind the spatial features of the places they were going to be played. However, all these compositions tend to be limited to a closed space were also the spatial possibilities are limited to that space. Suddenly, a high range of possibilities spreads when the closed concert hall is opened to the openness of the squares and public spaces. Coupled volumes, streets, galleries, balconies or even stairs belong now to this new stage for music that can be explored in infinite ways.

5   Conclusion

The present study concludes that a good acceptation of the spatialized sound in VR environments has been recorded by the students. Moreover, they value sense of presence, that is, the possibility to "feel like being in the place" that the spatialized sound gives to them. These results encourage us that the inclusion of localized sound into VR can be valuable in urban acoustic design applications.

Nevertheless, some important limitations need to be considered. Firstly, a wider range of possibilities in the VR environment in order to serve as an urban design tool. In fact, the decision on the position of the sound sources is one of the aspects that will make this tool more useful for architects. Also, the definition of the materials and position of the architectural elements of the scene could be one of the focus for further research.

Acknowledgments. This research was supported by the National Program of Research, Development and Innovation aimed to the Society Challenges with the references BIA2016-77464-C2-1-R & BIA2016-77464-C2-2-R, both of the National Plan for Scientific Research, Development and Technological Innovation 2013–2016, Government of Spain, titled *"Gamificación para la enseñanza del diseño urbano y la integración en ella de la participación ciudadana* (ArchGAME4CITY)", & *"Diseño Gamificado de visualización 3D con sistemas de realidad virtual para el estudio de la mejora de competencias motivacionales, sociales y espaciales del usuario* (EduGAME4CITY)". (AEI/FEDER, UE).

References

1. Shen, C.-X., Liu, R.-D., Wang, D.: Why are children attracted to the Internet? The role of need satisfaction perceived online and perceived in daily real life. Comput. Hum. Behav. **29**, 185–192 (2013)
2. Pelzer, S., Aspöck, L., Schröder, D., Vorländer, M.: Integrating real-time room acoustics simulation into a CAD modeling software to enhance the architectural design process. Buildings **4**, 113–138 (2014)
3. Fonseca, D., Redondo, E., Villagrasa, S.: Mixed-methods research: a new approach to evaluating the motivation and satisfaction of university students using advanced visual technologies. Univ. Access Inf. Soc. **14**, 311–332 (2015)
4. Weinzierl, S., Vorländer, M., Zotter, F., Maempel, H.-J., Lindau, A.: A real-time auralization plugin for architectural design and education. In: Proceedings of the EAA Joint Symposium on Auralization and Ambisonics 2014. Universitätsbibliothek Technische Universität Berlin (2014)
5. Kowaltowski, D.C.C.K., Bianchi, G., de Paiva, V.T.: Methods that may stimulate creativity and their use in architectural design education. Int. J. Technol. Des. Educ. **20**, 453–476 (2010)
6. Stromquist, N.P.: The impact of information and communication technologies on university students: a tentative assessment. Cult. y Educ. **21**, 215–226 (2009)
7. Tondeur, J., De Bruyne, E., Van Den Driessche, M., McKenney, S., Zandvliet, D.: The physical placement of classroom technology and its influences on educational practices. Cambridge J. Educ. **45**, 537–556 (2015)
8. Mercier, E.M., Higgins, S.E., Joyce-Gibbons, A.: The effects of room design on computer-supported collaborative learning in a multi-touch classroom. Interact. Learn. Environ. **24**, 504–522 (2016)
9. Sipilä, K.: Educational use of information and communications technology: teachers' perspective. Technol. Pedagogy Educ. **23**, 225–241 (2014)

10. Rawlins, P., Kehrwald, B.: Integrating educational technologies into teacher education: a case study. Innov. Educ. Teach. Int. **51**, 207–217 (2014)
11. Mumtaz, S.: Factors affecting teachers' use of information and communications technology: a review of the literature. J. Inf. Technol. Teach. Educ. **9**, 319–342 (2000)
12. Hodis, F.A., Hancock, G.R.: Introduction to the special issue: advances in quantitative methods to further research in education and educational psychology (2016). https://doi.org/10.1080/00461520.2016.1208750
13. Grimm, K.J., Mazza, G.L., Mazzocco, M.M.M.: Advances in methods for assessing longitudinal change. Educ. Psychol. **51**, 342–353 (2016)
14. Truax, B.: Handbook for Acoustic Ecology CD-ROM Edition Version 1.1 (CD) at Discogs. Cambridge Street Publishing, Burnaby (1999)
15. Brown, A.L.: Soundscapes and environmental noise management. Noise Control Eng. J. **58**, 493 (2010)
16. Cerwén, G.: Urban soundscapes: a quasi-experiment in landscape architecture. Landsc. Res. **41**, 481–494 (2016)
17. Yang, W., Kang, J.: Soundscape and sound preferences in urban squares: a case study in Sheffield. J. Urban Des. **10**, 61–80 (2007)
18. Fonseca, D., Navarro, I., de Renteria, I., Moreira, F., Ferrer, Á., de Reina, O.: Assessment of wearable virtual reality technology for visiting world heritage buildings: an educational approach. J. Educ. Comput. Res. (2017). https://doi.org/10.1177/0735633117733995
19. Simpson, P.: Street performance and the city. Space Cult. **14**, 415–430 (2011)
20. McNamara, L., Quilter, J.: Street music and the law in Australia: busker perspectives on the impact of local council rules and regulations. J. Musicol. Res. **35**, 113–127 (2016)
21. Los músicos callejeros de Barcelona denuncian el acoso que sufren por el ayuntamiento (2017)
22. Gehl, J.: Cities for People. Island Press, London (2010)
23. Gehl, J.: Life Between Buildings: Using Public Space. Island Press, London (2011)
24. Pifarré, M., Tomico, O.: Bipolar laddering (BLA): a participatory subjective exploration method on user experience. In: Proceedings of the 2007 Conference on Designing for User eXperiences - DUX 2007, p. 2. ACM Press, New York (2007)
25. Walliman, N.: Research Methods, 2nd edn. Taylor and Francis, New York (2017)
26. Pfeil, U., Zaphiris, P.: Applying qualitative content analysis to study online support communities. Univ. Access Inf. Soc. **9**, 1–16 (2010)
27. Wiley, J., Griffin, T.D., Jaeger, A.J., Jarosz, A.F., Cushen, P.J., Thiede, K.W.: Improving metacomprehension accuracy in an undergraduate course context. J. Exp. Psychol. Appl. **22**, 393–405 (2016)
28. Lee, M., Kim, H., Kim, M.: The effects of Socratic questioning on critical thinking in web-based collaborative learning. Educ. as Change **18**, 285–302 (2014)

Check for updates

Identification of Significant Variables for the Parameterization of Structures Learning in Architecture Students

Carles Campanyà[1(✉)], David Fonseca[1(✉)], Núria Martí[1(✉)], Enric Peña[1(✉)], Alvaro Ferrer[1(✉)], and Josep Llorca[2(✉)]

[1] La Salle, Universitat Ramon Llull, c/Sant Joan de La Salle 42, 08022 Barcelona, Spain
{ccampanya,fonsi,nmarti,enricp,ls31968}@salleurl.edu
[2] AR&M, Barcelona School of Architecture, BarcelonaTech, Catalonia Polithecnic University, Av/Diagonal 649, 08028 Barcelona, Spain
josep.llorca@upc.edu

Abstract. The present work can be included in a much broader research related to an improvement on the learning of structural concepts and their practical application for architecture students, and consists in identifying significant variables to predict the effect of different teaching practices using an academic analytics approach. This work gathers data from surveys answered by architecture students – from La Salle Architecture School in Barcelona - to confirm the hypothesis that motivation is a key aspect to focus on. The results confirm it, and configure the working basis to check the efficiency of the teaching practices to be analyzed next academic years and for defining a predictive model on structural learning for architectural students.

Keywords: Learning analytics · Structures for architecture students
Architecture studies · Learning indicators

1 Introduction

This article is based on an investigation aimed at identifying - using the learning analytics [1] approach –significant variables to predict the efficiency of different teaching practices, in order to improve the learning of structural knowledge and its application in architecture students, so that a predictive model can be defined based on these variables.

As a first stage for the research, motivation of students on the structural area is being analyzed, and also students' perception about the need of a deep structural knowledge on structures to become an architect.

This article presents results from surveys carried out on students close to finishing the architecture degree on one side, and to students of 1st, 2nd, 3th and 4th course on the other side. All respondents are architecture students in ETSALS (La Salle Architecture School in Barcelona, originally: Escuela Técnica Superior de Arquitectura La Salle). The main aim pursued with this study is to support or refute the following hypotheses:

© Springer International Publishing AG, part of Springer Nature 2018
Á. Rocha et al. (Eds.): WorldCIST'18 2018, AISC 747, pp. 298–306, 2018.
https://doi.org/10.1007/978-3-319-77700-9_30

- Students perceive that a deep knowledge in structures is not really needed to become an architect.
- Immersion in the atmosphere of schools of architecture boosts these students' perception.
- Generally, students close to finishing their studies do not know how to implement in practice their structural knowledge.
- Generally, motivation of architecture students in creative aspects is higher than in technical aspects.

On the other side, the surveys outcome is analyzed to find significant indicators that could be related to the representative values of the aspects that are intended to be improved, mainly the degree of learning and the degree of implementation of knowledge.

2   Context: Architecture and Structures

Even though architecture is broadly defined as a synthesis of arts and sciences from ancient times to the present [2–5], the huge amount and complexity of knowledge and competences needed has led most of the architects to specialization.

Today, architects have different professional attributions in each country, and that has led to different teaching approaches and school curricula, especially on competences required on technical subjects. Most European countries do not require deep structural knowledge to architecture students, for they have not professional attributions, while in other countries, like Spain, architects are considered to be the last responsible of building structures [6] and this professional attribution is automatically recognized when architect's title is obtained.

Even though these differences exist, European Directive 2005/36/CE [7] establishes automatic recognition of the Architect title in all EU countries, with the only condition of the possession of an Architecture title obtained in any state member of the EU, with no mention of the difference in professional attributions for each state. That leads to the fact than the same title has different professional attributions depending on the state where architecture is practiced, although titles come from diverse architecture training programs where technical aspects are treated differently.

2.1   Structures Subjects in the Spanish System

Spanish architecture study programs include a complete training in structures – design and calculation, in coherence with architectural concept, and execution surveying -, provided that architects are responsible for building structures, according to current regulations [8, 9]. With some difference between the different architecture schools, all curricula include subjects on mathematics and physics, on resistance of materials and on application of current structural codes in Spain, mainly the Building Technical Code, the Concrete Structural Code and the Steel Structural Code.

2.2 Structures Subjects in La Salle

The ETSALS was founded in 1997. The following subjects related to the structural area have been always included in the studies programs: mathematics, physics, introduction to structures, concrete and steel structures, and geotechnics and foundations. All these subjects can be found in the current syllabus [10] that meets the requirements according the Bologna process.

Currently, two study programs are mainly ongoing in ETSALS. Architecture degree is developed in five academic years in both programs. Structures area subjects are: Mathematics, Physics (1st year), Introduction to Structures (2nd year), Concrete and Steel Structures (3rd year) and Geotechnics and Foundations (4th year). Last academic program approved, dated in 2015, of 300 ECTS (European Credit Transfer and Accumulation System), requires to be completed with a Master's course of one academic year and 60 ECTS, that includes the Applied Structures subject.

Structural knowledge acquired is supposed to be applied in other subjects included in other areas, mainly in Construction (3rd, 4th and 5th years) and Projects (1st to 5th years), and evidently in the Final Degree Project (FDP).

2.3 Assessing Training: Learning Analytics

Assessing training is always a difficult task, since it is closely connected to the context where (formal, non-formal and informal) training to be assessed has taken place and to the type of assessment approach that has been used. Generally, in the literature on training evaluation, two major theoretical approaches can be found: evaluation training and effectiveness training. The former is based on the evaluation of learning outcomes achieved at the end of training, which is based on the effectiveness of training that was provided. In the former approach, objectives, content and design of training become the object of evaluation; in the latter approach, however, the training process is examined in all its stages (pre, ongoing and post) considering the variables that might have influenced the effectiveness of training activities [11].

Assessment supports and fosters the quality development of an education and training system because it:

- Identifies the strengths and weaknesses of an education and training system and actions
- Observes and analyses how resources are used
- Involves and empowers the stakeholders engaged in the training system and actions
- Ensures that a change has indeed occurred with effects on the institutional and social context
- Allows to identify critical issues in a primary phase using Pre and Profile tests, and using mixed methods (combining quantitative and qualitative approaches) for a better interpretation of the results [12].

When we try to incorporate new educational methods, we need to incorporate them into teaching in a controlled manner; there are some risks that need to be controlled before one can improve not only the curriculum but also student skills and knowledge.

For example, in the case of new technologies, the professor must be trained and capable of providing full-time support to students: he or she must be capable of offering a good and precise explanation of the practice and methodology, must correctly select the applications, and must provide clear final objectives. Previous studies describe "critical mistakes" in the implementation of these types of methods, mistakes that can generate negative perceptions among the students and which need to be avoided [13].

For all these reasons, it is necessary to develop educational decision support systems, in order to improve technical education. In this framework, Learning Analytics addresses the management and analysis of the educational data for the improvement of learning: "Using analytics we need to think about what we need to know and what data is most likely to tell us what we need to know" [14].

We can affirm that Learning Analytics has become the main topic of educational conferences, and it is more than a simple trend in education [15, 16]. Ferguson defined Learning Analytics as "the measurement, collection, analysis and reporting of data about learners and their contexts, for purposes of understanding and optimizing learning and the environments in which it occurs" [17]. Related with this concept, we can find other ideas; for example, the Educational Data Mining and the Academic Analytics. The first one is concerned with developing methods for exploring the unique types of data that come from educational settings, and the second idea, Academic Analytics, is a mixture process for providing higher education institutions with the data necessary to support operational and financial decision making [18, 19].

3   Method

As already stated, several surveys have been conducted on Architecture Students at ETSALS.

On one side, students of last year were consulted some weeks before the date of their FPD presentations, and therefore, only weeks before they became architects. This survey has been conducted four times between July 2016 and May 2017. The overall sample is comprised of 131 students, which were requested to answer questions on a scale of 1 to 5 with the aim of:

- Assess students' motivation on structural topics
- Know their perception about what they know on structures and about what should be known
- Know their perception about their capacity of being responsible of structures calculations
- Know their degree of satisfaction about their knowledge and about how structures are taught.

Students' grades obtained in subjects of structures area have been incorporated to complement this information, as has been done with the number of calls needed to pass each subject, the FDP grade, and other data like the number of structures corrections during the FDP development and the work experience of the students in architecture and/or structural studios.

On the other side, different surveys have been conducted in first weeks of the academic year 2017/2018 to undergraduate students of 1st, 2nd, 3th and 4th year. The total number of samples is 197: 55 from 1st year, 52 from 2nd year, 51 from 3rd year and 39 from 4th year students. These students were also asked to answer on a scale of 1 to 5 questions that search to evaluate the same three aspects described and to identify when significant changes take place.

Correlations between data obtained from surveys have been analyzed, especially those useful to parameterize the degree of knowledge of structures and the degree of application of this knowledge.

4   Results

This section summarizes the data analysis obtained from surveys. The numerical results presented here are the average of the answers on a scale of 1 to 5. Correlations are shown in percentage, according to Pearson correlation coefficient. Results are arranged according to the four hypotheses to check as described in the introduction.

4.1   Students' Perception About the Degree of Structural Knowledge Needed for Architects

Survey responses confirm the first hypothesis, that students perceive that a deep knowledge in structures is not really needed to become an architect. This is in contradiction with Spanish professional attributions that set architects responsible of building structures.

FDP students admit a low degree of structural knowledge, while consider not necessary to improve it. So, the average answer when requested about their capability of calculating buildings is 2.77, and 3.3 is the average score about their perception about pre-dimensioning capabilities. The students consider themselves not able to be responsible of building calculations (scoring an average of 2.43). Note that all FDP students consulted obtained the Architecture title some weeks after the date of the survey.

Even with better scores, students are critical of their degree of learning and of how structures are taught in ETSALS, as can be deducted from the answers to questions, i.e. 'if they think if they have enough structural knowledge to become architects' (average score is 3.5), if they are satisfied with what they have learnt (3.4) and the degree of satisfaction about how structures are taught (3.6).

We find especially significant that even students who answer with 4 or 5 when asked about if their structural knowledge is enough to be architects, admit not to have enough knowledge to calculate structures (scoring 2.8 and 3.6), not to be capable of being responsible of structures (averages of 2.6 and 2.75), and are the students who score lowest when asked about considering the idea of working in structures (1.8, close to the average of 1.9 got from all responses).

FDP students score with an average of 3.75 their interest in broadening structural knowledge, with 3.4 the need of deepening in it in undergraduate programs, and with 3.2 the idea of deepening it in master's programs. In this vein, an average score of 3.5 is obtained when asked for the idea of strengthening the teaching of applying structural knowledge in projects design, and 2.9 is the average answer to if it was good a more demanding assessment on structures subjects. 3.45 is the average score when asked if structural skills of architects should be improved.

It is significant that students who are more likely to work professionally in structures score higher when consulted about the need of improving architects' skills in structures. This need is also more perceived by the students that are more satisfied with their knowledge and with how structures are taught. It could be said that students who are more interested in structures think that architects' structural skills should improve.

Another interesting finding is the lack of relationship between the grade obtained in the FDP and the average grade of structures subjects got during the undergraduate studies. Correlation between these two variables is just 12%. Answers of the five students who got the maximum FDP grade of 10/10 are in line with the hypothesis being checked that students perceive that architects do not need to have deep structural knowledge. They score their ability to calculate with 2.4 (average is 2.8), their capability of being responsible of calculations with 1.5 (average is 2.4) and they are not thinking of working professionally in structures (1.6 when average is 1.9). They think it is not necessary to deepen structural knowledge (2.4 when average is 3.4) and are the ones who least want to improve their knowledge (3.2 in front of the average score of 3.8).

4.2   Influence of the Atmosphere of the School of Architecture in This Perception

Students are aware that many architects do not know how to calculate structures, and this awareness takes place in the first academic year. That can be deducted from the answers of undergraduate students when asked about it. At the beginning of the 2nd year, the average answer score is 3.2, similar to the score got from 3rd and 4th year students, while the students who begin 1st year mark it with an average of 4.2. Therefore, this change of perception occurs in one academic year, which is also reflected when asked if they think they will know how to calculate when they finish the degree, going from 4.55 (1st year students) to 3.7 (3rd and 4th year). The hypothesis seems confirmed, as this perception changes significantly in only one academic year.

The fact that the grade of the FPD has virtually no relation to the grades obtained in the subjects of the area of structures increases this perception that structures are not an essential part of the knowledge necessary to be an architect.

4.3   Perception of the Students About Their Capacity for Practical Application of the Knowledge of Structures

The answers of the students on this perception give rise to results that are thought to be different from those that teachers would have. It would therefore be necessary to contrast with the perception of teachers, to confirm whether both perceptions are similar or, as we suspect, differ. Teachers of subjects on structures at ETSALS have an average teaching experience of more than 15 years, and an average professional experience in structures of more than 25 years.

FDP students consulted think they are able to adequately select the structural typology for their buildings (average score of 4.1 for residential buildings and 3.8 for public buildings), which contradicts the teachers' experience. As has been said, the perception of the application of knowledge is low when asked if they are able to pre-dimension (3.4) and calculate (2.8) their structures.

When the FDP students are asked if they have applied the knowledge of structures to the workshops and projects they respond with a 3.7, and when they are consulted about the application in the FDP they score with a 4.0.

In the case of similar questions, students who begin 3rd year score 2.9 on being consulted about whether they apply the knowledge to projects and workshop, while in 4th grade this value rises to 3.8. 3rd year students respond with 2.82 when consulted about whether they know how to pre-dimension, and those of 4th respond with a 3.36.

As can be seen, values are generally low, and it would be confirmed (in the absence of contrasting data obtained from teachers) that there is a considerable margin for improvement.

4.4 Motivation of Students in Technical and Structural Aspects of Architecture

Responses reveal a lack of motivation in the technical field in general, and in structures in particular, as can be deducted from different answers: in students from 1st to 4th year, the motivation for the technical part (3.85 on average) is always below the motivation for the creative side (4.55 on average). FDP students declare that they do not have interest in working professionally in structures (1.9 on average), a score much lower than that declared by students from 1st to 4th year when asked on the interest in working in a technical branch (3.24) and even lower than the same students when asked if they would like to work in a creative field (4.5).

In addition, students who have not yet started studying any subject of structures perceive that structural knowledge is not completely necessary for architects and believe that structural subjects will be difficult for them. The finding of this predetermined thinking is found in the following answers: before taking any subject on structures, students think that they will not like them very much (score of 3.6) and that those subjects will be difficult to pass (3.4 of average score where 1 is easy to pass and 5 is difficult to pass). Later, when they have already taken one or two subjects of structures, they confirm that they do not like them (3.5 on average) and they find them even more difficult than they expected (2.85).

4.5 Other Significant Relationships Between Data

The relationship between data has been analyzed to identify variables that significantly affect students' perceptions.

It has been found that there is no significant correlation (16%) between the average grade in structural subjects and the perception of the students that they know enough. It is remarkable the fact that the average grade in structural subjects makes no difference on the score to the question about knowing how to calculate (around 2.8 on average for all the studied grade ranges). Students who have the best average grade of structures score lower than the rest on their ability to be responsible for the structures of the build-ings. Finally, the 5 people who achieved a higher than average grade in structures stand out to be poorly satisfied with the acquired knowledge (2.3 when the average is 3.4), and are also unhappy with how structures are taught (3 when the average is 3.6). At the same time, they are the ones who most believe that the level of structures of architects needs to be improved (4.2 when the average is 3.5).

5 Conclusions and Future Lines

The following hypotheses are considered to been proven:

- Architecture students have the perception that a deep knowledge in structures is not really necessary to be an architect.
- Immersion in the atmosphere of an architecture school boosts this vision.
- In general, students close to finishing the degree of architecture do not know how to apply the structural knowledge acquired in practice.
- In general, the motivation of architecture students is lower in the technical aspects than in the creative ones.

In order to achieve the learning objectives, and to parameterize the results as far as possible, identifying significant indicators, the following future lines of action are considered:

- Check if the students' motivation improves when focusing structural subjects on practical application of structural knowledge.
- Check if there is any improvement in results when motivation improves.
- Check if a change in the perception that structural subjects are difficult will result in an improvement of the results.

For this last point, as shown in various studies such as those collected by Ken Bain [20], perceptions of students may have an influence on the results they obtain. It may well be that students' performance decreases if they perceive that they do not need to know about structures and that it will be difficult to achieve a satisfactory grade in structural subjects.

References

1. Learning and Knowledge Analytics course blog. http://www.learninganalytics.net
2. Vitruvius, P., Morgan, M.H.: Vitruvius: The Ten Books on Architecture. Dover Publications, New York (1960)
3. Alberti, L.B., Tory, G.: Leonis Baptistae Alberti ... Libri De re aedificatoria decẽ. Opus integrũ et absolutũ: diligenter q recognitum. Distinctum est autẽ nuper opus ipsum totum. B. Rembolt, Parrhisijs (1512)
4. Viollet-le-Duc, E.: Entretiens sur l'architecture. A. Morel, Paris (1863)
5. Encyclopedia Britannica: "Architecture". https://www.britannica.com/topic/architecture
6. «BOE» núm. 266, 06/11/1999. Ley 38/1999. Ley de Ordenación de la Edificación. Departamento: Jefatura de Estado (1999)
7. Official Journal of the European Union: Directive 2005/36/EC of the European Parliament and of the Council (2005)

8. ANECA: Libro blanco de grado en Arquitectura. Agencia nacional de evaluación de la calidad y acreditación. http://www.aneca.es/media/326200/libroblanco_arquitectura_def.pdf
9. «BOE» núm. 184, 30/06/2010. Acuerdo de Consejo de Ministros por el que se establecen las condiciones a las que deberán adecuarse los planes de estudios conducentes a la obtención de títulos que habiliten para el ejercicio de la profesión regulada de Arquitecto. Departamento: Ministerio de Educación (2010)
10. La Salle Academic Guide: Syllabus of the degree in Architecture Studies. http://www.salleurl.edu/en/education/degree-architecture-studies/syllabus
11. Fonseca, D., Climent, A., Vicent, L., Canaleta, X.: Learning4Work. designing a new evaluation system based on scenario centered curriculum. Methodology: the pre-test, vol. 9753. Lecture Notes in Computer Science, July 2016, pp. 1–11 (2016). https://doi.org/10.1007/978-3-319-39483-1_1
12. Fonseca, D., Redondo, E., Villagrasa, S.: Mixed-methods research: a new approach to evaluating the motivation and satisfaction of university students using advanced visual technologies. Univers. Access Inf. Soc. **14**(3), 311–332 (2015). https://doi.org/10.1007/s10209-014-0361-4
13. Redondo, E., Giménez, L., Valls, F., Navarro, I., Fonseca, D., Villagrasa, S.: High vs. low intensity courses: student technological behavior. In: Proceedings of the 3rd International Conference on Technological Ecosystems for Enhancing Multiculturality, pp. 77–82 (2015). https://doi.org/10.1145/2808580.2808593
14. Long, P., Siemens, G.: Penetrating the fog: analytics in learning and education. EDUCAUSE Rev. **46**, 31–40 (2011)
15. Conde, M.A., Hernández-García, A.: A promised land for educational decision-making?: present and future of learning analytics. In: García-Peñalvo, F.J. (ed.) Proceedings of the First International Conference on Technological Ecosystem for Enhancing Multiculturality (TEEM 2013), pp. 239–243. ACM, New York (2013). http://dx.doi.org/10.1145/2536536.2536573
16. Johnson, L., Adams, S., Cummins, M.: The NMC Horizon Report: 2012 Higher, Education edn. The New Media Consortium, Austin (2012)
17. Ferguson, R.: The state of learning analytics in 2012: a review and future challenges. Technical report KMI-12–01, Knowledge Media Institute, The Open University, UK (2012)
18. Goldstein, P.J., Katz, R. N.: Academic analytics: the uses of management information and technology in higher education. EDUCASE, vol. 8 (2005)
19. Peña, E., Fonseca, D., Martí, N.: Relationship between learning indicators in the development and result of the building engineering degree final project. In: Proceedings of the 4th International Conference on Technological Ecosystems for Enhancing Multiculturality, pp. 335–340 (2016)
20. Bain, K.: What the Best College Teachers Do. Harvard University Press, Cambridge (2004)

Analysis of Different Approaches of Elementary School Students to Working with Personal Learning Environment

Tomas Javorcik[(✉)]

Department of Information and Communication Technologies,
Pedagogical Faculty, University of Ostrava, Frani Sramka 3,
709 00 Ostrava, Czech Republic
Tomas.Javorcik@osu.cz

Abstract. The diversity of students can be seen in all of their school and extracurricular activities. Addressing the diversity of students was part of the research aimed at integration of Personal Learning Environment and mobile technology into elementary school instruction. The paper describes the different approaches of upper primary school students to working with a PLE-based environment, with emphasis put on gender differentiation. One of the main research objectives was to determine the possible gender differences in the use of storage, the nature of saved files or their organization. The presented results show that there are significant differences between boys and girls. As far as the PLE is concerned, the differences can be understood at different levels. The results of analysis of working methods used by groups of students can be used in a number of fields. Identification of gender, age and other differences may not only help integrate the PLE into the education process, but also serve as a springboard for designing and/or implementing PLE software solutions.

Keywords: Personal Learning Environment · Gender · E-R-R-A model
Student diversity · Web 2.0 tools · ICT

1 Introduction

Since each student approaches the education process in a different manner (to a greater or lesser extent), each student is considered unique. The main characteristic in which students differ (physical predisposition aside) is a preferred learning style which determines how the student acquires, processes and remembers new information. The process of knowledge acquisition goes hand in hand with the diversity and number of information sources used during the process. While some may consider a pencil and a notepad to be suitable tools, others (especially today's students and teachers) may prefer computers, laptops, cellular phones, tablets and other devices that can be connected to the omnipresent Internet. These devices, together with the Internet services enable the creation of the so-called Personal Learning Environment (PLE), which can lead an individual through their student, professional and personal life.

© Springer International Publishing AG, part of Springer Nature 2018
Á. Rocha et al. (Eds.): WorldCIST'18 2018, AISC 747, pp. 307–316, 2018.
https://doi.org/10.1007/978-3-319-77700-9_31

2 PLE as an Appropriate Method for Using Ict in Schools

The positive effect of information and communication technology (ICT) on the education process, regarding the development of key competencies, knowledge and skills has been proved by numerous studies and researches. However, it needs to be stated that the vast majority of those researches used ICT for a single purpose (e.g. in one course only, in one part of the lesson or for selected thematic units). Since there is no use in only using a tablet as an electronic presenter of information when its potential is much greater, the use of ICT in instruction is now evaluated with regard to its use in humanities and natural science subjects during the entire course of instruction (Fig. 1).

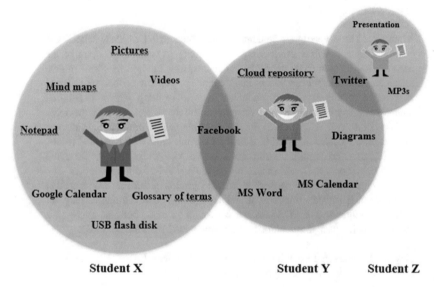

Fig. 1. Different nature and structure of PLE in students.

The Personal Learning Environment is a complex solution. This environment represents technology, Web 2.0 and social network tools which help people manage their learning process [1].

Downes [2] takes a closer look at the PLE elements. According to him, a PLE should contain:

- Profile management tools (change of password, arrangement of icons and files, etc.).
- Tools for editing and publishing materials (text editor, painting, tool for creating mind maps, etc.).
- Tools for saving external sources and websites (website directory, glossary of terms, etc.).

Aside from the aforementioned tools, a PLE can also contain *Tools for learning, (and other activities) management and planning* (notebook, calendar, task pad, etc.) and enable sharing of information (and knowledge) and cooperation of students.

It is important that students learn about the possibilities of PLE in elementary school so they can use it in their further studies (at higher levels of education) when they need to process large amounts of information in the most effective way possible. Systematically working with this environment, students will be able to use ICT in a complex manner. Moreover, it will help them realize that there is a plethora of ways of working with information.

3 Designing PLE

Upper primary school appears to be the best time to introduce students to the PLE. The initial contact with the PLE environment needs be controlled, however. The controlled and systematic use will ensure:

- Organized instruction – neither instruction nor the use of ICT is chaotic. It enables students to better understand when to pay attention to the teacher and when to use the PLE.
- Quality instruction – students use quality and appropriate PLE services provided by the school.
- Focused students – students do not disturb each other.

All of the above can be ensured by choosing an appropriate instruction model. The four-step E-R-R-A model, which is based on the three-step E-R-R model, was designed for the purposes of research (for more detailed information [3].

E-R-R-A-based instruction is divided into four stages:

1. Planning and evocation (E)
2. Realization of meaning, Contact with resources (R)
3. Reflection (R)
4. Aggregation of knowledge (A)

The E-R-R-A model is described in detail in a 2015 ICETA Conference paper [4]. In the first stage, the teacher informs students about the lesson plan, with the students marking the important dates (tests, seminar papers, laboratory exercises, field trips, etc.) in the calendar, which is an integral part of the PLE. In the *evocation* part, students use the notes they made earlier, which are related to the current thematic unit.

The second stage, *Contact with resources*, is influenced by the chosen organizational form of instruction. In this stage the student processes new information. In this stage, students have a certain degree of freedom as they themselves decide which of the available tools they use to process the information. The student can choose from a variety of tools, such as a note-taking tool, cloud storage, glossary of terms, pictures, videos, audio, etc.

In the *Reflection* stage, students share their knowledge with their peers. Knowledge sharing and discussing the processed information helps deepen, retain, systematize and expand knowledge.

In the *Aggregation of knowledge* stage, students use tools to create a concept map (for which they use the processed information). The aim of this stage is for students to revise the key terms they have learned in the current thematic unit and understand cross-curricular and interdisciplinary relations.

Each of the four stages may reveal different students' characteristics that may help determine how different groups work, how they will use the PLE, in which subjects they will use it and how often and when they will use it.

4   Differences Between Students

As far as the PLE and other ways of using information and communication technology are concerned, there are a number of differences between students (hence a large number of studies and researches). In the majority of cases, the main parameter for creating individual groups is:

- Age
- Socio-cultural or socio-economic environment
- Gender
- ICT type
- Student's learning style
- Grades

Groups created on the basis of the above criteria can differ in:

- Extent (the number of tools used in the PLE)
- Types of tools
- Organization of files
- Nature and length of notes
- Number of accesses
- Way of file sharing

Part of our research was dedicated to these and other PLE characteristics.

Secondary analysis of the results of international research activities [5] is aimed at describing the situation regarding this issue in the Czech Republic (CSI, 2016). The analysis results support the author's assumption that the academic society should take basic differences between students into account in their research activities. The aforementioned analysis looks at differences between students from multiple points of view. As far as the technical point of view is concerned, the results show that the students who use tablets and laptops for school-related purposes use these devices more often than those who use desktop computers. When comparing students from different schools, one needs to take the quality of the school into account. The indirect relationship between the quality of school and the time that the student spends on the Internet outside of school is worth mentioning, as is the relationship between the school's use of information and communication technology and its success rate. It can be deduced that students use their time meaningfully – they have mastered learning methods, methods of processing information and possess knowledge about information resources.

The aforementioned gender diversity in age-matched groups is often mentioned in studies aimed at all forms of education. From the psychological point of view, this is an obvious choice as it allows the researchers to learn more about the differences between male (boy) and female (girl) personalities. Those differences are especially evident in behavior, memory capabilities, motivation, learning, etc. All the aforementioned characteristics can be used in the creation and subsequent use of PLE in and outside of school. This is why the research part of this paper is aimed at this issue.

The comparison of previous researches with the contemporary ones conducted in a different socio-cultural environment than the Czech Republic shows that some of the gender differences have not changed in years. According to Shashaani's study [6], men use the computer more than women. At the time of the study, men considered the computer to be a useful tool.

A newer study [7] shows that – despite the greater availability of ICT and the Internet – the results have not changed much. According to the study results, boys use ICT more often than girls, but use it to play games and for entertainment. Girls, on the other hand, use technology to search for and process information and for communication. The results of this study were confirmed by the international ICILS study [8] – girls statistically outperformed boys by 18 points on average (the difference between Czech boys and girls was one of the lowest – 12 points on average). Thailand and Turkey were the only exceptions. Compared to the other participating nations, the Czech students achieved the best results.

The emergence of portable touch ICT (which are used ever more frequently for educational purposes) can influence the results of future studies. Further research will be aimed at determining whether or not this technology will deepen gender differences.

5   Research Part

The described analysis of different approaches of students toward PLE is part of research aimed at the development of key learning to learn competence through PLE. 83 of upper primary school students (7th and 8th grade students) participated in the research (Fig. 2).
The PLE was used in the following subjects:

- Mathematics
- Art education
- Informatics
- History
- Russian language
- English language
- Civic education

In these subjects, students accessed the PLE via laptops, tablets, desktop computers and cellular phones. The instruction was based on the E-R-R-A model. The experiment lasted one-half of the academic year.

The research part is aimed at analyzing the files which students saved into the PLE storage and the ways in which the individual PLE tools were used. The objective is to

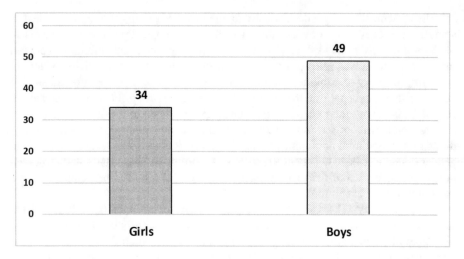

Fig. 2. Research sample

determine whether or not there are significant differences between boys and girls. The following are the results of the analysis, which was divided into 4 parts for better accuracy.

5.1 PLE Scope with Regard to the Number of Saved Files

The number of files saved in the PLE storage is one of the main parameters determining how frequently students use this type of environment. During the course of the experiment, 1,622 files of different data types were structured in the students' storage folders.

The entire number of files is divided into two almost identical parts, depending on whether they belong to boys or girls. The following table provides detailed information about the number of saved files by ownership (Table 1).

Table 1. Files in the PLE storage owned by boys and girls.

	Number of files	%	Average number of files per student
Girls (n = 34)	551	33.97%	16.21
Boys (n = 49)	1,071	66.03%	21.86
Total	1,622	100%	19.54

The table proves that – despite the even distribution of files – there is a significant difference between the average number of files per boy/girl. On average, boys had more files saved than girls. Therefore, it can be said that boys used the PLE more often than girls. The types and characteristics of saved files (see below) may confirm this hypothesis.

5.2   Types of Files Saved in the PLE Storage

The types of files used for working with the PLE will help determine whether or not the PLE tools are used evenly. Students' preferences may be influenced by a number of factors:

1. Preferred style and note editing may be influenced by the student's learning style,
2. Tool management,
3. Circumstances under which the note is taken.

The following are the file types the students used most often for storing their notes:

- Text files (*.doc, *.docx, *.pdf)
- Images, drawings and pictures (*.jpg, *.png)
- Audiovisual files (*.mp3, *.mp4, *.wav, *.avi)
- Concept maps
- Simple text notes

During the course of the experiment, the aforementioned file types were not used evenly. The following graph shows some of the file types as used by boys and girls (Fig. 3).

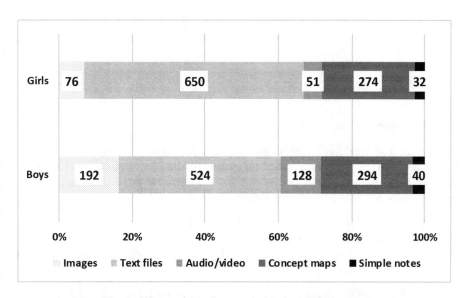

Fig. 3. Files saved by boys and girls in the PLE storage

In both groups, text files formed the largest share of saved files. Boys also saved images, drawings, schemes, pictures and audiovisual materials. The audiovisual materials were either created by the students themselves (using their cellular phones or tablets) or found on the Internet. Girls mostly saved texts with no images or other additional material. However, girls' texts were better structured than boys', and the key terms were highlighted. The number of concept maps and short text notes saved was similar.

5.3   Organization of Files

The organization of saved files plays an important role in being able to find them in the future. In the PLE storage, the files were organized in a directory structure. The analysis of directory structures showed the following:

1. All students (boys and girls) divided their storage space into directories (subjects in which experimental instruction took place). Files unrelated to those subjects, files related to more than one subject, and short-term files (e.g. homework assignments) were left out of the directories.
2. While the girls maintained the organization in the folders within the Level 1 directories (see above), the boys did not create any new folders within the Level 1 directories. Therefore, the content of the folder was not further divided. The author assumes that this type of organization is suitable for short-term use (and a relatively low number of files) as it makes the files easier to access. The question is whether or not the students would have used this type of organization if the number of files had been higher (Table 2).

Table 2. Number of created folders.

	Number of folders	Average number of files per student
Girls (n = 34)	233	6.85
Boys (n = 49)	194	3.96
Total	427	5.14

5.4   PLE Access Frequency

Monitoring user (students') logins helps determine a number of important characteristics which enable quantification of PLE-related work in terms of time. Of all online data, we have decided to include the number of accesses to the PLE. Apart from the acquired data, the following data could also be retrieved:

1. How long the student was logged in and used the PLE,
2. Most common time for users to login,
3. From where users logged in (i.e. from inside or outside of school) – this is important for determining whether or not students use the PLE in their leisure time.

Since the research was not primarily aimed at the above three points, they are yet to be processed. However, they will be addressed soon, and the results will be published.

At the conclusion of the experiment, 22,927 PLE accesses were recorded, which corresponds to approximately 312 logins per student. The boys were more active than the girls. The data is presented in the following table (Table 3).

Table 3. Number of PLE accesses during the course of the experiment.

	Number of login	Percentage	Average number of files per student
Girls (n = 34)	9,889	38.14%	290.85
Boys (n = 49)	16,038	61.86%	327.31
Total	25,924	100%	312.37

6 Interpretation of Results

The described part of our research shows that boys and girls have different approaches. However, none of the monitored characteristics confirms that one group uses the PLE more effectively than the other. Both groups (boys and girls) used the PLE in a specific manner.

As far as the number and nature of the saved files are concerned, the boys used more file formats than the girls. However, the girls' storage space was better organized and included mostly text notes (with the key terms being highlighted). There were a number of complementary files the majority of which had been downloaded from the Internet. Less variety of files and the number of downloaded notes proves that the girls did not exploit all the PLE possibilities. Therefore, it can be assumed that the less active approach of the girls may have been caused by their generally more conservative attitude toward information and communication technology. The author assumes that if the girls could have spent more time with the PLE, the differences would have been far less significant.

7 Conclusion

The paper presents part of the research aimed at the development of the key learning to learn competence through the Personal Learning Environment. During the course of the six-month long, PLE-based instruction, boys used the PLE (especially the electronic storage, which is an integral part of the PLE) in a different manner than girls. Identification of differences can help improve our software by making it more individually adjustable. That is why the E-R-R-A model used in the PLE-based instruction may be modified in the future. The results of our research can be used by other researchers, teachers and software developers who are interested in integrating the PLE into instruction or designing similar software solutions.

References

Haworth, R.: Personal learning environments: a solution for self-directed learners. TechTrends. **60**, 359–364 (2016). http://link.springer.com/10.1007/s11528-016-0074-z

Downes, S.: Elements of a Personal Learning Environment. Hämeenlina. http://www.slideshare.net/Downes/design-elements-in-a-personal-learning-environment

Steel, J.: Handbook I.: What is Critical Thinking, 2nd edn. Critical Thinking, Prague (2007)

Javorcik, T., Homanova, Z.: Proposition of model for use of personal learning environment in primary schools and possibilities of knowledge sharing. In: 3th IEEE International Conference on Emerging eLearning Technologies and Applications, Proceedings. IEEE Press, New York (2015)

Students and ICT: Secondary analysis of the results of the ICILS 2013 and PISA 2012 international studies. http://www.csicr.cz/html/SA_icils_pisa/html5/index.html?&locale=CSY&pn=10

Shashaani, L.: Gender differences in computer attitudes and use among college students. J. Educ. Comput. Res. **16**, 37–51 (1997)

İlban, M., Yildirim, B.: Students' preferences on computer using and a survey on variables impacting the preferences. Turkish Online J. Educ. Technol.(TOJET). **5**, 34–40 (2006)

Fraillon, J., Ainley, J., Schulz, W., Friedman, T., Gebhardt, E.: Preparing for Life in a Digital Age: The IEA International Computer and Information Literacy Study International Report. Springer International, Switzerland (2014)

Pervasive Information Systems

Advanced Open IoT Platform for Prevention and Early Detection of Forest Fires

Ivelin Andreev[✉]

Interconsult Bulgaria Ltd., Sofia, Bulgaria
ivelin.andreev@icb.bg

Abstract. The primary goal of the proposed architecture is to develop advanced architecture for early detection of forest fires that integrates sensor networks and mobile (drone) technologies for data collection and processing. Unmanned air vehicles (UAVs) will allow coverage of larger areas to raise the percentage of forest fires detections, monitor areas with high fire weather index and such already affected by forest fires. All information is forwarded and stored in cloud computing platform where near real-time processing and alerting is performed.

Keywords: Open platform · Edge gateway · Cognitive computing · Forest fire

1 Introduction

Forest fires, are uncontrolled accidents occurring in wild areas and causing considerable damages to natural and human resources. Unintentional events such as disposed cigarettes, short circuits, explosions and high temperatures can induce fires leading to a disaster. Therefore, early detection, prevention and timely dealing with forest fires becomes essential and drive the need for IoT platform tailored with usage scenarios in mind from which monitoring offices and crisis management centers would benefit.

The primary goal of the architecture is to enable flow of data from intermittent line-of-business devices. Video and sensor information from drones, on-site thermal cameras and ambient sensors are just some of the data sources supported by the platform. Once ingested by the cloud-based back-end system, data will be used for performing analysis, control and process automation. Most important, alarms and readings could be transferred from the field in fast and reliable manner and without human intervention.

The state of the art characteristics of the platform are mainly the utilization of open source and free subscription components to achieve proven performance, flexibility and extensibility. In addition the platform utilizes the benefits from cutting edge technologies such as artificial intelligence, cognitive algorithms and UAVs. The architecture provides sufficient level of resilience features to allow deployment in either public cloud, on-premises or a hybrid environment.

© Springer International Publishing AG, part of Springer Nature 2018
Á. Rocha et al. (Eds.): WorldCIST'18 2018, AISC 747, pp. 319–329, 2018.
https://doi.org/10.1007/978-3-319-77700-9_32

2   IoT Platform Architecture

The IoT cloud portal architecture is neutral, with regards to the use case scenario with rich customization capabilities in mind on each level. Its module architecture allows it to fit multiple scenarios and undergo customizations by requiring only a minimal set of modules to be engineered or refactored to fit from full on-premises, through hybrid to full in-cloud deployment. The main modules that determine platform flexibility are: *Adapters* to connect new devices and systems as data sources, using large number of protocols; *Analysers* to perform generic or specific processing of raw data; *Visual instruments* to present data in intuitive and easy to manage and perceive form; *Machine learning models* to provide domain knowledge for specific use case (Fig. 1).

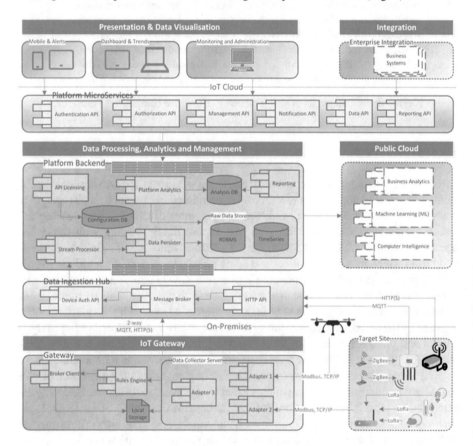

Fig. 1. Conceptual architecture at cloud level

Gateway scope is situated on-premises, physically close and communicating with the data source devices. These could be: intelligent sensors, field gateways that collect data directly from sensors, programmable logic controllers, etc. The main purpose of the on-premises setup is to collect, aggregate, pre-process and buffer raw data on-site and subsequently forward data in a reliable and secure manner to the cloud platform core.

Cloud scope encompasses the core of the platform. It ingests data securely via few protocols over the Internet, stores raw data, analyses and prepares reporting data sets. In addition, the platform core is also responsible for establishing secure communication with the source devices and licensing the usage of the service layer. At the same time it relies on external services for more advanced functions like computer intelligence, machine learning prediction and business intelligence visualization.

Public services scope refers mainly to advanced services offered by some of the major cloud providers. Development and maintenance of such services typically require vast amount of knowledge and experience and these are to be consumed on Software as a Service basis for specific needs like computer intelligence and business analytics.

Consumer scope is the layer at which end-user applications are. These consume data and perform actions on the cloud platform by using the cloud service layer. These could be mobile applications, dashboard applications for real-time monitoring, config-uration application or even 3-rd party systems that would consume raw or processed data. In case a system would supply data to the cloud platform, it would have to be connected via the ingestion hub. The main components of the architecture, their desig-nation and challenges they solve are explained below.

3   IoT Gateway and Data Ingestion Hub

A field gateway is a specialized device-appliance or general-purpose software that acts as a communication enabler, a local device control system and device data processing hub. It can perform local processing and control functions toward the devices. On the other side, it can filter or aggregate device telemetry and reduce the amount of data transferred to the cloud. [1] A cloud gateway is the part of the cloud-based architecture that enables remote communication to and from devices or field gateways, which poten-tially reside at several different sites. Both devices and field gateways may implement edge intelligence and analytics capabilities. This enables aggregation and reduction of raw telemetry data before transport to the backend and local decision-making capabil-ities (Fig. 2).

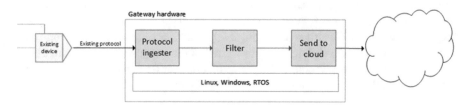

Fig. 2. IoT Edge data flow

The technology behind the gateway is Azure IoT Edge (previously known as Azure IoT Gateway SDK) which allows building modules that can be composed with great flexibility and serve the needs of the exact scenario. It is both a set of software components and architectural pattern that has no dependency on particular cloud service. IoT Edge gateway is a software gateway that can be deployed on a broad range of operating systems (OS) and respectively a wide range of hardware devices, supporting the OS mentioned [2]. There is a very large number of communication protocols available for device scenarios nowadays and the number is rapidly growing. IoT gateway ingests data by using a dedicated adapter-loader module which allows a specific (client or server) protocol endpoint to be published. Extension and support could be performed by implementing IoT Edge protocol adapters.

The platform data ingestion hub takes the main role in connecting IoT devices directly or via the cloud gateway to the platform core. The hub is capable to connect hundreds of thousands devices, provide 2-way communication for receiving telemetry and sending back commands to be executed on the edge. It also includes a built-in device identity store that stores a registry of device identities and manages security credentials.

The gateway supports 2-way communication with the cloud platform for interchanging data and configuration. The communication relies on a lightweight MQTT client module which utilizes the benefits of the current most preferred IoT related protocol. MQTT is publish-subscribe messaging protocol specially designed for constrained low-bandwidth, high-latency or unreliable networks. It provides sufficient multi-layer safety features [3]: Network security – using VPN secure connection as a foundation between broker and clients; Transport security – SSL/TLS is used for transport level encryption and confidentiality; Application security – client ID and user-password pair and x509 certificates for enhanced authentication. MQTT is often called "the messaging protocol for Internet of Things" [4] and covers out of the box a multitude of failure scenarios to assure reliable communication [5].

After in-depth comparison of capabilities and requirements, the platform description suggests Mosquitto MQTT as among the most suitable. Mosquitto MQTT is an open source software message broker, offered under Eclipse Public License (EPL) that implements MQTT protocol v.3.1 and v.3.1.1. [6] It provides a lightweight server implementation of the MQTT protocol that is suitable for all situations from full power machines to embedded and low power machines.

Message broker main designation is to facilitate the information exchange between the field and the cloud platform, via kind of messaging protocol. Multiple broker implementations exist and it does not appear necessary to develop an alternative solution. Brokers also have bridge capabilities, which allow them to connect to other MQTT servers, including, but not limited to, other broker instances. This allows networks of MQTT servers to be constructed, passing MQTT messages from any location in the network to any other, depending on configuration of the bridges. [7] The minimal hardware requirements make Mosquitto MQTT an ideal choice for deployment on IoT Cloud gateway for bridging and implementing some application level features.

4   Cloud Platform Core

The main responsibilities for the cloud platform core are to collect data in efficient manner and perform pre-processing and analytics.

Stream processor module designation is to perform some basic real-time data analytics after data ingression through the cloud gateway. Stream processor is intended to support multiple data streams concurrent and modify or determine the path of data. Typical analysis tasks in the scope of the stream processor are: detecting threshold limits and anomalies, or generating alerts.

Data persister is a data collector service with a built-in message broker client (MQTT) that receives raw data in the form of MQTT messages from the ingestion hub broker. The data collector provides an abstraction on the top of a set of supported database management systems to store and retrieve raw time series data.

Raw data (measurements sourced from devices via the ingestion hub) are stored in either relational DB management system or in a time series database. The supported engines in mind are: Microsoft SQL Server 2016 with in-memory OLTP, OSI Soft PI DB Time Series, Influx DB Time Series, MySQL Server 5.6 and later.

Time series data storage and management is one of the most critical elements of every IoT architecture. This task is typically solved by using Time Series (TS) systems that are optimized especially for handling such data types. Points are always tracked and aggregated over time with the key difference to relational data being that writes are more than 95% of operations, data are rarely updated and reads are typically sequential [9]. Some TS DBs support only measurements gathered from sensors at regular intervals while Influx is optimized for also irregular measures, which is a significant difference from other solutions like Graphite, RRD or OpenTSDB. Time series databases typically over perform relational DB systems in a number of scenarios: high I/O rate, number of tags, data volume, aggregation, continuous queries and data compression [8].

InfluxDB is maintained by a single commercial company following the open-core model, offering premium features. Influx DB is the top ranked time series DB as of the score of DB-engines initiative. From the ranking chart it is visible that InfluxDB is three times more popular than the second competitor. The key features making Influx DB so popular are [9]: easy setup, no external dependencies, implemented in Go; comprehensive documentation; scalable, highly efficient; REST API; support client libraries for .NET, Go, Java, JavaScript, Node.js, Perl, PHP, Python, R, Ruby; SQL-like syntax; on-premises and cloud offering.

Analyzers are a key element of the cloud platform flexibility and extensibility. Platform analytics are implemented in the scope of analysis service. The service allows analyser implementations to be plugged in a flexible manner only by configuration. Each analyser could be either generic and processing all data of certain origin or specific with programming logic tailored for a given use case. Analyzers typically operate in asynchronous manner on the raw persisted data with the outcome of their calculation being persisted for performance reasons. Analyzers are also intended to communicate with external services to enrich the platform with business intelligence capabilities, cognitive computing algorithm processing and machine learning for prediction. During prototyping phase Cortana Suite with Cognitive Services and Azure Machine Learning have been evaluated with very promising results.

5   Public Cloud

A set of services are intended to be used from public cloud vendors under SaaS. The motivation behind is mainly towards algorithms that require vast amount of knowledge and experience in science branches that are not in scope of the present platform.

Computer Intelligence is a set of nature-inspired computational methodologies and approaches to address complex real-world problems to which traditional programming approaches could not be as useful as expected [10]. There could be multiple underlying reasons: too complex process for formal description and execution, multiple uncertainties in the process; the process may have stochastic nature. One good candidate for the particular platform needs is the use of computer vision for detection of forest fires from images without human interaction. Computer Vision as a concept provides state-of-the-art algorithms to process images and return information. It is capable of analysing visual content and tagging it with the most relevant terms. Its algorithms output a number of tags based on the objects, living beings and actions identified in the image [11]. Tagging is not limited to the main subject, such as a person in the foreground, but also includes the setting (indoor or outdoor), furniture, tools, plants, animals, accessories, gadgets etc.

Machine learning (ML) originates from pattern recognition and from the theory that computers can learn without being programmed to perform specific tasks and that this learning could happen only on the basis of data. Machine learning also implies an iterative aspect and independent adaptation when models are exposed to new pieces of data. Although the concept is not new, the ability to automatically apply complex mathematical calculations to big data – over and over, faster and faster – is a recent trend. Machine learning shall not be at any time considered as a magic black box but rather than this, its relation with real life shall be understood together with its deep roots in data science [12]. The decision on the type of ML algorithm to be most appropriate depends on multiple factors like size, quality and nature of data. There could as well be multiple appropriate algorithms for the same type of tasks, but one could fit better than another. There are two major types of learning: supervised learning where known training data set is used to make predictions and unsupervised learning which identifies structure in the data given (cluster analysis). A high level comparison of the features of machine learning services in some of top vendors is available in Table 1 below.

For the particular predictive analytics nature of the cloud platform, an algorithm from the binary classification family shall be considered.

Business analytics solution building blocks are: datasets, dashboards and reports. Datasets are imported or connected and are the basis for subsequent interaction and presentation. Reports could be either predefined or presented by self-service capabilities. Power BI has been found appropriate for the following reasons [13]: consumes data from 60+ solutions in a centralized dashboard; software as a Service offering with pay-as-you-go; robust access control and security; best in class and continually updated dashboard visualizations; Suitable for analysis on data sets of up to 1 M rows.

Service level consists of multiple individually scalable micro- services instead of one monolithic and general service. The designation of these is to present data to the presentation layer of applications where mobile applications, dashboards and management portals could be developed. The service layer, on the other side, is also used to

Table 1. Comparison of the top machine-learning vendors

Characteristic	Azure ML	BigML	Amazon ML	Google prediction	IBM Watson ML
Flexibility	High	High	Low	Low	Low
Usability	High	Med	High	Low	High
Training time	Low	Low	High	Med	High
Accuracy	High	High	High	Med	High
Cloud/On-premises	±	+/+	±	±	±
Algorithms	Classification Regression Clustering Anomaly Recommend	Classification	Classification	Classification Regression	Semantic mining Hypothesis rank Regression
Customization	Parameters R-script Evaluation support	Own models C#, R, Node.js	Regression	N/A	N/A

provide data to 3rd party systems that are interested in the data collected, managed and analysed by the IoT cloud platform like regional, national, international crisis management centres and agencies. The data could be mapped contextually with other data which will allow post-processing and definition of new services for the end-users. Finally, creation of open virtual platform allows data integration from multiple other platforms and better processing and analysis in the future.

6 Validation

A substantial number of experiments were performed to verify the efficiency of individual components, the vitality of the overall concept and its suitability for the particular use cases. A special attention was paid to the data storage, computer intelligence and the data ingestion capabilities.

The research team identified the data storage component to be the most critical part. With the key limitations to support on-premises setup and reasonable popularity among developers the engines compared were from different categories. For benchmarking there was built a dataset of metrics reported by machines and servers. The largest exceeded 300 million data points. The first benchmarking was performed comparing the latest engine versions: Influx DB v1.3 and SQL Server 2016 Standard. Identical hardware and software setups were used (Windows 10 Ent. x64, Intel i7-2600 K 3.40 GHz quad, 16 GB, SSD Samsung 850 EVO 250 GB, HDD 7200 RPM).

Database benchmarking started with SQL server. As illustrated in the Figs. 3 and 4, Influx DB over performs 40 times SQL Server 2016 with Influx DB being able to operate at about 200,000 writes per second and SQL just 5,000. SQL Server 2016 implements in-memory OLTP technology that significantly improves transaction processing and data ingestions. Enabling it with delayed write to disk improves speed with 45% to about 7,800 writes/sec but this still leaves SQL Server far behind.

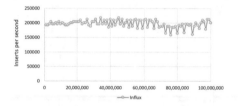 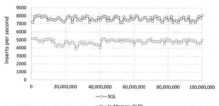

Fig. 3. Influx DB write performance   **Fig. 4.** SQL Server 2016 write performance

Storage space required is another important parameter which illustrates the data compression capabilities and determine the number of required disk I/O operations. The particular test was performed using a subset of 38 million points which proves to be sufficient to provide credibility of the overall tendency. As identified during benchmarking, Influx DB is capable to achieve 26 times better compression (98 MB vs 2580 MB) for the particular scenario. SQL Server provides standard compression mechanisms on both page and row level. However, enabling them improved the storage requirements with just 27% to 1884 MB. Another major characteristic of database engines is the capability to respond to queries. To benchmark this a query that aggregates data chunks of 10000 points was used with the result averaged over 1000 executions. The test was executed in parallel from 2, 4 and 8 threads to measure the abilities for parallel processing. The overall conclusion is that SQL Server mean query response time is 78 ms (nearly 13 queries/sec), while Influx DB can achieve 60 times better throughput of 769 queries/sec and response time of 1.3 ms.

All tests were performed on SSD as well as on HDD and another important finding is that while SSD is officially recommended for Influx DB and affects its performance dramatically, SQL Server was not affected in such extent due to its policy of caching. The findings from other tests for ingesting, managing and storing time series data provide similar confidence in the advantages of Influx DB and are summarized in Table 2 below [14–18]:

Table 2. Influx DB performance for time series data compared

Influx DB	Mongo DB	Elasticsearch	Cassandra	OpenTSDB
Write (val/sec)	27× Slower	8× Slower	5.3× Slower	5× Slower
Query (qry/sec)	Equal	7.5× Slower	168× Slower	4× Slower
Storage (GB)	84× Larger	4× Larger	9.3× Larger	16.5 Larger

MQTT broker, being the heart of the data ingestion layer, was benchmarked using an open source MQTT stress tool. Scalability of the test setup was evaluated by incrementally increasing the number of publishers with the goal to reach maximum number of publishers, rather than message throughput. During the stress test the broker was easily capable to ingest the target requests from 30'000 publishers simultaneously. It was also identified that transmission latency appears somewhere above 10'000 messages.

Cognitive algorithms are another critical concept to be verified for potential in the target scenario to replace the necessity of human operators and expensive specialized hardware. Experiments show with high success rate the capabilities of modern computer vision engines and prove that artificial intelligence is a promising candidate in raising reliable warnings from just processing an image. A test on both positive samples (visually clear that a fire is burning, Fig. 5) and negative samples (clouds or sunset colours mimic the appearance of a fire, Fig. 6) was performed. The response is provided in machine-readable JSON for both a positive and a negative sample. Each tag is given a corresponding confidence score in the range (0:1) with 1 being the highest degree of certainty for relevance. The tested AI engine was capable to successfully tag all of the provided images with the tags of highest importance being: Smoke – the higher is the rating, the greater is the likelihood for a fire been captured; Clouds, Steam – often fire or smoke could be visually confused with clouds or steam.

{"name":"sky", "confidence":0.99},
{"name":"outdoor", "confidence":0.99},
{"name":"grass", "confidence":0.98},
{"name":"mountain","confidence":0.77},
{"name":"smoke", "confidence":0.54},
{"name":"clouds", "confidence":0.23}

{"name":"outdoor", "confidence":0.96},
{"name":"clouds", "confidence":0.81},
{"name":"nature", "confidence":0.71},
{"name":"cloudy", "confidence":0.59},
{"name":"dark", "confidence":0.48},
{"name":"day", "confidence":0.11}

Fig. 5. Fire positive sample and response   **Fig. 6.** Fire negative sample and response

Machine learning predictive experiment was created in addition to computer vision on the basis of test dataset from more than 500 real forest fires. The dataset was normalized and the experiment was created using multiple binary classification modules to compare their performance and select the best candidate. The performance was evaluated based on area under curve (AUC) using two-class locally deep SVM, two-class boosted decision tree, two-class neural network, two-class SVM and two-class decision forest. The parameters of the learner were additionally tuned to improve the performance of the predictor. The best result of 74% was achieved by utilizing boosted-decision tree. Although the result could be considered generally good, it must be pointed out that there is a strong subjective human factor which is involved in a great number of forest fires.

However, it is evident that further research is necessary for extending the used set of features and measurements, but this is out of scope of the present work.

7   Conclusions and Further Development

Unintentional factors such as lighted cigarettes, short circuits, explosions, elevated temperatures or storms can cause fires leading to disasters with severe impact on social, business and environmental areas. Sensors integrated into structures, machinery, and the environment, coupled with the efficient delivery of sensed information, could provide tremendous benefits to society. Forest fire prevention is based mainly on risk analysis, weather forecast and sensors deployed in the forest. Whenever the risk is high drones could be applied for faster fire confirmation.

The present paper proposes an advanced, extensible and open IoT platform to address exactly the challenges faced by crisis management centres but at the same time allows deployment for wide range of application areas. The crucial elements in the approach are flexible and reliable technologies, the utilization of advanced machine learning and cognitive algorithms and the ability to extend sourcing of input to modern IoT devices such as drones. The key platform elements and their integration have been experimented and proven to correctly operate as an integrated solution at TRL (Technology Readiness Level) 5 with its modules having technology maturity from TRL 5 to TRL 9.

Acknowledgements. A major part of the research work is performed in the scope of Horizon 2020 project Advanced Systems for Prevention and Early Detection of Forest Fires (ASPires), funded by European Civil Protection and Humanitarian Aid Operations 2016/PREV/03 (ASPIRES).

References

1. Dotchkoff, K.: Microsoft Azure IoT services ref. architecture, pp. 13–29, October 2015
2. Azure IoT Edge on GitHub, August 2017. https://github.com/Azure/iot-edge
3. Introducing the MQTT Security Fundamentals, December 2016. http://www.hivemq.com/blog/introducing-the-mqtt-security-fundamentals
4. Five Things to Know About MQTT – The Protocol for Internet of Things, September 2014. https://www.ibm.com/developerworks/community/blogs/5things/entry/5_things_to_know_about_mqtt_the_protocol_for_internet_of_things
5. Are your MQTT applications resilient enough?, May 2016. http://www.hivemq.com/blog/are-your-mqtt-applications-resilient-enough/
6. Mosquitto MQTT Broker Home Page, July 2017. https://mosquitto.org/
7. Eclipse Mosquitto, May 2017. https://projects.eclipse.org/projects/technology.mosquitto
8. Schwartz, B.: TS Database Requirements, June 2014. https://www.xaprb.com/blog/2014/06/08/time-series-database-requirements/
9. InfluxDB Properties, August 2017. https://db-engines.com/en/system/InfluxDB
10. Siddique, N., Adeli, H.: Computational Intelligence, Synergies of Fuzzy Logic, Neural Networks and Evolutionary Computing. Wiley, Chichester (2013)

11. Computer Vision API Version 1.0, August 2017. https://docs.microsoft.com/en-us/azure/cognitive-services/computer-vision/home
12. Evolution of machine learning, July 2017. https://www.sas.com/en_us/insights/analytics/machine-learning.html
13. Any data, anywhere, any time, July 2017. https://powerbi.microsoft.com/en-us/features/
14. InfluxDB Markedly Outperforms Elasticsearch in Time Series Data & Metrics Benchmark, May 2016. https://www.influxdata.com/influxdb-markedly-elasticsearch-in-time-series-data-metrics-benchmark/
15. InfluxData: Benchmarking InfluxDB vs MongoDB for Time-Series Data, Metrics & Management, p. 12, February 2017
16. InfluxData: Benchmarking InfluxDB vs Cassandra for Time-Series Data, Metrics & Management, p. 16, September 2016
17. InfluxData: Benchmarking InfluxDB vs OpenTSDB for Time-Series Data, Metrics & Management, p. 14, November 2016

On Green Scheduling for Desktop Grids

Thanasis Loukopoulos[1]([✉]), Maria G. Koziri[2],
Kostas Kolomvatsos[2], and Panagiotis Oikonomou[2]

[1] Computer Science and Biomedical Informatics Department, University of Thessaly,
2-4 Papasiopoulou st., 35100 Lamia, Greece
luke@dib.uth.gr

[2] Computer Science Department, University of Thessaly, 2-4 Papasiopoulou st.,
35100 Lamia, Greece
{mkoziri,kolomvatsos,paikonom}@uth.gr

Abstract. Task scheduling is of paramount importance in a desktop grid environment. Earlier works in the area focused on issues such as: meeting task deadlines, minimizing make-span, monitoring and checkpointing for progress, malicious or erroneous peer discovery and fault tolerance using task replication. More recently energy consumption has been studied from the standpoint of judiciously replicating and assigning tasks to the more power efficient peers. In this paper we tackle another aspect of power efficiency with regards to scheduling, namely greenness of the consumed energy. We give a formulation as a multi-objective optimization problem and propose heuristics to solve it. All the heuristics are evaluated via simulation experiments and conclusions on their merits are drawn.

Keywords: Scheduling · Green computing · Desktop grids
Volunteer computing

1 Introduction

A desktop grid is comprised of volunteers sharing their computational resources when in idle mode, usually common PCs, in order to jointly accomplish some computationally intensive job. Due to their scalability, desktop grids are used in a plethora of scientific projects, e.g., [1, 2] as an alternative to parallel and cluster computing, but also have the potential of tackling non scientific, yet time consuming tasks, e.g., video coding, with minimal financial cost [3] to job issuers.

Realizing the scalability potential of desktop grids involves solving a number of problems, such as interoperability, splitting and scheduling tasks, monitoring and checkpointing. Here, we focus on the scheduling problem in desktop grids, the general statement of which is as follows: *given a set of volunteer peers with their characteristics and a set of tasks to be executed, allocate tasks to peers so that some performance metric, usually related to the total make-span, is optimized.* The problem (see related work) has attracted significant interest during recent years with researchers tackling various statements of it, differing both in the environment assumptions and the performance parameters considered.

© Springer International Publishing AG, part of Springer Nature 2018
Á. Rocha et al. (Eds.): WorldCIST'18 2018, AISC 747, pp. 330–340, 2018.
https://doi.org/10.1007/978-3-319-77700-9_33

Furthermore, a rising interest exists on energy efficient computation spanning areas varying from chip design [4] to the data center level [5]. Naturally, energy efficiency has also attracted the interest of research on desktop grids with most works focusing on reducing energy consumption. Nevertheless, only reducing the total energy consumed by peers doesn't suffice as a greenness criterion since it doesn't always translate directly to carbon dioxide emissions, e.g., the power source of a high energy consuming peer might be a solar or wind plant.

In this paper we tackle the problem of scheduling tasks in desktop grids with regards to the cleanness of the energy sources used by peers. We focus on static scheduling of independent tasks. Our contributions include the following: (a) we model the problem as a two function constrained optimization problem with the first target being the cleanness of the used energy and the second one fault tolerance and with task deadlines considered as constraints; (b) we present heuristics for each of the optimization targets that combined together produce promising results; (c) we evaluate the heuristics experimentally and conclude upon their merits.

The rest of the paper is organized as follows. Section 2 presents the related work. Section 3 gives the problem formulation. Section 4 illustrates the heuristics, which are experimentally evaluated in Sect. 5. Finally, Sect. 6 concludes the paper.

2   Related Work

Cycle stealing systems that harness the idle cycles of desktop PCs and workstations enjoy large success with popular projects such as SETI@home [2] sustaining the throughput of over one million CPU's and 100's of Teraflops/second [6].

Concerning the necessary mechanisms for task-peer assignment, researchers in [7] studied dynamic component deactivation. In [8] the authors presented a framework based on three components: an on/off model using an energy-aware resource infrastructure, a resource management system adapted for energy efficiency, and a trust delegation component to assume network presence of sleeping nodes. An energy-aware framework to manage different resources on Grid, Cloud and dedicated networks was proposed in [9]. The generic framework includes scheduling algorithms to optimize reservation placements, algorithms to switch resources into sleep mode, prediction algorithms to anticipate workload and workload aggregation policies to avoid frequent on/off cycles for the resources. A survey on reservation schemes for computational grids can be found in [10].

In [11] the authors investigated whether grid volunteer systems have an overall green advantage over service grids and data centers. They discussed several approaches that can be used by green volunteer systems, including the use of sleeping states. An energy-aware approach for opportunistic grids based on virtualization and consolidation was presented in [12]. The authors proposed to consolidate virtual machines into the resources already executing process of physical users. In [13] sleeping and wake-up strategies in opportunistic grids were studied. Sleeping strategies were employed to reduce the energy consumption of the Grid during idle periods, while energy-aware scheduling was used for allocating tasks to the available machines. In [14], the same authors extended their work by analyzing the break-even time and the maximum potential of sleeping states to save energy in P2P grids.

In [15] a cooperative game theoretical methodology was used to jointly optimize energy consumption and response time in computational grids. In [16] the problem of independent batch-scheduling in grids to optimize makespan and energy consumption was addressed. The problem was formulated as a two-objective global minimization problem. The authors proposed an evolutionary-based scheduling algorithm with dynamic voltage and frequency scaling for the management of the cumulative energy utilized by grid resources. In [17] the problem of scheduling tasks on grid systems was investigated with the aim of minimizing both the schedule length and energy consumption subject to tasks' memory requirements and deadline constraints. A set of eight heuristics was introduced, including two evolutionary algorithms and six greedy list-based scheduling algorithms. Scheduling both task execution and the related data transfers was the aim of [18] for independent tasks, while task dependencies were considered under the concept of operator placement in [19].

Replication has also attracted research interest. A comparative study of power aware non preemptive scheduling algorithms with replication was done in [20]. Fault tolerance issues were discussed in e.g., [21, 22], while in [14] the authors developed methods to give probabilistic guarantees on result correctness using a peer credibility metric. Some of these methods have been deployed in existing projects. For instance, in BOINC [23] tasks are recomputed to minimize false positives.

Our approach differs from the above works primarily with regards to the optimization target, since the aim (among others) is to favor green peers that use energy from renewable sources.

3   Problem Formulation

Assume N tasks denoted by T_i, with $1 \leq i \leq N$. Let L_i denote the load for T_i. We assume L_is are known (or can be estimated) and express the amount of time required to complete T_i on a base machine. Let there be M peers, whereby P_j denotes the jth peer assuming a total ordering of them. Each peer is associated with a speedup factor SP_j measured as the ratio of its processing power to the processing power of the base machine used to estimate L_is.

Depending on the country it resides, each peer is assigned a greenness value $g(P_j)$ that corresponds to the percentage of energy in this country coming from renewable sources. We also denote with E_j the energy consumed by each peer when working close to 100% utilization, i.e., when achieving speedup of SP_j.

Let X be an $N \times M$ matrix encoding the assignment of tasks to peers as follows: $X_{ij} = 1$ iff T_i is allocated to P_j, 0 otherwise. Each task must be allocated in at least one peer and replication of tasks is allowed, presumably for purposes of fault tolerance. In the paper we assume that all tasks have the same maximum degree of replication (let r), which is a common approach in many systems [6, 23]. In other words, for a placement to be valid, the following must hold:

$$\sum_{j=1}^{M} X_{ij} \leq r, \forall 1 \leq i \leq N \tag{1}$$

In case where $rN > M$, each peer must execute more than one tasks. This might also happen as a result of favoring powerful green peers in task assignment. We assume that the tasks assigned on the same peer arrive sequentially, i.e., the peer receives the second task once it has finished the first one. We encode such ordering in the following way. We assume that the maximum number of tasks assignable to a peer, may never exceed $O = c\lceil rN/M \rceil$ for some constant c. Notice that an upper bound exists on O namely, $O \leq N$, meaning that in the worst case a replica of every task will be assigned to a peer. Having defined the maximum number of tasks per peer in the above way, we can extend the matrix X to encode both task-peer assignment and the order with which it is performed as follows: X is now an $N \times M \times O$ matrix, whereby $X_{ijk} = 1$, iff T_i is assigned at P_j as its kth task to be executed. Clearly, the following must hold for a placement to be valid:

$$\sum_{j=1}^{M} \sum_{k=1}^{O} X_{ijk} \leq r, \forall 1 \leq i \leq N \tag{1'}$$

$$\sum_{k=1}^{O} X_{ijk} \leq 1, \forall 1 \leq i \leq N, \forall 1 \leq j \leq M \tag{2}$$

Equation 1' replaces Eq. 1 in counting the number of replicas per task. Equation 2 says that a task cannot be assigned to a particular peer more than once (recall that the values on the X matrix are binary).

Peers are not always available for task execution. Let $p(P_j)$ be the percentage of time the peer P_j is available over a time period TP. This information can be gathered by the monitoring tool of the desktop grid. In case of a first comer, the mean value of all peers for which historical information exists, can be used as an estimate.

A peer P_j that commences execution of a task T_i, will finish it in time equal to: L_i/SP_j, provided that the execution will not be interrupted. Assuming P_j has other tasks to execute as well and by taking into account the fact that only a fraction of time is devoted by the peer to task execution, we can calculate the completion time of a task T_i assigned in P_j as its kth such task using the following:

$$C_{ijk} = \frac{\sum_{x=1}^{k} \sum_{y=1}^{N} X_{yjx} L_y}{p(P_j) SP_j}, \forall 1 \leq i \leq N, \forall 1 \leq j \leq M, \forall 1 \leq k \leq O \tag{3}$$

By taking into account the fact that some portion of the consumed energy comes from renewable sources, the environmental unfriendly energy consumed by P_j for executing all the assigned tasks can be given by:

$$G_j = \left(1 - g(P_j)\right) E_j \frac{\sum_{k=1}^{O} \sum_{i=1}^{N} X_{ijk} L_i}{SP_j} \tag{4}$$

Furthermore, we assume that each peer has a failure probability of $f(P_j)$, which is the probability of producing an erroneous result. Such probabilities can be calculated by the monitoring/checkpointing mechanism. Given a placement matrix X, the probability that the computation of T_i fails in all the peers it is assigned to, is given by:

$$F_i = \prod_{j=1}^{M} (f(P_j) \sum_{k=1}^{O} X_{ijk} + (O - \sum_{k=1}^{O} (1 - X_{ijk}))) \qquad (5)$$

which means that in case T_i is allocated at P_j the failure probability is taken into account, otherwise the product remains unaffected (a value of 1 is used).

Let each task T_i have a deadline D_i within which it should be completed. The greenness-aware scheduling problem with deadlines, replication and fault tolerance can be stated as a two function optimization problem as follows: Given L_is, SP_js, $p(P_j)$s, $f(P_j)$s, O and r, find an X matrix such that the following two functions are minimized:

$$f_1 = \sum_{j=1}^{M} G_j \qquad (6)$$

$$f_2 = \max\{F_i : 1 \leq i \leq N\} \qquad (7)$$

subject to the constraints imposed by Eqs. 1', 2 and the following (Eq. 8) that captures the fact that tasks should finish before their deadlines:

$$X_{ijk} C_{ijk} \leq D_i, \forall 1 \leq i \leq N, \forall 1 \leq j \leq M, \forall 1 \leq k \leq O. \qquad (8)$$

Notice, that Eq. 8 doesn't take into account the failure probability but only the potential execution time of a task. Thus, Eq. 8 ensures that task allocations (assuming no failures) are such so that deadlines are met and together with the minimization of Eq. 7 leads to maximizing the number of finished tasks (within deadlines).

4   Scheduling Algorithms

The algorithms proposed in the paper follow into two categories. First, are algorithms that aim to assign tasks without replication, therefore taking into account only f_1. Second, are algorithms that create extra replicas in order to optimize (among others) f_2. In the sequel we present our heuristics.

4.1   Algorithms for Assigning Tasks (No Replication)

Random (Rand): In the random approach each task will be assigned once to a random peer such that its deadline will not be violated. If it is, the random peer choice is discarded and a new one is made. In case all possible peer assignments lead to deadline violation the task will remain unassigned.

Completion Time (CT): Starting with the heaviest task, CT checks all possible (*M* in number) peer assignments and assigns the task to the peer where it will be completed at the earliest possible time. It then proceeds with the remaining tasks assigning each in decreased order of load (L_i). Notice that this step of the algorithm assigns each task to exactly one peer.

Greenness (G): The algorithm sorts tasks according to their load and peers according to: $E_j(1 - g(P_j))/SP_j$, meaning in order of the energy consumed per computational speedup (taking into account the energy percentage from renewable sources). Then it assigns iteratively the heaviest task to the most energy efficient peer, provided the completion time doesn't violate deadlines.

4.2 Algorithms for Creating Replicas

Random (Rand): The algorithm selects a task at random and creates to a random peer a replica in case the deadline is not violated. If it is, the random peer choice is discarded and a new one is made. In case all possible peer assignments lead to deadline violation for the particular task, the task is taken out of the list and no other replicas for it are made.

Replication for Fault Tolerance (RFT): First the list of tasks is sorted in descending order of their failure probability. Then starting with the one most likely to fail an extra replica is created at the peer that has the minimum failure probability and can complete the task within the required time. The sorted task list is updated with the new failure probability of the task. The algorithm iterates until no eligible assignments exist, i.e., all possible assignments violate task deadlines, or the maximum number of replicas per object or the maximum number of assigned tasks per peer is reached. Another possibility is to have a threshold for the desired task failure probability.

Green Aware Replication (GAR): GAR is similar to RFT algorithm with the exception being that instead of assigning a task to the peer with the minimum failure probability, it assigns it to the most energy efficient peer.

5 Experiments

5.1 Simulation Setup

We considered the case where a total of 1,000 tasks must be assigned to 1,000 peers (both with and without replication) and conducted simulation experiments with settings defined as follows.

Tasks: Task load depicted the expected execution time over a baseline machine and varied uniformly from 100 to 10,000 time units. The deadline for each task was set as a multiple of its load (×2, ×5 and ×10 depending on the experiment). In all the experiments the maximum allowable number of tasks per peer was set to 20.

Computational Peers: The computational power of peers (speedup over the baseline machine) was set to vary uniformly between 0.1 and 10 and the failure probability of a peer was set to vary between 0.1 and 0.001. The percentage of time a peer devotes to volunteer computing tasks was set to vary between 0.001 and 0.3 while the greenness factor varied between 0.01 and 0.3. Finally, energy consumption was set to vary between 100 and 500.

In the experiments we measure the performance of the algorithms both as standalone and in combinations. For instance G_GAR means that the G algorithm decides the initial (single copy) task assignment then GAR is used for replica creation.

5.2   Results

First we evaluated the performance of the algorithms that create no replicas (Rand, CT and G). Figure 1 plots (average of 10 runs) the non-green energy consumption (f_1 optimization target) for various task deadline settings. CT's performance remains constant, which is expected since it decides based on task load and peer properties which remain unaffected by the deadline increase. Rand and G algorithms experience opposite trends. Specifically, as the deadlines increase G achieves better performance (low values are good) while Rand's performance deteriorates. This is due to the fact that as deadlines become less tight more tasks per peer can be assigned without violating the related constraint. G maps these tasks to green peers while Rand maps them randomly. On the other hand, when deadlines are shorter Rand is forced to assign some tasks to green peers in order to avoid deadline constraint violation. Summarizing, Fig. 1 shows that G achieves by far the best performance on greenness criterion compared to its competitors. Particularly for large deadlines (×10) the relative gains can be 3 and 6 times higher compared to CT and Rand respectively. Finally, it should be noted that all algorithms achieved the same value concerning the failure probability criterion (f_2), while no task deadlines were violated.

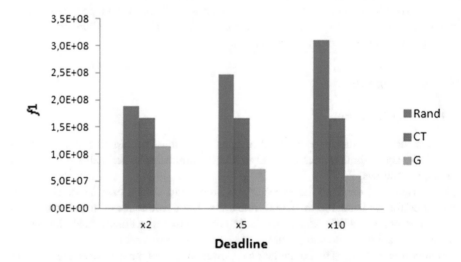

Fig. 1. Energy criterion (f_1) for different task deadlines (one replica per task).

Next, we evaluated the algorithmic combinations when 2 replicas per task could be assigned. Again we obtained results for different deadline values. Figures 2 and 3 depict the performance on $f1$ (dirty energy consumption) and $f2$ (failure probability) criteria respectively. From Fig. 2 it can be inferred that the best energy performance is offered by GAR variations, while RFT offers the best results for fault tolerance.

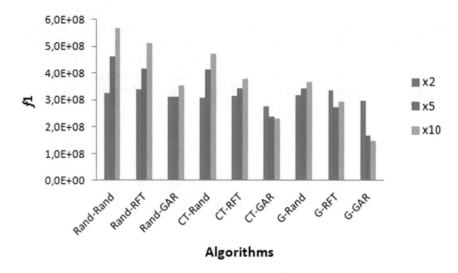

Fig. 2. Energy criterion (f_1) for different task deadlines (replicas per task = 2).

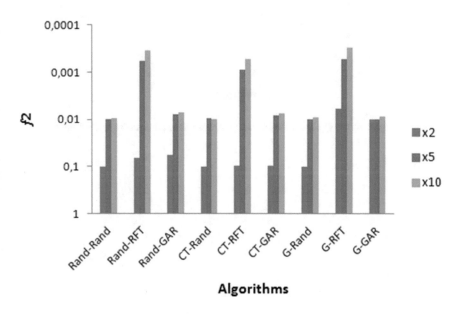

Fig. 3. Fault tolerance criterion (f_2) for different task deadlines (replicas per task = 2).

In a last experiment, the most promising combinations, i.e., CT, G together with RFT, GAR are evaluated as the number of allowable replicas increase. Figures 4 and 5 plot performance on the two optimization criteria energy and fault tolerance respectively. As expected, using more replicas leads to higher energy consumption as more computations run totally but also decreases the failure probability. Comparing results from Figs. 4 and 5 we can conclude on the following: (i) CT-GAR is inferior to G-GAR (the later achieves the best performance on energy criterion and comparable performance on failure probability); (ii) for similar reasons G-RFT is superior to CT-RFT and (iii) G-GAR and G-RFT are the top performers on energy and failure probability respectively. Finally, we should notice that G-GAR increases energy consumption less steeply compared to G-RFT (Fig. 4), while G-RFT achieves less failure probability by at least two orders of magnitude compared to G-GAR.

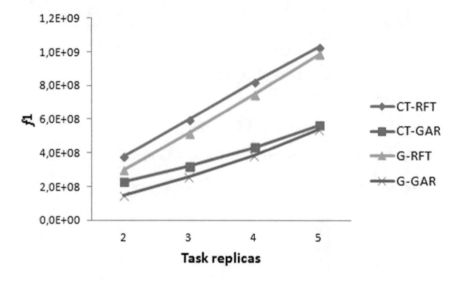

Fig. 4. Energy criterion (f_1) for different number of replicas per task (task deadline $= \times 10$).

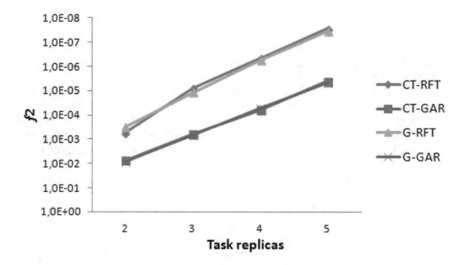

Fig. 5. Fault tolerance criterion (f_2) for different number of replicas per task (task deadline $= \times 10$).

6 Conclusions

In this paper we studied the task scheduling problem on desktop grids. We formulated the problem as a 2-objective optimization with power efficiency and greenness being the first objective and the second one being fault tolerance. We developed heuristics that perform both single copy and replica task assignment. In the experiments we evaluated the performance of heuristics with regards to power efficiency and fault tolerance. Among the alternatives G_GAR that uses the greenness criterion both when placing a task for the first time and when creating replicas was found to achieve the best performance energy wise, while G-RFT was the winner in fault tolerance criterion.

Acknowledgments. This work was supported by the "ENFORCE" project which is part of the SoftFIRE grant agreement no 687860, European Commission (Horizon 2020).

References

1. Rosetta@home. http://boinc.bakerlab.org/rosetta/
2. The SETI@home project. http://setiathome.ssl.berkeley.edu/
3. Koziri, M.G., Papadopoulos, P.K., Tziritas, N., Dadliaris, A.N., Loukopoulos, T., Stamoulis, G.I.: On planning the adoption of new video standards in social media networks: a general framework and its application to HEVC. Soc. Netw. Anal. Min. **7**(1), 1–32 (2017)
4. Dadaliaris, A.N., Oikonomou, P., Koziri, M.G., Nerantzaki, E., Hatzaras, Y., Garyfallou, D., Loukopoulos, T., Stamoulis, G.I.: Heuristics to augment the performance of tetris legalization: making a fast but inferior method competitive. J. Low Power Electron. **13**(2), 220–230 (2017)

5. Ahmad, R.W., Gani, A., Hamid, S.H.A., Shiraz, M., Yousafzai, A., Xia, F.: A survey on virtual machine migration and server consolidation frameworks for cloud data centers. J. Netw. Comput. Appl. **15**, 11–25 (2015)
6. Oram, A.: Peer-To-Peer: Harnessing the Power of Disruptive Technologies. O'Reilly & Associates, Sebastopol (2001)
7. Develder, C., Pickavet, M., Dhoedt, B., Demeester, P.: A power-saving strategy for grids. In: 2nd International Conference on Networks for Grid Applications, pp. 1–8 (2008)
8. Da Costa, G., Gelas, J.-P., Georgiou, Y., Lefevre, L., Orgerie, A.-C., Pierson, J.-M., Richard, O., Sharma, K.: The GREEN-NET framework: energy efficiency in large scale distributed systems. In: 23rd IEEE International Parallel & Distributed Processing Symposium, Rome, pp. 1–8 (2009)
9. Orgerie, A., Lefevre, L.: ERIDIS: energy-efficient reservation infrastructure for large-scale distributed systems. Parallel Process. Lett. **21**(2), 133–154 (2011)
10. Qureshi, M.B., Dehnavi, M.M., Min-Allah, N., Qureshi, M.S., Hussain, H., Rentifis, I., Tziritas, N., Loukopoulos, T., Khan, S.U., Xu, C.-Z., Zomaya, A.Y.: Survey on grid resource allocation mechanisms. J. Grid Comput. **12**(2), 399–441 (2014)
11. Schott, B., Emmen, A.: Green desktop-grids: scientific impact, carbon footprint, power usage efficiency. Int. J. Scal. Comput. Pract. Exp. **12**(2), 257–264 (2011)
12. Castro, H., Villamizar, M., Sotelo, G., Diaz, C.O., Pecero, J.E., Bouvry, P.: Green flexible opportunistic computing with task consolidation and virtualization. Cluster Comput. **16**(3), 545–557 (2013)
13. Ponciano, L., Brasileiro, F.: On the impact of energy-saving strategies in opportunistic grids. In: 11th ACM/IEEE International Conference on Grid Computing, pp. 282–289 (2010)
14. Ponciano, L., Brasileiro, F.: Assessing green strategies in peer-to-peer opportunistic grids. J. Grid Comput. **11**(1), 129–148 (2013)
15. Khan, S., Ahmad, I.: A cooperative game theoretical technique for joint optimization of energy consumption and response time in computational grids. IEEE Trans. Parallel Distrib. Syst. **20**(3), 346–360 (2009)
16. Kolodziej, J., Khan, S.U., Wang, L., Zomaya, A.: Energy efficient genetic-based schedulers in computational grids. Concurr. Comput. Pract. Exp. **27**(4), 809–829 (2015)
17. Linberg, P., Leingang, J., Lysaker, D., Khan, S., Li, J.: Comparison and analysis of eight scheduling heuristics for the optimization of energy consumption and makespan in large-scale distributed systems. J. Supercomput. **59**(1), 323–360 (2012)
18. Loukopoulos, T., Lampsas, P., Sigalas, P.: Improved genetic algorithms and list scheduling techniques for independent task scheduling in distributed systems. In: 8th International Conference on Parallel and Distributed Computing, Applications and Technologies, Adelaide, Australia, pp. 67–74 (2007)
19. Tziritas, N., Loukopoulos, T., Khan, S.U., Xu, C.-Z.: Distributed algorithms for the operator placement problem. IEEE Trans. Comput. Social Systems (TCSS) **2**(4), 182–196 (2015)
20. Tchernykh, A., Pecero, J.E., Barrondo, A., Schaeffer, E.: Adaptive energy efficient scheduling in peer-to-peer desktop grids. Future Gener. Comput. Syst. **36**, 209–220 (2014)
21. Sarmenta, L.: Sabotage-tolerance mechanisms for volunteer computing systems. Future Gener. Comput. Syst. **18**, 561–572 (2002)
22. Taufer, M., Anderson, D., Cicotti, P., Brooks, C.L.: Homogeneous redundancy: a technique to ensure integrity of molecular simulation results using public computing. In: 19th IEEE International Parallel & Distributed Processing Symposium (2005)
23. BOINC. https://boinc.berkeley.edu/

Towards a Pervasive Intelligent System on Football Scouting - A Data Mining Study Case

Tiago Vilela, Filipe Portela, and Manuel Filipe Santos[✉]

Algoritmi Research Center, University of Minho, Guimarães, Portugal
mfs@dsi.uminho.pt

Abstract. Football, which is a popular world-wide sport, has become one of the most practiced sports but also, with more study cases. Scouting and game analysis that is currently made has offered the possibility to improve the competition and increase the performance levels within a team. Taking this into account it emerged the term Scouting. The objective of this study is to streamline the Scouting process in Football, through Data Mining (DM) techniques and following the Cross Industry Standard Process for Data Mining (CRIPS-DM) methodology. The goal of DM was to develop and evaluate predictive models capable of forecasting a score of a football player's performance. Based on this target, 2808 classification models and 936 regression models were developed and evaluated. For the classification, the maximum accuracy percentage was centered at 94% for the Forward player position, while for the regression the minimum error value was 0.07 for the Forward position. The results obtained allow to streamline the Scouting process in Football thus enhancing the sporting advantage.

Keywords: Data mining · Football · Scouting
Knowledge discovery in databases

1 Introduction

Currently, it is recurrent clubs investing in training and gaming management software [1] that using sensors and video system, generate a huge amount of statistic data referent to the involved players, such as number of shots with each foot, number of penalties scored to the right side, kilometers run, scored goals, disarms made, among others. These technologies enrich and empower football, since they allow to analyze, in a more correct way, the available data, offering a set of more reliable solutions facing the desired profitability [1]. However, the amount of obtained data is huge, making the decision process quite empirical, depending on the technician experience or on some technical team member.

A football club Scouting department has as main purpose to discover the best players that can empower a club, either on sports and/or on financial level. Morais [1] refers that "Scouting is, for me, one of the more important tools from modern football, since is consists in detecting talents, in rigorous observations, analysis and after evaluations from players, games or teams".

© Springer International Publishing AG, part of Springer Nature 2018
Á. Rocha et al. (Eds.): WorldCIST'18 2018, AISC 747, pp. 341–351, 2018.
https://doi.org/10.1007/978-3-319-77700-9_34

In order to achieve this mission, Scouting accomplishes, in most times, slow process of analysis and observation of a player until his hiring, quite a few games are analyzed, sometimes in person, which together with the spent time to make a detailed analysis to each of those games, makes it harder to gain competitive advantage. Therefore, regarding Scouting, one of the most common problems is the inability of a scout to evaluate quickly and effectively a player, due to the immense volume of existing data.

It is within this context that derives the motivation of this study, in which was carried out the statistical data treatment of a player, resulting as output a mark that reflects a player's classification in a specific game.

Our main goal is developing a Pervasive Intelligent Decision Support in the scout football area. The first step is the induce of Data Mining models able to find patterns.

The motivation to this study awakened from the opportunity to work with a large volume of football data, applying Data Mining (DM) techniques in order to obtain precious information to the people who, direct or indirectly, are evolved in the game. Thus, this study allowed to speed up the process of Scouting. Based on the predicting models developed, it can be possible to quickly identify which players had a better performance in one or more games. Furthermore, the scouts have a new tool that will facilitate the analysis of data from the observed players, thus becoming this study's result an improvement for Scouting.

This article is divided in 6 sections: Introduction; Background; Methods and Tools; Case Study; Discussion; Conclusion. In Background, it is presented the state of the art for this study's main topic, by analyzing the fundamental concepts which are inherent to it. In Methods and Tools it is proceeded the identification of the methodological approaches that support the dissertation project (Design Science Research Methodology and Cross Industry Standard Process for Data Mining) and the used tool for DM developing (R language). On the Study Case, it is detailed the practical developed work, organized by each of CRISP-DM methodology phase (business comprehension, data comprehension, data analysis, modeling, evaluation and implementation). In Discussion proceeds the analysis and discussion of the obtained results on this work. Finally, in Conclusion is performed a project conclusive synthesis, giving emphasis to obtained results and its contribution. Furthermore, it is identified the main future possible researches and the study's limitations.

2   Background

2.1   Football

Football is a sport that fits in collective games, occupying an important place in contemporary sports culture. It is, as stated by [2], the world most respected and practiced sport. In the last years, in this sport has been given "excessive importance to euros (millions) and to petrodollars" [1] whereby it is possible to verify that this modality is more and more a business and less and less a sport [1, 3]. In the perspective of [1], the best of the revenues would be to empower players so that they would also generate revenues and, consequently, bet on resources monetization. Santos [2] agrees with [1] in this regard, by saying that "the initiation levels are already understood as a source of resources, from where will be extracted quality, to allow to obtain competitive teams with players adjusted to the desired profile". According to Santos [2], the evolution and search for young talents is, currently, an activity that achieved unpaired levels of growth and improvement. It quickly became apparent that the importance given to elite football has spread to younger layers and to club's formation teams. The search for talent in early ages has been an opportunity for clubs' economical, human and social growth. Analysis to arising new players should have as start point the identification of skills and potentialities so that they can achieve success in football. Therefore, the demand by these individuals initiated, developed, appreciated and specialized in observation, converting in what nowadays is called Scouting [2].

2.2 Scouting

According to Santos [2], Scouting consists in planning games to follow, identifying lacks in team structures, elaborate reports, made throw observations. Thus, Scouting can be seen as a means for minimizing losses and monetize resources [1], being able to offer information about the team itself, the opponent team, the acting style of referee teams, the ecological environment external to the game, among others.

In the perspective of Mendes [1], Scouting can decompose in three parameters: observation and analysis of team itself, individual observation and analysis and opponent observations and analysis, as it can be seen in Fig. 1.

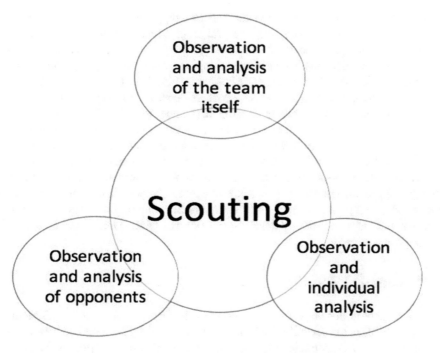

Fig. 1. Scouting Perspectives (adapted of [1])

To ensure the observation, it has emerged the called game observation systems, facilitating the Scouting process. This software allows to create and explore observation results; organize data, focusing performance data, features and statistics from all observed players; perform searches in databases from the kind of needed player; elaborate an agenda of games to observe; develop evaluation models to different kind of players and to conceive dynamical reports about one or more players and share it with the remaining team. After this data gathering, scouting departments used to apply manual methods to follow players' performance and, therefore, use a traditional decision-making approach taking into consideration its intuition. However, more and more, the quantity of collected data is huge, bringing great troubles to analysis and decision. Hence, sports organizations have started to adopt an advanced analysis based on digital technologies to evaluate players' technical skills [4].

This way, new talent identification using technologies would be improved with the application of DM techniques. Examples of application of DM techniques would be a decision support system development to be used during a game, as well as the use of data to help selecting referee or place for the game. DM techniques are based mainly on the search for existing patterns, relating events in a non-trivial way [5].

DM enters, thereby, in this field with the intention to find more precise information, avoiding unneeded ones [6].

2.3   Relate Work

Relatively to the carried-out research for understanding approached state of the art themed, has denoted that already exist several articles that point to an approach for applying DM techniques associated with multiple sports. Football, basketball, volleyball, tennis and American football have been using DM, fundamentally, to predict future game results [7, 13–15]. In Scouting world has been verified that the amount of obtained data is getting bigger, letting the decision process dependent on experience from the technician or from another Scouting department member [6]. Therefore, has emerged proposals for using DM techniques to help team's Scouting departments improving its performance in sports such as volleyball, basketball and baseball. Advanced Scout program, developed to support NBA coaches in identifying hidden patterns in game statistics and data can be taken as an example. This program, since mid-ninety, uses DM techniques to provide knowledge to coaches so they can prepare their team for the next game [8]. Regarding research about DM related to Scouting in football is possible to notice that, until now, was not yet developed any kind of solution that can be compared to the present case study.

3   Methods and Tools

The development of this project was supported by the Design Science Research Methodology (DSRM) in the scientific field, since this methodology is composed by techniques and policies that guide scientific research in information systems [9]. For the practical component, the choice relied on the cross-industry standard process for data mining

(CRISP-DM). This methodology presents a flexible life cycle composed by six stages, in which its sequence is not rigid and allows to move back and forward between phases [10]. The CRISP-DM life cycle is represented in the context of this study.

- **Business Understanding:** Understand the projects objectives and requirements to define the problem and the work plan.
- **Data Understanding:** Initial collection of data and activities that allow to check its quality.
- **Data Preparation:** Activities to build the final data set.
- **Modeling**: Application of regression and classification techniques, in which, for each task, were selected thirteen techniques and created four scenarios.
- **Evaluation:** Review of the obtained results to verify if the objectives were achieved.
- **Implementation:** This project was not implemented in a final customer, as it was not part of this study's scope.

For the development of DM tasks, R language was used due to its simplicity of development and extensibility. For model development and evaluation, Rminer package was used. Throughout the project, a total of thirteen packages were used, which allowed the manipulation of Excel files, graphics construction and application of Feature Selection algorithms.

4 Case Study

4.1 Business Understanding

According to the background of this study, it is possible to understand that, until now, there are no DM techniques to improve the process of Scouting in football. This means that, nowadays, one of most common problems in Scouting is to evaluate a player in a fast and efficient way, due to the enormous amount of available data. This evidence brought, therefore, the need to carry out this case study, which has, at its business goal, to offer a relevant contribution to Scouting in football, in order to streamline the process of data analysis of players and, consequently, enable their better identification and recruitment. Concerning to DM objectives, the case study was based on obtaining prediction models through two tasks of DM: Classification and Regression. The DM tasks were chosen to explore some of existent predictive tasks, followed by the results analysis in each one.

4.2 Data Understanding

The present data is relative to of football players' performance at a given time and the dataset records are related to:

- Six different seasons (2010–2011, 2011–2012, 2012–2013, 2013–2014, 2014–2015 and 2015–2016);
- Four main leagues of European football (Serie A, La Liga, Premier League and Bundesliga)

This forms a total of 119 teams and 2888 players. Although the dataset contains only 6870 records, each record contains 35 attributes. As already mentioned, the main goal is to forecast a player's future rating.

First, it was necessary to divide the data by the position attribute, so that the DM study was differentiated by each position, with these being Defender, Midfielder and Advanced. The Table 1 below shows the target (rating attribute) for each position, indicating the maximum, minimum, mean, standard deviation and number of records.

Table 1. Analysis of rating values

Function	Positions		
	Forward	Midfielder	Defender
Maximum	8.88	7.89	8.01
Minimum	5.99	6.12	6.23
Mean	6.84	6.90	6.94
Standard deviation	0.39	0.30	0.25
Row count	2199 (32%)	1655 (24%)	2548 (37%)

Regarding the Forward position, it should be noticed that it holds the higher dispersion between player's classification (Rating) values - higher standard deviation value - and, therefore, the global maximum and minimum values are present in this position. It is in Defender position that player's mean rating achieves higher value (6.94). However, this position also contains a greater number of records.

Based on the presented data exploration, and in order to complete its analysis, we will then identify changes made for data improvement, listed in the following points:

- Transformation of age attribute (Age), because it presented player's current age instead of his age at the time in question.
- The replacement of "-" character by the digit 0 (zero).

Disaggregation of number of goals attribute to the holder and the alternate (Apps). Currently it is constituted by x (y), where x represents the number of games to the holder and y the number of games in the condition of substitute.

4.3   Data Preparation

At this stage, it was intended to correct the problems identified in the previous phase of CRISP-DM. These problems were identified based on the assumption that, in this project, it was intended to perform the prediction of a given player's rating, using two DM approaches: Classification and Regression. These approaches were divided in 4 tasks:

- **Task 1:** The initial change was based on the transformation of age attribute (Age). It was felt the need to change this attribute, as age (Age) represented the player's current age and not his age referring to the time of data recording. Thus, age attribute (Age) became the player's age at the time referring to the remaining data.

- **Task 2:** The next fundamental change to use the mentioned DM tasks, classification and regression, was the substitution of the character "-" with the digit 0 (zero). This change occurred in order to analyze the attributes as being of the numeric type and not as text.
- **Task 3:** In this task, the change to be made was centered on the breakdown of the attribute number of games to holder and substitute (Apps) in two distinct attributes (number of games to holder (Apps_titular) and number of games to spare (Apps_suplente). The attribute number of games for the starter and the substitute (Apps) consists of x(y), where x represents the number of games to be held and y the number of games as a substitute.
- **Task 4:** Besides the solved problems identified in the previous phase, data preparation also consisted in the preparation of classes (value intervals) to be possible to make predictions using Classification task. Class construction was achieved using a random method (4 classes), using the mean (2 classes) and also using quartiles (4 classes).

4.4 Modeling

In this stage, two prediction tasks (classification and regression) were studied in order to predict the performance of a given player. Concerning target value, it differs in both tasks. In classification, the target consists on classes, which were built in the data preparation phase. In the regression, target was formed by the original values, being discrete and decimal values.

Regarding sampling methods, three were used: Cross Validation, Holdout Simple and HoldOut Order, in order to divide data for training and testing multiple models. For Cross Validation, the dataset was divided into 10 folds. For each model, the Holdout were set to ten runs and its result was consisted in the joined results of each run. It was chosen to calculate the ten runs' mean, in order to ensure that evaluation was not represented by one training only, resulting, this way, on an increase in model's confidence. For the method of representation, only the conventional one was used.

Regarding prediction scenarios, for each position, eight different scenarios were developed. In a summary, we obtained:

- Two DM tasks:
  - Classification
  - Regression
- Three positions addressed
  - Defender
  - Midfielder
  - Forward
- Three approaches in class construction
  - Random
  - Mean
  - Quartiles
- Three sampling methods

- – Cross Validation
- – HoldOut Simple
- – HoldOut Order
- • 19 DM techniques
  - – ctree; rpart; kknn; ksvm; mlp; mlpe; randomForest; bagging; boosting; lda; lr; naiveBayes; qda; mr; mars; cubist; pcr; plsr and cppls
  - – 8 scenarios: C1, C2,…, C8

Finally, for the classification task 3 * 3 * 3 * 13 * 8 = 2808 models were obtained and for the regression task 3 * 3 * 13 * 8 = 936 models, totaling 3744 prediction experiments.

4.5   Evaluation

As an evaluation metric for the classification models, accuracy (ACC) was selected because it represents the calculation of the proportion of correctly classified cases in a given model [11]. This means that it translates the percentage that the model accurately predicts the class. The selection of the ACC metric was chosen since a great majority of the experiments are multiclass (construction of classes by the mean, random method and quartiles) (Table 2).

Table 2.   Predict results of classification

Scenario	Class construction method	Accuracy(%)		
		Forward	Defender	Midfielder
C1	Aleatory	78,17	66,76	74,62
	Mean	92,28	82,69	91,30
	Quartiles	79,54	59,30	73,91
C2	Aleatory	78,58	71,94	64,13
	Mean	90,86	85,39	86,05
	Quartiles	76,44	64,68	62,68
C3	Aleatory	84,45	60,83	78,31
	Mean	94,00	78,92	93,30
	Quartiles	83,81	52,73	76,98
C4	Aleatory	77,63	67,78	64,13
	Mean	91,13	82,34	86,05
	Quartiles	75,99	58,71	62,68
C5	Aleatory	77,40	61,34	67,07
	Mean	91,41	79,08	88,77
	Quartiles	78,35	53,62	66,30
C6	Aleatory	77,22	65,03	63,77
	Mean	91,27	79,96	87,14
	Quartiles	76,13	56,43	62,68
C7	Aleatory	74,62	75,27	59,42
	Mean	90,18	86,57	82,07
	Quartiles	75,58	66,33	56,52
C8	Aleatory	71,08	68,41	63,44
	Mean	89,09	83,39	86,96
	Quartiles	70,94	60,32	63,59

For the regression evaluation, there are several metrics that are based on the construction of an error estimate [12]. Mean Absolute Error (MAE) and Root Mean Squared Error (RMSE) were the ones selected.

According to Witten et al. [12], MAE calculates the mean magnitude of errors in a set of predictions, without considering its direction, that is, it corresponds to the mean, on test sample, of the absolute differences between the forecast and the actual observation; whereas RMSE refers to a quadratic scoring rule that evaluates the error mean magnitude. It reflects the square root of the mean square differences between the prediction and the actual observation. Therefore, as they deal with error estimation, their values are both better and closer to the zero value (Table 3).

Table 3. Predict results of regression

Scenario	Forward		Defender		Midfielder	
	RMSE	MAE	RMSE	MAE	RMSE	MAE
C1	0,10	0,08	0,13	0,10	0,10	0,08
C2	0,10	0,08	0,11	0,08	0,14	0,11
C3	0,07	0,06	0,15	0,12	0,08	0,07
C4	0,10	0,08	0,13	0,10	0,14	0,11
C5	0,10	0,08	0,15	0,12	0,13	0,10
C6	0,10	0,08	0,14	0,11	0,14	0,11
C7	0,11	0,09	0,10	0,08	0,16	0,13
C8	0,14	0,11	0,12	0,09	0,15	0,12

5   Discussion

In general, it is possible to verify that the created models help improving Scouting process in Football, once the obtained results of the predictions have presented very positive results. These results allowed the creation of models that help to predict the grade of a given player (Rating) with high precision, as it was verified in the evaluation chapter. It is important to discuss the results by analyzing the two Data Mining (DM) techniques used: classification and regression. In classification models, the best prediction resulted in an accuracy (ACC) of 94%, being relative to the Forward position. For the Midfielder position, the best prediction resulted in 93.3% accuracy and for the Defender position resulted in 86.6% accuracy.

In regression models, the predictions were evaluated using two metrics: Mean Absolute Error (MAE) and Root Mean Squared Error (RMSE). The obtained results with the RMSE metric presented a low error value for the three player positions. The lowest error value for the Forward position was 0.07, for the Defender position 0.10 and for the Midfielder position 0,08. It could be seen that the lowest RMSE value occurred for the Forward position. However, the highest value also occurred for this position. Regarding the mean value of RMSE, it was found that the lowest value refers to the Forward position and the highest to the Midfielder position. It should be noticed that the overall error value was centered at 0.15, a globally satisfactory value.

6 Conclusions

The designed predictive models help to improve Scouting process in Football, helping scouts in the treatment of players' data and the evaluation of the player's performance. Based on certain data of players, preliminary predictive models were built, which were capable of obtaining the rating grade of player's performance.

In this way, DM allows to accelerate the process of decision making in players' performance evaluation. Based on the analysis of players' performance characteristics, the developed models allow to predict their evaluation without the need of observation from Scouting elements. Thus, it is possible to say that the development of DM allows to speed up the Scouting process in Football, when it comes to players evaluation. Although indirectly, it can also be stated that, by accelerating the analysis of players' performance through the DM, a competitive advantage is obtained in the decision-making process for hiring new talents, in the identification of opposing players that may constitute a threat to the team and in the individual analysis of the team itself.

Although developed work proves to have a great importance for improving Scouting process in Football, only an initial approach to this theme was made, since there is no representative study on the theme. Therefore, in the future, there is a need to acquire different data from those who worked on this project, including new analysis attributes, in order to explore new approaches in Football Scouting. In addition, considering that this is an initial approach, it will be possible in future investigations to prepare scenarios and techniques with the best results obtained in this project, in order to optimize and deepen their study. In the future all the Pervasive System will be designed and the Data Mining models will be integrated.

Acknowledgements. This work has been supported by COMPETE: POCI-01-0145-FEDER-007043 and FCT – Fundação para a Ciência e Tecnologia within the Project Scope: UID/CEC/00319/2013.

References

1. Mendes, A.: Scouting, o Futebol (Re)nasce Aqui (2016)
2. Santos, P.: O modus operandi de um Departamento de Scouting de Futebol Estágio Profissionalizante realizado na Futebol Clube do Porto (2012)
3. Garganta, J.: Atrás Do Palco, Nas Oficinas Do Futebol. In: Futeb. muitas cores e sabores. Reflexões em torno do desporto mais Pop. do mundo, January 2004, pp. 227–234 (2004)
4. Radicchi, E., Mozzachiodi, M.: Social talent scouting: a new opportunity for the identification of football players? Phys. Cult. Sport Stud. Res. **70**(1), 28–43 (2016)
5. Nunes, S., Sousa, M.: Applying data mining techniques to football data from European championships. In: Actas da 1ª Conferência Metodol. Investig. Científica, December 2005 (2006)
6. Butzen, É.: Proposta de um Módulo de Data Mining para Sistema de Scout no Voleibol, p. 82 (2008)
7. Leung, C.K., Joseph, K.W.: Sports data mining: predicting results for the college football games. Procedia Comput. Sci. **35**, 710–719 (2014)

8. Solieman, O.: Data mining in sports: a research overview. Department of Management Information Systems, August 2006
9. Vaishnavi, V., Kuechler, B.: Design Science Research in Information Systems Overview of Design Science Research, Ais, p. 45 (2004)
10. IBM.: IBM SPSS Modeler CRISP-DM Guide, p. 53 (2011)
11. Gorunescu, F.: Data Mining: Concepts, Models and Techniques (2011)
12. Witten, I.H., Frank, E., Hall, M.A.: Data Mining: Practical Machine Learning Tools and Techniques (Google eBook) (2011)
13. Gomes, J., Portela, F., Santos, M.F.: Real-time data mining models to predict football 2-way result. Jurnal Teknologi, Penerbit UTM Press (2016). ISSN: 0127-9696
14. Gomes, J., Portela, F., Santos, M.F.: Pervasive decision support to predict football corners and goals by means of data mining. In: Advances in Intelligent Systems and Computing, vol. 445, pp. 547–556. Springer (2016). ISBN: 978-3-319-31306-1
15. Gomes, J., Portela, F., Santos, M.F., Machado, J., Abelha, A.: Predicting 2-way football results by means of data mining. In: ESM - 29th European Simulation and Modelling Conference. Leicester, UK, EUROSIS (2015)

Check for
updates

Step Towards a Pervasive Data
System for Intensive Care Medicine

Paulo Gonçalves[1], Filipe Portela[1(✉)], Manuel Filipe Santos[1],
José Machado[1], António Abelha[1], and Fernando Rua[2]

[1] Algoritmi Research Center, University of Minho, Guimarães, Portugal
cfp@dsi.uminho.pt
[2] Intensive Care Unit, Centro Hospitalar do Porto, Porto, Portugal

Abstract. The use of technologies that can facilitate and streamline the processes of those who constantly need to perform various actions or comply with the most varied procedures, requires constant adaptability, either by organizations or users. The focus of this research is precisely the adaptability of technologies and in this case the technology used in the Intensive Care Unit (ICU) of the Centro Hospitalar do Porto (CHP). The increasing use of different electronic devices, all with different characteristics and dimensions, requires the optimization of the platforms that transmit and manipulate all the information, so that it is possible to use it regardless of which device is being used. Through the introduction of new functionalities to the current system, it is intended with this artefact to show the optimization made on the INTCare platform, with the main purpose of increasing its responsiveness.

Keywords: INTCare · Pervasive Computing · Pervasive environments
Mobile Health · Pervasive health

1 Introduction

The use of computed technologies in healthcare, has been growing day by day. This new approach, combined with mobile technologies, brings a new era: the e-health. With the mobile technology, being adapted to heal-related areas, opens the improvement of the dissemination of public health information. Also facilitates remote consulting, diagnosis and treatment, patient management, public health monitoring, distribution of information to doctors and nurses and increased efficiency of administrative systems [1]. With these, comes the scan for intelligent systems or that are capable to support the processes required to a nurse or doctor, making their decisions easier and faster. One of the areas that needs a helpful application is the Intensive Medicine (IM). The possibility of having the information available anytime, changes completely, the performance of an Intensive Care Unit (ICU) and the intensivists capacity of the patients' needs [2]. INTCare is an Intelligent Decision Support System (IDSS), which fills the requirements of a web application developed to an ICU. It is implemented on Centro Hospitalar do Porto (CHP) and the focus of this system is the support to the decision-making process of ICU professionals and became a necessary

© Springer International Publishing AG, part of Springer Nature 2018
Á. Rocha et al. (Eds.): WorldCIST'18 2018, AISC 747, pp. 352–362, 2018.
https://doi.org/10.1007/978-3-319-77700-9_35

tool to agile process and tasks of IM [2]. The number of functionalities of INTCare is in continuous grow, according the needs and problems, with the objective of develop a highly capacitated system with requisites like: adaptability, security, privacy, pervasive and ubiquitous, real-time, data mining and decision models and secure remote access. To accomplish this, has been implemented a series of intelligent agents that autonomously execute some essential tasks. To a better data visualization, INTCare has a specific functionality: Electronic Nursing Record (ENR) [3]. ENR is a web based platform, optimized for touch screens, that allows the ICU medical staff to register electronically the paper-based nursing records. With the ENR, is possible to register various types of data, confirming if some therapeutic was performed or not and consult all the present and past data about the patients. This platform is the front page of INTCare system because congregate independent data sources and enable the access of all data always in the same place [4]. Motivated by the importance of ENR, this paper aims to show the difference between the before and now, of INTCare front page. The objective of this work was renewing the platform to be have a better data visualization and surpass some limitations that are hold up a better use of the remote access to the system.

The structure of this work is divided as follows: Sect. 1 reserved to the introduction of the content; Sect. 2 presents the background study; Sect. 3 is about the methodology applied; Sect. 4 defines the INTCare system; Sect. 5 refers to the ENR platform; Sect. 6 concludes the research presented.

2 Background

2.1 Intensive Medicine and Intensive Care Units

Intensive Medicine (IM) can be described as an area of medicine that focuses on the diagnosis and treatment of patients with serious health problems. These problems are usually considered to be a threat to the quality of life of the patients and, for this purpose, the IM seeks to reverse this debilitated life condition, to a state in which the patient can be accompanied in a hospital [5]. Intensive Care Units (ICUs) are specialized hospital units where IM treatments are applied. These units are known as critical environments where patients with very weak health status are hospitalized, usually in a state of organ failure or life-threatening [6]. ICUs are endowed with a large amount of resources, mostly technological, aimed at reversing the condition of patients through a constant monitoring of their health and vital signs [7]. The adaptability of technologies to these areas, is important because the fact that data related to a patient can be available at any time and place makes the ICU much more efficient, since they are environments dependent on the correct presentation of the information, so that decisions can be made as accurately as possible.

2.2 Mobile Health

Mobile Health is considered the strongest contribution when it comes to the use of health technologies. This tendency has shown a constant growth and brings the interaction between patients and health professionals to a higher level, making use of the mobile devices to support the decisions and tasks of the professionals of the different health areas [8]. The constant monitoring of the evolution of patients' health status is crucial, either after surgery or when transferred to another hospital unit, or in support of prevention/prediction of possible complications, thus being able to sustainably support the management of an hospital unit and consequent increase the quality of the service provided [9]. However, these systems are usually developed in isolation, hampering the task of interoperability, which implies more complex and heterogeneous processes [10].

2.3 Pervasive Computing and Technologies

Pervasive technologies imply that these are so integrated into our lives that they become tools that are influential and indispensable to our day-to-day life, without causing dependency and as intuitive as possible. These technologies have their implementation divided into four types: implantable, wearable, portable, and environmental.. As for wearable technologies, or technologies in intelligent environments, do not have invisibility as a mandatory feature. However, they should always be the less intrusive and inconvenient possible [11, 12]. Pervasive Computing is an addition to the Mobile Computing, focusing not only on mobility, but also on interoperability, scalability, adaptability/responsiveness and invisibility, so that users can take advantage of easier use when they need it [13].

Pervasive Healthcare, described as healthcare services to everyone anywhere and anytime, is considered a key factor in the reduction of expenses of the hospital units and in the advance of the treatment and cure of numerous health problems. The pervasive technologies adapted to health, seek to alleviate some of the main failures of hospital units, such as records in paper format and the failure of information always available, allowing their use remotely [14]. A pervasive environment is designed to allow the interaction of mobile devices, with sensors and servers, becoming important in the health area, since it is necessary the support to the doctors, so that the best decision is made according to the status of patients. To describe the points in favour or against and the opportunities or threats that arise from the implementation of pervasive environments, a SWOT analysis was constructed, which is a reference that allows to obtain an analysis of the environment under study, according to the context of the Intensive Care Units. Table 1 shows this SWOT analysis [3].

Table 1. SWOT analysis for pervasive environment in ICUs.

Strengths	Weaknesses
• Information always available;	• Secure networks;
• Possibility of remote access to data;	• Confidentiality and privacy of information;
• Access to specific information;	• Lack of technological infrastructures;
• Improvement and optimization of services;	• Need for technological training for users;
• Support to medical decisions;	• Problems in application maintenance;
• Processes innovation;	• Dependence of decision support systems
• Improvement of the quality of the services	
Opportunities	**Threats**
• High quality of decisions;	• Non-qualified staff in the technological area;
• Availability of information;	• Security and/or privacy failures;
• Use of new technologies;	• High costs in the implementation or extension
• Optimization of tasks and processes;	of infrastructures; Risk of attacking networks
• Fill hospital needs;	or servers
• Avoid problems in decision making	

3   Methodology

The Design Science Research (DSR) method goal, is create and evaluate the different artefacts' that aim to solve organizational problems [15]. This model is composed by six activities: first, "Identify the problem & Motivate", define which is the problem to investigate and present and justify the value of the solution; second, "Define Objectives of a Solution", interpret the objectives of a solution according to the definition of the problem and the knowledge of what is possible and feasible to do, these objectives can be quantitative or qualitative; third, "Design & Development", refers to the creation of the artefact, based on research, and may be models, methods or instantiations; fourth, "Demonstration", aims to demonstrate the use of the artefact when solving one or more phases of the problem and may involve the use of the same in appropriate simulations or case studies; fifth, "Evaluation", observe and quantify the quality of the artefact in the resolution of the problem, for which it is necessary to compare the objectives of the solution with the results obtained in the Demonstration phase; sixth, "Communication", stage of demonstration and presentation of the problem and its importance, the artefact, utility and innovation, the rigor of its design and its effectiveness in solving the problem [16]. In the Identifying the problem and motivation phase, we sought to understand how the INTCare's ability to adapt to the different characteristics of different devices would be beneficial for the decision-making process in the ICU of CHP. The main objective, is the optimization of the INTCare platform, ENR. The Design and Development phase encompasses all idealizations for the optimization of the artefact, which for this project was a responsive prototype, that is, able to adapt visually to any device. In the Demonstration phase, the solution was implemented to present the work done. The results of the work demonstrated were estimated in the Evaluation phase, quantitatively, to see if the solution developed is adequate to the identified problem. The Communication phase includes the writing of scientific articles and the final dissertation report, where the final results obtained are presented.

4 INTCare

INTCare is an Intelligent Decision Support System (IDSS). The main objective of this system is to automatically collect and analyse data of patients hospitalized in the ICU. This system can make predictions regarding the future condition of a patient such as a possible failure of vital organs [4]. Is considered a IDSS because in an autonomous way, in real time and using the predictive model, it can be a great support in the process of clinical decision making, since it allows the creation of important information, turning the processes within the ICU much more effective and efficient, reducing the uncertainty caused by the possible stress of decision-making [3].

4.1 System Features

The whole system is interconnected in order to ensure secure communication and access to data everywhere and also respects a number of essential characteristics [3], which are detailed below:

- **Real time:** all patient data are collected in real time using sensors that have been added to the ICU. The necessary information is entered into a database, immediately following the occurrence of an event;
- **Electronic mode:** all data must be available in an electronic form, otherwise should be registered, by the professionals, with an available application, the Electronic Nursing Record (ENR);
- **Online:** all information must be online, allowing it to be used regardless of where the data is accessed;
- **Autonomous:** the tasks must be done in the most automatic way possible, using the intelligent agents integrated in the INTCare, and the validation must be done manually;
- **Data mining models:** The success of IDSS depends, on the acuity of the DM models, which implies that the prediction models must be reliable. These models make possible to predict events and avert some clinical complications to the patients;
- **Decision models:** The achievement of the best solutions depend heavily on the decision models created. Those are based in factors like differentiation and decision that are applied on prediction models and can help the doctors to choose the better solution on the decision-making process;
- **Safety:** all patient data should be stored safely and securely. Access controls are also mandatory in the system;
- **Reliable:** the health professionals are responsible for the validation of the data inserted in the ENR and the system presents only the information that is required, guaranteeing the truthfulness of the same;
- **Accurate:** all operations must be approved prior to a final decision to avoid placing the patients' health at risk and possible attacks to the system;
- **Privacy:** data for any patient must be private and can't be collected outside the hospital unit. To this, it is necessary to ensure that all tasks performed are recorded and associated to a professional to prevent problems and establish responsibilities;

- **Adaptive:** the system must be able to adapt automatically and according to the needs, guaranteeing a constant and correct operation of the same;
- **Secure access:** access to the network that supports the system, must be highly protected and encrypted. Tools like VPNs with integrated access protocols are essential. Only authorized staff may obtain and use the information, either through local access or remotely;
- **Context awareness:** it is important to raise the awareness of those who use the system so that they know the value of the information used to avoid possible negligent acts;
- **Risk list and contingence plan:** it is necessary to define risks and possible impact, as well as the probability of occurrence, to predict future problems.

4.2   INTCare as a Pervasive System

The implementation of a pervasive system, implies some responsibilities as shown in the anterior section. As other systems in general, privacy, safety and secure access, are some of the prior aspects otherwise not only patients' life will be compromised, but hospital and ICU staff will be responsible for unexpected problems [3]. Another important characteristic is the system autonomous operations.

INTCare system is able to give the user what he needs, through the use of intelligent agents, that work behind all the features and simplify complex tasks i.e. data calculations, allows the ICU professionals to consult and validate data, no matter the electronic device used to access. With some improves made to the system, the patients information is always saved securely and can be accessed anywhere, although if the user is not at the ICU, validating the data is not allowed.

The INTCare system can be categorized as a pervasive system considering its ability of responsiveness and accessibility, anywhere and anytime.

5   Electronic Nursing Record

The most common obstacle to the development of Intelligent Decision Support Systems (IDSS) in medicine is the high number of paper data and the large number of data sources. The reality of ICU is not different and when the project started, about of 80% of information that were essential to the development of Data Mining Models were registered in paper and, only 8% were registered punctually in database, most of them in an offline mode, i.e., the date between the paper registration (date of collection) and database storing can have some days of delay [17].

The Electronic Nursing Record (ENR) is a platform developed for collecting and monitoring data and it is incorporated in the INTCare. ENR has the objective of collecting all clinical data, which will be made available to the medical teams.

5.1 Description

ENR is a web based touch screen platform that was developed with the objective to receive all medical data and put it available to Doctors and Nurses in real time and automatically, through an intuitive interface [4].

The ENR has a set of features that allows a complete patient data monitoring, i.e., with this platform the ICU staff can register, analyse, validate and consult all data associated to a patient in critical situation. The introduction of ENR in the ICU represents a good improvement when compared with earlier situation. Before the ENR the data of vital signs, medications, patient scores, fluid balance were always registered in the Nursing Record Paper Based (NRPB). A previous study [17] showed the impact of ENR after the implementation [17]. The information is available online and in real-time now.

5.2 Features

Throughout ENR evolution, some new features and functionalities were added to simplify the work in the hospital and promote dematerialization of processes. In complement, the application gives the possibility to have a set of graphics that can show the evolution of patient condition. Other possibility is the automatic/manual calculation/insert of the ICU medical scores (SOFA, SAPS II, SAPS III, Glasgow, TISS28 and MEWS) [18].

For that was very important the system interoperability that allows the data access from other data sources like, lab results, electronic health records, drugs systems, therapeutic attitudes and procedure, and others [19].

Complement application characteristics:

- Record, Store, Validate and consult the data from the previous day;
- Consult therapeutic plan of day after of the patient;
- Auto/Manual data save, refresh or validation;
- Auto start, according the PID, bed and monitors;
- Easy connection to other platforms;
- Show patient identification (PID);
- Other Charts (events, lab results …);
- Patient data is real, confidential and secure;
- Historical data consult;
- Lab Results comparative table;
- Automatic calculation of fluid balance and, Glasgow, pain scales and ICU patient Scores;
- Hourly data validation and block;
- Patient Alerts;
- INTCare System connectivity;
- Auto PDF creator and reader system.

5.3   Overview

The ENR is one of the most important tools of the entire INTCare system, making its correct operation vital. Being such an important platform, it becomes an essential work tool for doctors and nurses since it makes the decision-making process faster and safer. One of the main features of ENR is the responsiveness. With this, ENR adapts to every devices and characteristics', allowing the intensivists to access the information everywhere (Fig. 1).

Fig. 1.   ENR responsive table.

For a quick access of data and data processing, ENR a set of intelligent agents were developed. Those mechanism not only contributes to improve health care practices and patient care, but also provides information in an easily, secure and quick way. At the same time, it makes the data available to be used by an IDSS in real-time, anywhere and anytime. The digital nature of a nursing record allows data contained within it can be, easily, searched and retrieved [18]. With all of this ENR still has a good performance, the information is always available immediately, and without requiring any action or concern of the user, save and load all data, only once, so that each time the user needs to switch between different menus, the response is immediate (Fig. 2 - ENR therapeutics table). All variables used by these features are updated automatically when a change is made by the user, always guaranteeing the presentation of the most recent data. It was always a concern to develop an application that is easy to maintain and as dynamic as possible. ENR is an application directly dependent on defined variables, which when changed in no way will change its operation.

The INTCare system, as an IDSS, collects all the information from different sources. One of them is the ENR platform, that when connected with other systems like AIDA, turns possibly the automatically data processing and the online learning that provides the necessary information to data mining and decision models, i.e., this data collected automatically and in real time makes possible to obtain results about ICU Scores, Critical Events, Prediction organ failure and patient outcome, etc. [17].

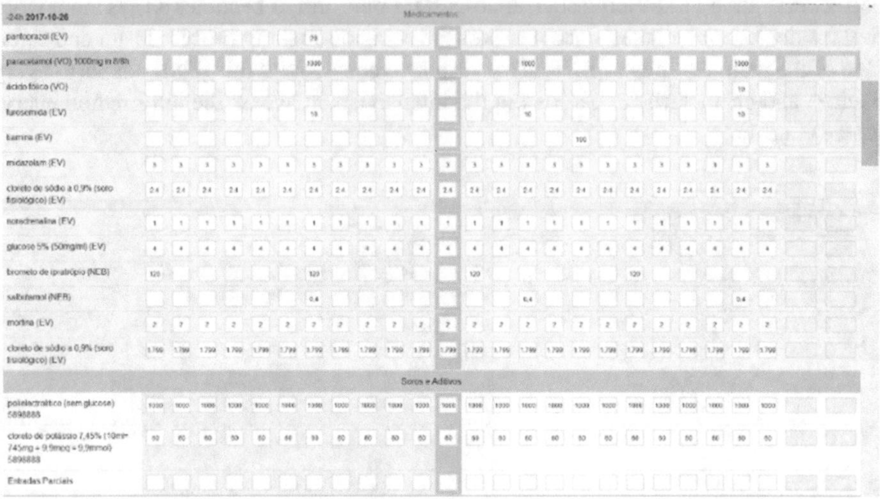

Fig. 2. ENR therapeutics table.

6   Conclusions and Future Work

The paper presented seeks to demonstrate the impact of pervasive systems in ICUs. In an area where information has a crucial impact, and decisions are taken within seconds, the implementation of a systems that is capable to the needs of the users is an important tool. The INTCare system turns out as a mechanism that offers the ICU professionals the support they need to their decision-making process. Despite the problem, was easy to understand that adding a feature that allow the user to consult the information in real time and with every technological device, knowing that the application response will always be adapted, is an important functionality of a system.

These changes support a bigger control of the patients monitoring and, doctors and nurses, have the possibility of accessing to cleaner and better visualization of the data, being that an important element to the service provided.

Resuming, the work done was a total optimization of an earlier version of INTCare system. The work done along with this paper, are two of the artefacts of the research project presented that was evaluated with a grade if eighteen out of twenty. INTCare has now new interfaces and functionalities, turning the access and use of ENR more user friendly and more capable to answer all the user needs.

Although, the work is not complete because there are some new features on the line to be implemented, such as, a bed map to the ICU. This feature works as a satellite view of the ICU facility, permitting the professionals a better overview of all beds and a quicker access and control to a specific bed in case of some alert triggered.

Acknowledgement. This work has been supported by COMPETE: POCI-01-0145-FEDER-007043 and FCT – Fundação para a Ciência e Tecnologia within the Project Scope: UID/CEC/00319/2013

References

1. Alnanih, R., Radhakrishnan, T., Ormandjieva, O.: Characterising context for mobile user interfaces in health care applications. Procedia Comput. Sci. **10**, 1086–1093 (2012)
2. Santos, M.F., Portela, F., Vilas-Boas, M.: INTCARE: multi-agent approach for real-time intelligent decision support in intensive medicine. In: 3rd International Conference on Agents Artificial Intelligence ICAART, vol. I (2011)
3. Portela, C.F., Santos, M.F., Silva, Á., Machado, J., Abelha, A.: Enabling a pervasive approach for intelligent decision support in critical health care. In: CCIS PART 3, vol. 221, pp. 233–243. Springer, Heidelberg (2011)
4. Portela, F., et al.: Knowledge discovery for pervasive and real-time intelligent decision support in intensive care medicine. In: MedicineKMIS (2011)
5. Braga, A., et al., Pervasive patient timeline for intensive care units. In: New Advances in Information Systems and Technologies, pp. 527–536. Springer International Publishing, Cham (2016)
6. Portela, F., Santos, M.F., Machado, J., Abelha, A., Rua, F.: Step towards pervasive technology assessment in intensive medicine. Int. J. Reliab. Qual. E-Healthc. IJRQEH **6**(2), 1–16 (2017)
7. Braga, A., et al.: Step Towards a patient timeline in intensive care units. Procedia Comput. Sci. **64**, 618–625 (2015)
8. Silva, B.M.C., Rodrigues, J.J.P.C., Lopes, I.M.C., Machado, T.M.F., Zhou, L.: A novel cooperation strategy for mobile health applications. IEEE J. Sel. Areas Commun. **31**(9), 28–36 (2013)
9. Rodrigues, R., et al., An intelligent patient monitoring system. In: International Symposium on Methodologies for Intelligent Systems, pp. 274–283 (2012)
10. Cardoso, L., Marins, F., Portela, F., Santos, M., Abelha, A., Machado, J.: The next generation of interoperability agents in healthcare. Int. J. Environ. Res. Public Health **11**(5), 5349–5371 (2014)
11. Varshney, U.: Pervasive computing and healthcare. In: Pervasive Healthcare Computing, pp. 39–62. Springer, Boston (2009)
12. Varshney, U.: Pervasive healthcare and wireless health monitoring. Mob. Netw. Appl. **12**(2–3), 113–127 (2007)
13. Saha, D., Mukherjee, A.: Pervasive computing: a paradigm for the 21st century. Computer **36**(3), 25–31 (2003)
14. Pereira, A., Portela, F., Santos, M.F., Machado, J., Abelha, A.: Pervasive business intelligence: a new trend in critical healthcare. Procedia Comput. Sci. **98**, 362–367 (2016)
15. Hevner, A.R., March, S.T., Park, J., Ram, S.: Design science in information systems research. MIS Q. **28**(1), 75–105 (2004)
16. Peffers, K., Tuunanen, T., Rothenberger, M.A., Chatterjee, S.: A design science research methodology for information systems research. J. Manag. Inf. Syst. **24**(3), 45–77 (2007)
17. Portela, F., et al.: Intelligent decision support in intensive care: towards technology acceptance. In: ESM 2012 - 26th European Simulation and Modelling Conference (2012)

18. Marins, F., Cardoso, L., Portela, F., Santos, M., Abelha, A., Machado, J.: Intelligent information system to tracking patients in intensive care units. In: Ubiquitous Computing and Ambient Intelligence. Context-Awareness and Context-Driven Interaction, pp. 54–61. Springer, Cham (2013)

19. Gago, P., Santos, M.F., Silva, A., Cortez, P., Neves, J., Gomes, L.: INTCare: a knowledge discovery based intelligent decision support system for intensive care medicine. J. Decis. Syst. **14**(3), 241–259 (2005)

Analysis of the Water Quality of the Monjas River: Monitoring and Control System

Marcelo León[1], Maritza Ruiz[2], Teresa Guarda[1(✉)],
Robert Montalvan[1], León Arguello[3], and Ana Tapia[1]

[1] Universidad Estatal Península de Santa Elena - UPSE, La Libertad, Ecuador
marceloleonll@hotmail.com, tguarda@gmail.com,
robertmontalvan@hotmail.com, lawyer_atb@hotmail.com
[2] Universidad de las Fuerzas Armadas - ESPE, Sangolqui, Quito, Ecuador
marilolita3@hotmail.com
[3] Universidad Técnica Estatal de Quevedo - UTEQ, Quevedo, Ecuador
larguello@uteq.edu.ec

Abstract. Running water available for human consumption inside Quito Metropolitan District began to be deficient by year 1960 due to an increase of its population. This water shortage caused several annoyances to its inhabitants up to the point that local government started on year 1990 new measures for a more effective management of its water resources. In 1992, ordinance 2910 ruled for the prevention and control of pollution produced by industrial liquid discharges and emissions into the atmosphere. Contaminating sources were manufacture of fabrics, metal products, food, beverages and tobacco with low biodegradability and high toxicity inorganic inputs and wastewater of Northern Quito majorly caused contamination of Monjas River. This research describes current water situation and proposes different alternatives of treatment plants based in a pervasive information systems.

Keywords: Water treatment · Water contamination · Territory organization
Pervasive information systems

1 Introduction

Currently, the law of territory planning requires increasingly personalized services, and more quickly. Public drinking water utilities are responding to these market needs through the use of comprehensive business intelligence, improving traditional business intelligence, with the ability to capture, interpret, and act on data immediately, to make faster decisions, in order to create a proactive and reactive interaction environment between the stakeholders in the business and providing the appropriate decision support information to managers based on the discovery of knowledge in the current data. These companies look for alternative ways to increase the value of their business intelligence initiatives. Increasingly, organizations are struggling to achieve Pervasive Business Intelligence (PBI) [1, 2]. The emergence of PBI is an natural evolution of business intelligence applications in organizations, with application from the strategic level to the operational level.

© Springer International Publishing AG, part of Springer Nature 2018
Á. Rocha et al. (Eds.): WorldCIST'18 2018, AISC 747, pp. 363–374, 2018.
https://doi.org/10.1007/978-3-319-77700-9_36

Ecuadorian water supply depends on conservation and stability of highlands, so deforestation, erosion and changes on rainfall causes a reduction of water source. At the beginning of 1990s this source began to be scarce in Quito and its water management considered possible solutions to protect it and manage the efficient use of its water resources through better practices.

The water resources available to Quito Metropolitan area (DMQ) are constituted by water from Esmeraldas river upper basin, surrounding groundwater and part of eastern subbasins and watershed of the Guayllabamba River [1]. The 2005 Water Quality Management Plan (PMCA), refers to additional water sources:

- San Pedro river: starts at 2.760 m above sea level and ends at Machángara at 2.080 m above sea level;
- Machángara river: starts at 2.180 m above sea level and it is nourished by several waterfall from South Quito; most of waste water from South Quito is discharged onto this river;
- Guayllabamba: starts at 2.080 m above sea level by mixture of San Pedro River and Machángara River; other sources are Chiche, Guambi, Uravia, Coyago, Pisque and Monjas Rivers;
- Monjas: starts at 2.470 m above sea level and joins to Guayllabamba River at 1.655 m above sea level. Most of waste water from North Quito is discharged onto this river.

Here four water treatment plants were considered: two Belgian plants (one for the treatment of Dyle River and the other one for treatment of Lasne River). Among the national ones we have considered Ucubamba (Cuenca) wastewater treatment plant and the drinking water treatment plant of Puengasi (Quito) in order to estimate the best plant layout required to solve the pollution problem of the Monjas River [2, 3].

The watershed of the Monjas River, is located north-west of the province of Pichincha, DMQ. The river Monjas begins at approximately 4440 m above sea level, and converges with the Guayllabamba River at 1655 m above sea level. It is part of the Guayllabamba river system, where the main river that crosses the whole basin is the Monjas (Fig. 1).

The Monjas River, which receives approximately 20% of the wastewater from Quito, is located in a sector with several changes in recent years due to the development of new neighborhoods and subdivisions with growthing population that has affected the conditions of the river and its environment, adding uploading urban and rainwater drainage from North Quito. The total area of the basin is 163.64 km^2 and crosses the urban area to the north of Quito. In Quito there are four main rivers: Machángara, Monjas, San Pedro and Guayllabamba that present an important level of contamination. This contamination does not only respond to mountains of rubbish, debris, furniture and even dead animals that exist along the rivers, but by the residual waters of the houses near this zone [4, 5].

Monjas River, considered the second most contaminated river in the capital because the sewer system of Quito Metropolitan area ends in the river (81% of pollution due to wastewater and 19% to industrial discharges). Pollution in Monjas River is quite

Fig. 1. Monjas river location.

visible. In different visits, there were found puddles of putrid and yellow water of about a meter deep. Its contamination affects surrounding inhabitants and vegetables crops and avocado trees that has been harvested for about 20 years. Affluent was small and clean but today has radically changed [6].

Area inhabitants' testimonies is that color and the smell of Monjas River usually becomes unbearable. This situation keeps citizens in fear for personal and their family well-being. A decontamination program for Quito's surrounding rivers plans the construction of interceptors to separate rainwater, residual and domestic waters, which are those that contaminate Monjas River affecting health of its population.

Three wastewater treatment plants (two in the south and one in the north) might be built to decontaminate the waters of Machángara, Monjas and San Pedro rivers. Other small ones will be built in the villages of the northeast Quito.

Quito's water company (EPMAPS) started in 2015 a campaign to prevent people from throwing rubbish or debris in riverbanks. Also, an analysis of water pollution and treatment plants construction studies, with the goal in mind that in the future the waters of the rivers may have recreational uses. This proposal, determines that remaining sludge will be treated for agriculture, livestock, forestation, soil recovery, landscape restoration, fertilizer or fuel for industrial furnaces. A theoretical basis that complements the selection of a treatment plant of the Monjas River was considered. Additionally, a standardized diagnosis on quality of water was proposed and finally, an analysis of four plants layout were presented to estimate best location according to the current regulations and permissible limits for discharges [2, 7].

2   Methodology

2.1   Monjas River Sampling

Three different samples were obtained and labeled from sources to analyze each parameter three times with a previous calibration of equipment.

> *PH:* approximately 25 mL of the water sample was taken, water pH was measured using the pH meter device and the obtained value was recorded.
> *Conductivity:* 25 mL of the water sample was taken, conductivity was measured using the conductivity meter device and the reported value was recorded.
> *Alkalinity:* in 50 mL of sample add 3 drops of indicator Phenolphthalein and 3 drops of methyl orange to finally add sulfuric acid (H_2SO_4), titrate the sample and record the volume spent
> *Hardness:* 10 mL of sample was placed, 1 mL of buffer solution formed by 16.9 g of NH_4Cl, 143 mL of NH_4OH and a flash of Helium Chromium was added. The sample takes on a pink color, but after shaking it takes a blue coloration

2.2   Morfometrics Parameters

Using the ArcGis 9.3 Calculate Geometry tool in the attributes table, the area in km^2 and the perimeter in km were calculated, both data are very important, delimiting the total volume that the basin receives during the precipitation and the surface and shape of the basin.

2.3   Monjas River Flow

In order to calculate flow of the Monjas River, an hydrological analysis was carried out with the information of the "Formulation and Implementation of the National Integrated and Integral Management Plan for the Water Resources of the Watersheds and Hydrographic Microbasins of Ecuador", elaborated by the Changjiang Institute of Survey Planning Design And Research - CISPDR, which contains an INAMHI database. After analysis of the available hydrological stations for the sector, it was established that the one with similar hydrogeomorphologic characteristics is the Alambi station in Churupamba H136, located in the coordinates 16226N, 757927E with an area of 442 km^2 and with registered data by a 16 years lapse. Factors of area variation and precipitation between the two basins.

2.4   Monjas River Map

The hydrographic maps of Monjas river basin were obtained and used for treatment plant location. In each map symbology shows the two points where the sampling was taken, the populated areas at scale 1: 120000.

2.5   Quality of Water at Monjas River

Water quality not only suffices with the latent appreciation of pollution, which for the Monjas River is estimable by the naked eye of experts and inexperienced people who pass through the river shores, but it was also necessary to determine Water Quality Index (WQI) or ICA (in Spanish) [8].

2.6   Technical Evaluation

Features of each of the four treatment plants models: Lasne treatment plant [9]; Dyle treatment plant; Treatment plant of Ucubamba and Puengasí treatment plant. An individual analysis of the treatment plants is presented to perform a compilation of the four plants for comparison. A detailed analysis of the flow diagram of the process of each of the treatment plants is carried out, as well as the detailed description of each unit operation that is part of the process and the general description of the equipment to then confront the data.

After comparing the evaluated technical data of the plants, the best option for the treatment of the residual waters of Quito is selected and later the location of the plant will be determined to avoid contamination of the river Monjas.

3   Results and Discussion

3.1   Sample Taken

Results of the water analysis (Table 1), the analysis was performed at the coordinates [utm]: coordinates $X = 17M0779663$; coordinates $Y = 9990022$.

Table 1.   Results of water sample at upper area.

#	Time	Temp. (°C)	pH	O.D. (ppm)	Temp. (°C) environment
1	7:40	18.9	7.35	2	19.7
2	8:40	19.7	7.52		19.9
3	9:40	20.3	7.70		20.9
4	10:40	20.9	7.68		21.5
5	11:40	22.1	7.82		22.7
Promedio		20.4	7.61	2	21.0

In Table 2 the results of the sensorial characteristics of the sample are observed. Equivalence is given by: + Scarce; ++ Medium; +++ Plentyful. In Table 3 the results of the laboratory analysis are observed.

Table 2. Lab analysis results upper water area

#	Time	Appearence	Color	Smell	Solids	Foam
1	7:40	+	+	+ +	+	+
1	7:40	+	+	+ +	+	+
2	8:40	+	+	+ +	+	+
3	9:40	+	+	+ +	+	+
4	10:40	+	+	+ +	+	+
5	11:40	+	+	+ +	+	+
Promedio		+	+	+ +	+	+

Table 3. Water results from upper water area

Parameter	Unit		Max permissible limit	Total value	Acceptance criteria
Oil and fat	Solubles hexano	mg/l	30	<20	Fulfill
Residual Chlorine	Cl2	mg/l	0.5	0.36	Fulfill
Color	Color	Color	0	119	Not fulfill
Electric Conductivity	CE	µS/cm	N/A	227	N/A
Biochemical Oxygen Demand	DBO	mg/l	100	32	Fulfill
Chemical Oxygen Demand	DQO	mg/l	160	46	Fulfill
Phosphates	PO4	mg/l	N/A	4.45	N/A
Nitrates	NO3	mg/l	N/A	12.0	N/A
Nitrites	NO2	mg/l	N/A	3.84	N/A
Dissolved Oxygen	OD	mg/l	N/A	2,0	N/A
Hydrogen Potential	pH	pH	6 a 9	7.6	Fulfill
Suspended Solids	SST	mg/l	80	155.0	Not fulfill
Total Solids	ST	mg/l	N/A	434	N/A
Sulfates	SO4	mg/l	1000	56	Fulfill
Temperature	T	°C	35	20.4	Fulfill
Tensoactives	MBAS	mg/l	0,5	0.332	Fulfill
Opacity	–	NTU	5	44	Not fulfill

Water analysis results (Table 4), coordinates [utm]: Coordinates X = 17M0784874; Coordinates Y = 9998789.

Table 4. Water results from down water area.

#	Time	Temp. (°C)	pH	O.D. (ppm)	% HR	Temp. (°C) environment
1	1:00	21.7	8.12	3	64	26.3
2	1:40	22.3	8.17		63	26.8
3	2:30	22.5	8.20		61	27.3
4	3:30	23.7	8.51		62	27.0
5	4:40	23.4	8.50		63	26.5
Promedio		22.7	8.30	3	63	26.8

In Table 5 sensory characteristics are upstream. Equivalence equal to the previous one. Table 6 shows the results of the laboratory analysis.

Table 5. Sensorial water results from upper water area.

#	Time	Appearence	Color	Smell	Solids	Foam
1	7:40	+++	++	+ +	+++	+++
1	7:40	+++	++	+ +	+++	+++
2	8:40	+++	++	+ +	+++	+++
3	9:40	+++	++	+ +	+++	+++
4	10:40	+++	++	+ +	+++	+++
5	11:40	+++	++	+ +	+++	+++
Promedio		+	+++	++	+++	+++

3.2   Gravelius Index

Relationship between the perimeter of the basin and the perimeter of the circle. The more irregular the basin, the greater its coefficient of compactness. There is a greater tendency to increase as this number closer to 1.

$$kc = \frac{Basin\ Perimeter}{Circle\ Perimeter} = 0.282 * \frac{P}{\sqrt{A}}$$

$$kc = 1.56$$

(1)

Table 6. Lab water results from down water area

Parametro	Unidad		Límite máximo permisible	Valor total	Criterio aceptación
Oil and fat	Solubles hexano	mg/l	30	28,4	Fulfill
Residual Chlorine	Cl$_2$	mg/l	0,5	0,88	Not fulfill
Color	Color	Color	0	374	Not fulfill
Electric Conductivity	CE	µS/cm	N/A	667	No aplica
Biochemical Oxygen Demand	DBO	mg/l	100	36	Fulfill
Chemical Oxygen Demand	DQO	mg/l	160	63	Fulfill
Phosphates	PO$_4$	mg/l	N/A	6,15	N/A
Nitrates	NO$_3$	mg/l	N/A	12,0	N/A
Nitrites	NO$_2$	mg/l	N/A	0,64	N/A
Dissolved Oxygen	OD	mg/l	N/A	3,0	N/A
Hydrogen Potential	pH	pH	6 a 9	8,3	Fulfill
Suspended Solids	SST	mg/l	80	388,0	Not fulfill
Total Solids	ST	mg/l	N/A	911	N/A
Sulfates	SO$_4$	mg/l	1000	42	Fulfill
Temperature	T	°C	35	22,7	Fulfill
Tensoactives	MBAS	mg/l	0,5	0,655	Not fulfill
Opacity	–	NTU	5	123	Not fulfill

3.3   Form Factor (Kf)

Relationship between the mean width and the axial length of the basin.

Average wide: $B = A/L$

$$B = \frac{163,64}{12,51} = 13,08$$

Form Factor: $Kf = B/L$ \hfill (2)

$$Kf = \frac{13,08}{12,51} = 1,04$$

B = Average wide; A = Basin area; L = Axial Length

3.4   Monja River Basin Flow

In Table 7 shows the minimum, average and maximum monthly flows for the point of interest, thus determining an average monthly flow of 2,60 m^3/s.

The average winter flow is approximately 3,9 m^3/s and summer time is 1,3 m^3/s.

Table 7. Determination of monthly flow data.

Month	Q min (m³/s)	Qaverage (m³/s)	Q max (m³/s)
January	1.5340	2.9565	5.5237
February	1.6917	4.1030	9.2900
March	2.8458	4.5585	8.1944
April	2.4857	4.9723	7.6895
May	2.3013	4.3634	7.4604
June	1.6597	2.2814	3.0141
July	1.2317	1.5867	2.4008
August	0.7803	1.1325	1.6410
September	0.5162	1.0966	2.3984
October	0.7077	1.2150	2.3065
November	0.5065	1.3308	5.1071
December	0.6217	1.6311	6.9227

3.5   Monjas River Maps

Figure 2 shows the hydrographic map of the river basin of the Monjas that will serve for the location of the treatment plant.

In each map the symbology is shown the two points where the sampling was taken, the populated areas at scale 1:120000.

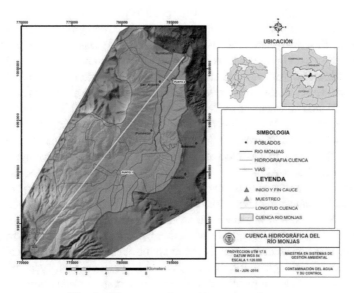

Fig. 2. Histogram of monthly average flows.

3.6   Water Quality of Monjas River

In the WQI water quality index, it is calculated with 9 parameters: fecal coliforms (NMP/100 mL), pH (pH units), biochemical oxygen demand in 5 days (BOD5 in mg/L), Nitrates In mg/L), Phosphates (PO4 in mg/L), temperature change (°C) turbidity (FAU), total dissolved solids (mg/L), dissolved oxygen (OD in % saturation). In Tables 3 and 6 the results of these parameters and Table 8 shows WQI interpretation to classify water quality [9].

The WQI result for upstream was 37.22, and the result for low waters of 27.54, the two bad waters, which can only support a low diversity of aquatic life and are probably experiencing problems with pollution.

Table 8. Water quality classification according to calculated WQI.

Quality of water	WQI value
Excellent	91 a 100
Good	71 a 90
Fair	51 a 70
Bad	26 a 50
Pésima	0 a 25

3.7   Selection and Location of Treatment Plant

According to EPMAPS, future demand for potable water will grow from 9000 L/s in 2010 to 14000 L/s by year 2040, assuming a high population growth. The supply of the flow rates to 95% of the systems is 7000 L/s. Drinking water sources for DMQ are limited, water demand requirements will grow to 233 L/s annually during the next 30 years [10]. Monjas River originates from eastern slopes of the Rucu Pichincha volcano and flows into the Guayllabamba River at an altitude of 1,660 m above sea level in San Antonio de Pichincha township, with an average winter flow of 3.9 m^3/s and 1,3 m^3/s and an average flow of 2,60 m^3/s per month. The Dyle plant has a capacity for a flow of 8.300 L/s; higher than is required for the Monjas River, this oversize excludes the selection of this plant to be selected as a treatment plant in this Monjas river. The Ucubamba plant treats 2,200 L/s, lower monthly flow (2,600 L/s) of the same source, without estimating an increase during the winter. In addition, the surface is extensive to carry out the biological treatment. The capacity of the Puengasi plant is 2400 L/s but it only treats 1800 L/s of raw water with a modern clarification system.

Based on the data collected the pervasive system provides the data in real time assisting the manager in making the decision on the choice of the most suitable plant, in this case the selected plant for this project could be the Lasne type, which treats wastewater with 2 plants in parallel considering the average monthly flow of 2.60 m^3/s, but with the average winter flow of 3.9 m^3/s it is necessary to have three plants which in summer (average flow of 1.3 m^3/s), each plant works at a third of its capacity. It takes 21 ha of land in the city of Quito. The proposed place of coordinates

−0.0844508, −78.4836454, 15.18 z, is located at Avenida Manuel Cordova Galarza Km 2 ½ via Mitad del Mundo. This selected place has 40 ha available to locate plants in parallel.

4 Conclusion

PBI empowers people at all levels of the organization with timely analysis, alerts and feedback tools. PBI systems have to reflect the real-time concept for a particular business, that is, the right time. The system provides the data in real time, helping the manager in the decision make process, to chosing the most appropriate plant choice. In this context the Lasne treatment plant was selected to treat waters from Monjas River. The chosen plant minimizes the impact on landscape, noise reduction and atmospheric. The three plants located in parallel will occupy an area of approximately 21 ha and the individual assessment of the Lasne treatment plant investment is around 26 million euros. The process followed to treat the effluent will report values within the latest European standards for phosphorus and nitrogen spills. The system will maintain a constant monitoring of the quality of the effluent discharged, complying with the current regulations regarding BOD and COD, thus improving aquatic life. The plant guarantees a physical pre-treatment process that will remove the thick and thin solids suspended in the water to be treated, facilitating the process of biological treatment in the system. The stabilization of the sludge will certify to eliminate the excess of activated sludge which is no longer reusable. These criteria allow to adapt to the Plan of Territorial Ordinance of Pichincha, the Water component and its identification as to the quality and future distribution.

References

1. Ortiz, S.: Taking business intelligence to the masses. Computer **43**, 12–15 (2010)
2. Portela, F., Santos, M.F., Machado, J., Abelha, A., Silva, Á., Rua, F.: Pervasive and intelligent decision support in intensive medicine – the complete picture. In: Bursa, M., Khuri, S., Renda, M.E. (eds.) Information Technology in Bio- and Medical Informatics. Lecture Notes in Computer Science (LNCS), vol. 8649, pp. 87–102. Springer, Cham (2014)
3. Camacho, D.: Esquemas de pagos por servicios ambientales para la conservación de cuencas hidrográficas en el Ecuador. Investigación Agraria: Sistemas y recursos forestales **17**, 54–66 (2008)
4. Programa de Saneamiento Ambiental para el Distrito Metropolitano de Quito (PSA): Estudios de actualización del plan maestro integrado de agua potable y alcantarillado para el DMQ, EPMAPS (2011)
5. Da Ros, G.: La contaminación de aguas en Ecuador: una aproximación económica. Editorial Abya Yala (1995)
6. Distrito Metropolitano de Quito: Perspectivas del ambiente y cambio climático en el medio urbano. ECCO (2011)
7. Campaña Lozano, R., Gualoto Kirochka, E.: Evaluación físico-química y microbiológica de la calidad del agua de los ríos Machángara y Monjas de la red hídrica del distrito metropolitano de Quito (DMQ) (2015)

8. EMAPS: Machángara, Monjas y San Pedro entraran en plan de descontaminación total. In: Noticias Quito. http://www.noticiasquito.gob.ec/Noticias/news_user_view/machangara_monjas_y_san_pedro_entraran_en_plan_de_descontaminacion_total–12391. Accessed 2014

9. Srebotnjak, T., Carr, G., Rickwood, C.: A global Water Quality Index and hot-deck imputation of missing data. Ecol. Ind. **17**, 108–119 (2012)

10. Asadollahfardi, G.: Water quality indices (WQI). In: Water Quality Management, pp. 21–39. Springer, Heidelberg (2015)

11. Ávila, L., Guerrón, M., Carolina, A., Flores, L., Andrade, L.: Programa para la Descontaminación de los Ríos de Quito, PDRQ (2014)

Pervasive Smart Destinations

Teresa Guarda[1,2,3(✉)], Lidice Haz[4], Maria Fernanda Augusto[1], and José Avelino Vitor[5]

[1] Universidad Estatal Península de Santa Elena - UPSE, La Libertad, Ecuador
tguarda@gmail.com, mfg.augusto@gmail.com
[2] Universidad de las Fuerzas Armadas - ESPE, Sangolqui, Quito, Ecuador
[3] Algoritmi Centre, Minho University, Braga, Portugal
[4] Universidad de Guayaquil, Guayaquil, Ecuador
victoria.haz@hotmail.com
[5] Instituto Politécnico da Maia, Maia, Portugal
javemor@iiporto.com

Abstract. In the current competitive environment, the maturation of worldwide markets, lead companies to rethink and align new business strategies that allow them to be more competitive. To be competitive in contemporary society is to be hostage to the use of adopted information and communication technologies. Technological advances lead to the creation of new opportunities in all areas, including tourism. These transformations require managers to position themselves against the current reality. The current challenges point at the evolution of tourism, more specifically to the path from the evolution of touristic destinations to smart destinations. In this study, we propose a framework for a system that obtains tourist trends and provides touristic information through the use of beacons and a cloud service.

Keywords: Smart destinations · Mobil technologies
Pervasive smart destination · Proximity Based Technology · Beacons
Cloud service

1 Introduction

Nowadays, any organization to be competitive is hostage to the right technology to meet the needs of customers. In tourist activity, the need for ICT (Information and Communication Technologies) is even more relevant, in order to sell a tourism product it is necessary to disclose the information that characterizes it, organize and manage that information, so that it can be sold in accordance with the customer's expectations. After consumption of the tourist product, it is important to share the experience gained with other travelers. The process of buying and consuming a tourism product is only possible if tourists and professionals in the sector have access to the information they want, and for this it is necessary to use systems that can planning the trips, manage flights and accommodation, retain customers, among others. All these systems have in common the management of tourist information, which allows concluding that they are indispensable to the touristic activity.

© Springer International Publishing AG, part of Springer Nature 2018
Á. Rocha et al. (Eds.): WorldCIST'18 2018, AISC 747, pp. 375–382, 2018.
https://doi.org/10.1007/978-3-319-77700-9_37

Domestic tourism in Ecuador generated 12.3 million trips during 2016, according to the General Coordination of Statistics and Research of the Ministry of Tourism. 40% of these trips were made during the different holidays, which represented more than 4.9 million trips. All this movement facilitated to boost the national economy in USD 285.5 million, highlights the Secretary of State through its website. Only the End of the Year holiday, which was held between December 31, 2016 and January 2, 2017, generated 618,972 trips, with an economic movement that exceeded USD 37.6 million. The most visited province was Santa Elena, followed by Manabí and Esmeraldas. In 2015, 6.3 million trips were registered, which meant an economic movement of USD 531.2 million [1].

The tourist sector has undoubted influences in Ecuador, with a growing influx of visitors searching of everything that all provinces have to offer. The constant evolution and rapid acceptance of mobile technologies allow not only to customize these experiences but also to improve them.

Tourism promotion is an important regional activity to revitalize regional exchange and intergenerational communication. The province of Santa Elena has many tourist resources with much potential that are not well known due to lack of information. As part of our research, we intend promoting local tourism in Santa Elena, Ecuador, and for that we present the local touristic tracking and promoting system (LTTPS) framework for the Peninsula of Santa Elena, projecting the results achieved by collecting information through the beacons.

The paper is organized as follows: the second section discusses the areas related to the topic of study; the third is about beacon real time tracking system; the fourth section presents the LTTPS framework; to conclude the fifth section gives the conclusions.

2 Pervasive Smart Destinations

Tourist destinations are booming supported by the use of ICT; new modes of communication; and new ways of collecting and analyzing data, both on a personal and business level. In this context, new opportunities for management and value creation arise [2].

Technologies such as computing cloud computing and the Internet of Things (IoT) have unleashed a new perspective and new opportunities in bringing innovative services to tourists, organizations and businesses linked to tourism [3].

A tourist destination is the place visited which is the central point for the decision to take a trip. An Smart Destination is an innovative tourist space, accessible to all, consolidated on an avant-garde technological infrastructure that guarantees the sustainable development of the territory, facilitates the interaction and integration of the visitor with the environment and increases the quality of its experience in the destination and the quality of life of the residents [4–6].

Although intrinsic and created resources are the main motivators for tourists' choices, prosperous tourist destinations also depend on resources and support services. Resources/support services have a side effect on the motivation of tourists and provide a basis on which a destination can be established, such as infrastructure, transport, communications, security, and so on. However, the availability of resources is not enough for a destination to establish its position in the market.

A destination that establishes a vision for tourism, shares this vision among all stakeholders, has a management that develops a proper marketing strategy and a government that supports the tourism industry, with an efficient tourism policy, tends to be more competitive than those who have never asked themselves what role tourism plays in their economy [7]. Thus, the competitiveness of destination is mediated by management, which can increase the appeal of key resources, enhance the quality and effectiveness of support resources, and better adapt fate to the constraints imposed by situational conditions.

The concept of Smart Destination emerges within the context of Smart City. The incorporation of technology into the environment has the potential to enrich the experiences of competitiveness of the destination [8].

Smart Destination can be defined as an urban tourism platform integrated with ICT. This platform dynamically links the entities involved in tourism and information technology to collect, create and exchange information that can be used to enrich tourism experiences in real time [3, 7, 8]. For Wang et al., a Smart Destination incorporates the use of ICTs in the development and production of tourism processes [9].

A Smart Destination is a complex infrastructure of systems, including a wide range of technologies that provide direct support to tourism, such as decision support and recommendation systems; systems that detect the context; autonomous agents that search and collect data from the Web; and also systems that create increased realities. Intelligent systems seize the resources generated by these technologies to provide intelligent applications for tourism, such as the use of monitoring devices, services based on visitor location, and recommendation services [10–12].

The number of European and Asian countries where the concept of Smart Destination appears as an integral part of national economic development policy is increasing [2].

In an extremely competitive market of tourist destinations, competitiveness is based on the resources of the destination and on the ability to mitigate them [7]. The IoT is full of opportunities to produce and deliver new products and services through the interconnection of electronic devices and various wearable's carried by people. Sensors, RFID (Radio Frequency Identification), NFC (Near Field Communication), and smart mobile networks provide pervasive Apps and services based on the user's location and context [12]. These technologies, devices and objects enable the creation of intelligent systems in tourism, which is a major application domain for intelligent systems due to the overall complexity of decisions to be made in moving tourist contexts. Intelligent tourism systems represent next-generation information systems that promise to provide tourism consumers and service providers with more relevant information, greater decision support, greater mobility, and ultimately, more enjoyable tourism experiences. They cover a wide range of relevant travel and tourism technologies, such as recommendation systems, context-aware systems, autonomous agent research, and mining network and environmental intelligence capabilities.

3   Beacon Real Time Proximity Based Technology

Proximity Based Technology (PBT), are all the technologies that have a fixed location, have some kind of possibility of data transmission, and in which we can limit the area where they can be associated with other entities. There are many technologies that meet these criteria to make them technologies based on proximity, being Beacon the selected technology for this work.

The development and standardization of low-power wireless technologies, short-range wireless technologies, have led to new concepts of pervasive computing. There are currently several options available for the development of a pervasive computing environment [13]. These wireless sensors have several advantages, such as adaptability, flexibility, adaptability and low power consumption.

Beacons are small transmitters or low-power unit that broadcasts its identification using Bluetooth that enables communication with the context of the user's device (smartphones) present in a predefined radius [13], usually 70 m, although it is possible to make the signal stronger and cover areas greater than 1,000 m perimeter. Beacons allow smartphones to know the context it is part of. These communication devices, which are more commonly seen in the retail industry, are increasingly being used in various service industries, especially those dealing with large audiences such as tourism and hospitality, education and sport. Beacons work through a push or emission of data made through Bluetooth technology and that must correspond in an application installed on a smartphone inside the covered area [14]; it is possible to send text (small messages, information about products or services, proximity to stores, promotions or other) or media (ads, images, discount coupons).

Beacons are extremely accurate and relevant because they are only available within a certain perimeter or radius of the store for which they are intended [15].

Imagine that you are sightseeing in a foreign city, and along the most interesting areas there are a few beacons stuck to the walls. These can issue signals with historical building data, facilitate their opening hours, offer discount coupons in the corresponding entries.

4   LTTPS Framework

Use of the IoT in the field of culture and tourism in a city\region that starts from the user, whether resident or tourist, as a pillar on which to evolve and towards which to focus these technologies, detecting in it two potential cases of use for Smart Points of Interest: as a co-creation tool to ask the user about their interests, locating the questionnaires at strategic points; and as an online tool to include multimedia information on cultural heritage, museums, interesting activities, quick ticket purchase, restaurant location, hotel location, etc. that in a physical way could not be provided to the user. In this way, the use of the IoT allows opening a door to a philosophy of co-creation with the visitor and citizen, a differential value for projects of cultural and tourist nature that contributes to the creation of Smart Destinations, thus allowing a reality trans-media that combines the physical cultural experience as well as the multimedia possibilities of the online world.

Santa Elena is the youngest province of the current 24 and has a holiday infrastructure and a rich variety of archaeological, historical, natural, cultural attractions, extensive beaches and fishing villages.

Salinas and Montañita are the two great icons of the province and true magnets to attract national and foreign tourists. The economy is based on tourism and fishing. The peninsula has important fishing ports: Santa Rosa, San Pedro and Chanduy.

The province of Santa Elena is very popular in holiday season and holidays, among the attractions include: Salinas, Montañita, Chocolatera, Museo los Amantes de Sumpa, Baños de San Vicente, Acuario de Valdivia, Malecón la Libertad, Manglaralto, Colonche, Playa Punta Blanca, Ayangue, Museo Real Alto, Ballenita, Ancón, Playa San Pablo, Dos Mangas and others.

In this work we propose a framework to create intelligent points of interest that contribute to the development of Smart Destinations in Peninsula de Santa Elena.

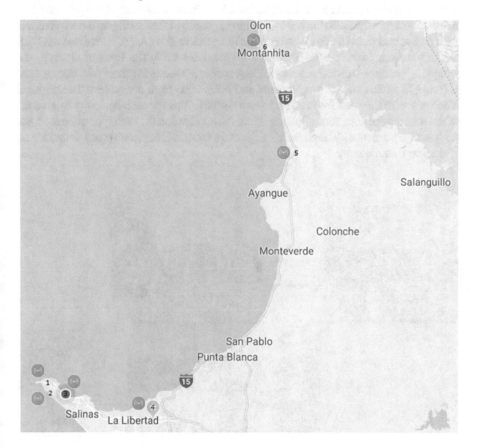

Fig. 1. SmartPOIs localization: ● 1. Chocolatera; ● 2. La Loberia; ● 3. Malecón de Salinas; ● 4. Museo los Amantes de Sumpa; ● 5. Acuario de Valdivia; ● 6. Playa de Montañita.

To achieve this goal, different technologies are analyzed and understood, and IoT can contribute to the sector, thus discovering its strengths and weaknesses as well as its different lines of action in the field of tourism and cultural heritage. Beacons are the key technological tool for the creation of smart points of interest (SmartPOI). These Bluetooth Low Energy (BLE) Beacons are the instrument that allows creating the connection between the online and offline world in a specific point of interest (POI).

Several SmartPOIs were placed in different points of interested in Peninsula of Santa Elena, to send to the user\tourist a notification that directs them to a Web App with a suggestion box related to the place (Fig. 1).

This research is supported in turn by various technological advances such as the Physical Web, a concept that represents a before and after in the use of beacon technology. Before Chrome appearance, the user had to download a beacon tracking app or one specifically designed for it. Now, with the Chrome App, the emission of any URL (Uniform Resource Locator) that is being broadcast in the environment is detected, when the Bluetooth and the GPS (Global Positioning System) of the intelligent device is turned on. Another tool used in this research is the Progressive Web Apps, with these the URL that the Beacon emits can be directly a responsive web in the form of an App, not occupying space on the device and avoiding access problems for users who do not want to install it. In this way it can be made accessible also from computers. The physical Web ends with the need for users to have to install Apps for each company to give the public beacons, or companies that use them. An application for each type of beacon and use is not very practical. The physical Web works with the Eddystone protocol and works by finding URLs nearby.

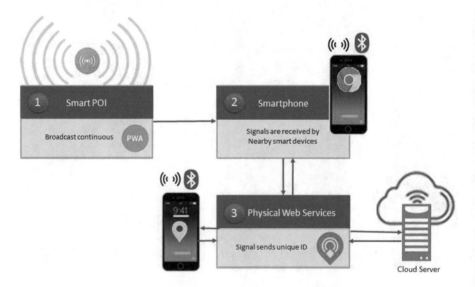

Fig. 2. Framework for Smart Points of Interest.

The SmartPOI have direct connection via Bluetooth Low Energy with devices as SmartPhones, IPods, or Tablets in multiplatform environments. There are two roles assigned to use the SmartPOI, the system administrator, and the user. In system administration the following processes are performed: configuration of the Bluetooth enabled device in the real world; make SmartPOI, providing the visible to all, providing the context and positioning for its application; and engage users. Regarding the user, it goes through the following steps: going through the physical world, with Bluetooth enabled; receive notifications with information; and choose to interact with the notification.

In the proposed framework (Fig. 2) configured BLE beacons localized in the points of interest are broadcasting in continuous with PWA. The signs will be received by the nearby smart devices using BLE (Smartphones, Tablets, Ipods), providing it's unique ID number to the smart device. After that user will decide accept or not the notification. In case of acceptance of the notification the smart device sends the ID number to the cloud server, which checks the action to be assigned to that ID number and responds.

5 Conclusions

The growing interaction between visitors and destinations through ICT is something that deserves to be better explored by public, private and aided by the academy before the undeniable change of behavior of people who rely more and more on digital media for research, use and memories of their travels.

Currently technology allows users to select what they need at any given moment, be it information, location of places, among others; thus creating a network that gathers everything necessary to live an optimal experience. By locating Smart POIs at strategic points, from a tourist point of view, is possible send specific information on a Progressive Web App, via Bluetooth, to nearby intelligent devices that inform tourists.

Because of these transformations the cities have perceived this change and begin to adapt to its residents and visitors. While cities have become smart by improving interaction with their citizens through ICT, especially in the governance aspect that is fundamental to the development of the whole (economy, quality of life, environment, mobility and society), the Smart Destination are destinations where accessibility, innovation, sustainability and technology are their main pillars.

References

1. MT: Ministerio de Turismo. In: El Turismo dinamizó en 285.5 millones de dólares la economía el (2016). http://www.turismo.gob.ec/el-turismo-dinamizo-en-285-5-millones-de-dolares-la-economia-el-2016/
2. Gretzel, U., Reino, S., Kopera, S., Koo, C.: Smart tourism challenges. J. Tourism 16(1), 41–47 (2015)
3. Boes, K., Buhalis, D., Inversini, A.: Conceptualising smart tourism destination dimensions. In: Tussyadiah, I., Inversini, A. (eds.) Information and Communication Technologies in Tourism 2015, pp. 391–403 (2015)
4. Benckendorff, P., Sheldon, P., Fesenmaier, D.: Tourism Information Technology. Cabi, Wallingford (2014)

5. Buhalis, D.: eTourism: Information Technology for Strategic Tourism Management. Pearson Education, Harlow (2003)
6. Trépanier, M., Tranchant, N., Chapleau, R.: Individual trip destination estimation in a transit smart card automated fare collection system. J. Intell. Transp. Syst. **11**(1), 1–14 (2007)
7. Oye, N., Okafor, C., Kinjir, S.: Sustaining tourism destination competitiveness using ICT in developing countries. Int. J. Comput. Inf. Technol. **2**(1), 48–56 (2013)
8. Buhalis, D., Amaranggana, A.: Smart tourism destinations. In: Xiang, Z., Tussyadiah, I. (eds.) Information and Communication Technologies in Tourism, pp. 553–564. Springer, Cham (2014)
9. Wang, D., Li, X., Li, Y.: China's "smart tourism destination" initiative: a taste of the service-dominant logic. J. Destination Market. Manag. **2**(2), 59–61 (2013)
10. Zhao, X., Wang, S.: Application of smart technology in the integrated environmental management of urban wetland park. Int. J. U E Serv. **8**(7), 243–250 (2015)
11. Xiang, Z., Tussyadiah, I., Buhalis, D.: Smart destinations: foundations, analytics, and applications. J. Destination Market. Manag. **4**, 143–144 (2015)
12. Lamsfus, C., Martín, D., Alzua-Sorzabal, A., Torres-Manzanera, E.: Smart tourism destinations: an extended conception of smart cities focusing on human mobility. In: Spring, C. (ed.) Information and Communication Technologies in Tourism, pp. 363–375 (2015)
13. Gubbi, J., Buyya, R., Marusic, S., Palaniswami, M.: Internet of Things (IoT): a vision, architectural elements, and future directions. Future Gener. Comput. Syst. **29**(7), 1645–1660 (2013)
14. Portela, F., Santos, M., Silva, Á., Machado, J., Abelha, A., Rua, F.: Pervasive and intelligent decision support in intensive medicine – the complete picture. In: Information Technology in Bio- and Medical Informatics. Lecture Notes in Computer Science (LNCS), vol. 8649, pp. 87–102. Springer, Cham (2014)
15. Haartsen, J., Mattisson, S.: Bluetooth-a new low-power radio interface providing short-range connectivity. Proc. IEEE **88**(10), 1651–1661 (2000)
16. Raza, S., Misra, P., He, Z., Voigt, T.: Building the Internet of Things with bluetooth smart. Ad Hoc Netw. **57**, 19–31 (2017)
17. Ott, J., Kärkkäinen, L., Walelgne, E.A., Keränen, A., Hyytiä, E., Kangasharju, J.: On the sensitivity of geo-based content sharing to location errors. In: 2017 13th Annual Conference Wireless On-demand Network Systems and Services (WONS), pp. 25–32. IEEE (2017)

Technologies in the Workplace - Use and Impact on Workers

Social Media Use in the Workspace: Applying an Extension of the Technology Acceptance Model Across Multiple Countries

Samuel Fosso Wamba[(⊠)]

Toulouse Business School, Toulouse, France
s.fosso-wamba@tbs-education.fr

Abstract. Social media technologies and tools are emerging as important source of firm business value creation. In this study, an extended version of the technology acceptance model (TAM) that integrates perceived risk and security is used to assess the acceptance of social media within the workspace across multiple countries and to test for user's behaviors homogeneity. In addition, the study investigates the moderating effects of user computer experience on the relationship between user's attitude and intention to use social media. To test the proposed model, the study uses data gathered from the US, Australia, the UK, Canada, and India. In the data analysis process, the study uses the full data and data from each country. The results detect the existence of user's behaviors heterogeneity across the countries under study; confirm the robustness of the TAM in the context of social media within the workspace. Finally, implications for research and practice are proposed.

Keywords: Social media · Adoption and use · Intention · TAM
Perceived risk · Perceived security · Computer experience

1 Introduction

Social media technologies and tools are emerging as important source of firm business value creation. They have been considered as the driving forces behind the growth of social commerce [1]. Social media offer new ways to interact with a given firm key players [2, 3], including real-time one-to-one communication with online consumers. In addition, online consumers can now use social media to communicate in real-time with their peers before their final purchasing decisions [4]. The social media high business value has attracted firms across various sectors, as "social media positively influences most companies' revenue and sales" (p. 1) [5]. Indeed, it is estimated that about two billion people are current active users of social media. This number will reach about 2.5 billion in 2018 [5]. In addition, advertising revenues generated by social media are quite impressive. They went from about 17.85 billion U.S. dollars in 2014 to reach the notable amount of 41 billion U.S. in 2017 [6].

While the high operational and strategic value of social media has been acknowledged by prior studies [7–11], very few studies have explored key factors of social media adoption and use within workspace across many countries. Therefore, the

© Springer International Publishing AG, part of Springer Nature 2018
Á. Rocha et al. (Eds.): WorldCIST'18 2018, AISC 747, pp. 385–392, 2018.
https://doi.org/10.1007/978-3-319-77700-9_38

main objective of this study is to examine the main factors that influence social media adoption and use by organizations across multiple countries. To achieve our research objective, the study aims at examining the following research questions:

1. Is the extended TAM suitable to study social media adoption and use within organizations across multiple countries?
2. What is the moderating effect of user computer experience on the relationship between user attitude and the behavioral intention to use social media?

To address our research questions, the study draws on the emerging literature on social media, an extended version of TAM that integrates two new constructs namely: perceived risk (PR) and perceived security (PS). In addition, the study tests the moderating effect of user computer experience on the relationship between user attitude and the behavioral intention to use social media. Then, the proposed model is tested using the collected data from each country and the pooled or global data. After the introduction, we first present our theoretical development. Then, in the next section, our research methodology is discussed. The subsequent section presents and discusses our results. Then, the discussion, implications, limitations and future research perspectives section is presented. In the final section, we present the conclusion of the study.

2   Theoretical Development

Our proposed research model (Fig. 1) was developed based on the emerging literature on social media, an extended version of TAM that integrates two new constructs namely: perceived risk and perceived security.

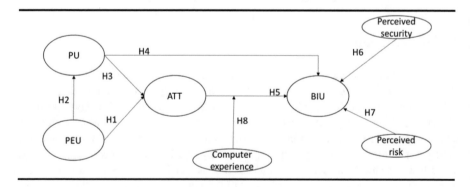

Fig. 1. Proposed research model

The TAM is probably the most used research framework in the information systems field [12–15]. The model was first developed by Davis [12] to explore the reasons behind the acceptance or the rejection of information technology (IT) by potential users using a small set of constructs. TAM argues that perceived usefulness (PU) or the "degree to which a person believes that using a particular system would enhance his or her job performance" (p. 320) [12] and perceived ease of use (PEU), which is defined as "the degree to which a person believes that using a particular system would be free of effort" (p. 320) [12] are the two particular beliefs that are of "primary relevance for computer acceptance behaviors" (p. 985) [15]. The model also identifies a positive relationship between PEU and PU. Furthermore, TAM posits that IT usage is determined by the user's behavioral intention, which in turn is jointly explained by the user's attitude toward (ATT) using the computer system and PU.

Drawing on the above discussion, we propose the following hypotheses in the context of social media use within organizations (Fig. 1):

H1: PEU has a significant positive effect on ATT.
H2: PEU has a significant positive effect on PU.
H3: PU has a significant positive effect on ATT.
H4: PU has a significant positive effect on the user's behavioral intention to use (BIU).
H5: ATT has a significant positive effect on BIU.

In addition to these five hypotheses that are the founding blocks of the TAM, we argue that in the context of social media PS or "the subjective probability with which users believe their sensitive information (business or private) will not be viewed, stored, and manipulated during work sessions by unauthorized parties in a manner consistent with their confident expectations" (p. 165) [16] and PR or "the extent to which a functional or psychosocial risk a user feels he/she is taking when using a product" (p. 165) [16] are key determinants of the behavioral intention to adopt social media within organizations. Indeed, in the current digital world where a massive amount of data is generated and use by users as well as exchanged between various firm stakeholders, ensuring that private users information is stored safely is an important step toward facilitating the acceptance of social media within organization.

Therefore, we propose the following hypotheses (Fig. 1):

H6: PS has a significant positive effect on BIU.
H7: PR has a significant negative effect on BIU.

Computer experience which is defined as "a level in which someone have ever used a technology to ease his/her work" (p. 27) [17], has been viewed as an important adoption factor in the diffusion theory [17]. It has been used as a moderator for the relationships between effort expectancy and behavioral intention and social influence and behavioral intention in the context of consumer acceptance of personal information and communication technology services [18]. Drawing on this discussion, we hypothesize the following:

H8: Computer experience moderates the relationship between user's attitude and its behavioral intention to use social media within workspace.

3 Methodology

Our study uses a web based-questionnaire to collect data from 2,556 social media users in their workplaces in various countries including: the US, Australia, the UK, Canada, and India. The data collection was realized in January 2013 by a market research firm called *Survey Sampling International (SSI)*. Our study uses items derived from the existing literatures. They were adapted to the social media adoption and use in firms context [12, 16], and measured with a seven-point Likert scale.

For the data analysis, a partial least squares (PLS) structural equation modeling (SEM) called SmartPLS tool version 3.0 was used to assess the measurement and structural model [19]. The study assesses the reliability and validity of all the items. More precisely, the study looks at the item loadings values, the composite reliability value, and the average variance extracted (AVE). To meet the minimum requirement, these values should be respectively higher than 0.70, 0.70 and 0.50 [20]. Finally, the study uses the two stage approach proposed by [21] and embedded into SmartPLS 3.0.

4 Results and Discussion

Table 1 presents the outer loadings for full data and each country. As we can see, they all have a value higher than 0.7. Table 2 presents all Cronbach's alpha values, composite reliability and AVE values of our constructs.

As showed in Table 2, all displayed values are meeting the suggested acceptable threshold values of respectively, 0.7, 0.7 and 0.5 [20, 22], and thus justifying the use of all constructs included in our research model.

The study also tested for discriminant validity [23–25] by looking at all correlation matrixes with the square root of the AVEs in the diagonals. All the values were exceeding the inter-correlations of the construct with the other constructs in the model, and thus ensuring the discriminant validity.

Table 3 displays the results of our structural models. From the table, we can observe that all the standardized path coefficient of the core relationships in the TAM model in the context of social media adoption and use within organizations are significant at a level of 0.001. Therefore, all hypotheses derived from TAM (H1, H2, H3, H4, and H5) are supported for the full data, Australia, Canada, the UK, the USA and India. The standardized path coefficient related to the relationship between PS and BIU is only significant (at the level of 0.05) for the full data and for the UK (at the level of 0.1), and thus supporting our hypothesis H6 only for the full data and for the UK. This hypothesis is not supported for Australia, Canada, the USA and India, and therefore suggesting differences in adoption behavior across the countries under study when looking at the impact of perceived security on the behavioral intention to use social media within organizations.

Table 1. Outer loadings

	Full data	Australia (C1)	Canada (C2)	UK (C3)	USA (C4)	India (C5)
ATT1	0.970	0.974	0.965	0.976	0.971	0.929
ATT2	0.956	0.957	0.945	0.966	0.961	0.912
ATT3	0.964	0.966	0.959	0.962	0.970	0.939
ATT4	0.954	0.953	0.936	0.962	0.958	0.934
BIU1	0.847	0.832	0.813	0.859	0.827	0.819
BIU2	0.917	0.919	0.930	0.927	0.911	0.867
BIU3	0.845	0.820	0.812	0.835	0.829	0.863
BIU4	0.912	0.915	0.911	0.926	0.907	0.876
PEU1	0.933	0.937	0.939	0.934	0.923	0.913
PEU2	0.841	0.828	0.815	0.864	0.852	0.828
PEU3	0.940	0.946	0.931	0.948	0.934	0.929
PR1	0.995	0.986	0.961	0.979	0.995	0.961
PR2	0.751	0.750	0.831	0.791	0.718	0.958
PS1	0.972	0.977	0.954	0.966	0.977	0.967
PS2	0.973	0.976	0.962	0.968	0.977	0.966
PU1	0.767	0.754	0.735	0.760	0.770	0.760
PU2	0.921	0.918	0.884	0.913	0.930	0.904
PU3	0.929	0.923	0.901	0.926	0.934	0.902
PU4	0.938	0.940	0.919	0.941	0.935	0.907
PU5	0.929	0.930	0.906	0.923	0.930	0.904

Table 2. Cronbach's alpha values, rho and average variance extracted

	α						Rho_A						AVE					
	Full data	C1	C2	C3	C4	C5	Full data	C1	C2	C3	C4	C5	Full data	C1	C2	C3	C4	C5
ATT	0.972	0.974	0.965	0.977	0.975	0.947	0.973	0.974	0.966	0.977	0.976	0.947	0.923	0.927	0.905	0.934	0.931	0.862
BIU	0.903	0.895	0.890	0.910	0.892	0.879	0.907	0.901	0.904	0.916	0.897	0.882	0.776	0.761	0.754	0.788	0.756	0.734
PEU	0.891	0.890	0.880	0.905	0.889	0.870	0.923	0.923	0.930	0.932	0.918	0.888	0.820	0.820	0.804	0.839	0.817	0.794
PR	0.809	0.771	0.783	0.788	0.784	0.914	3.762	2.001	1.066	1.563	3.896	0.915	0.777	0.767	0.806	0.792	0.753	0.920
PS	0.942	0.952	0.911	0.930	0.952	0.929	0.942	0.952	0.916	0.930	0.952	0.929	0.945	0.954	0.918	0.935	0.954	0.934
PU	0.939	0.937	0.920	0.937	0.941	0.924	0.939	0.937	0.921	0.940	0.941	0.923	0.809	0.802	0.760	0.801	0.814	0.769

Also, we can observe a negative non-significant effect of PR on the BIU for the full data, Australia, Canada, the USA and India. Surprisingly, we found a positive non-significant effect of PR on the BIU for the UK data, and thus reinforcing the existence of differences in adoption behavior across the countries under study for social media adoption and use within organizations. H7 is not supported for the full data and none of the country under study.

When looking at the moderating variable, we can see that computer experience has a direct positive significant effect only on the BIU for the full data and data collected from Canada. However, the standardized path coefficient related to the moderating effects is only significant for the full data (at the level of 0.1) and data collected from

Table 3. Results of the structural model for the full data and each country

	Full data	C1	C2	C3	C4
	Beta(sig.)				
ATT -> BIU	0.563****	0.584****	0.597****	0.600****	0.528****
CEXP -> BIU	0.017*	0.012	0.042**	0.030	0.016
Moderation	0.013*	−0.002	0.027	0.033**	0.000
PEU -> ATT	0.397****	0.455****	0.385****	0.454****	0.350****
PEU -> PU	0.573****	0.537****	0.514****	0.512****	0.577****
PR -> BIU	−0.009	−0.003	−0.039	0.025	−0.010
PS -> BIU	0.033**	0.046	0.022	0.059*	0.001
PU -> ATT	0.531****	0.493****	0.520****	0.476****	0.566****
PU -> BIU	0.352****	0.323****	0.309****	0.309****	0.423****

****$P < 0.001$; ***$P < 0.01$; **$P < 0.05$; *$P < 0.1$

the UK (at the level of 0.05), and therefore, we must accept H8 only for the full data and for the UK, and reject H8 for Australia, Canada, the USA and India.

5   Conclusion and Future Research Directions

This study starts with the aim of using an extended version of the TAM that integrates perceived risk and security to assess the acceptance of social media within the workspace across multiple countries and to test for user's behaviors homogeneity. Also, the study investigates the moderating effects of user computer experience on the relationship between user's attitude and intention to use social media. To test the proposed model, the study uses data collected from the US, Australia, the UK, Canada, and India. In the data analysis process, the study uses the full data and data from each country.

The study found that all core relationships in the TAM model in the context of social media adoption and use within organizations are significant at a level of 0.001 for the full data, and data collected from each country under study (Australia, Canada, the UK, the USA and India), and thus confirming the robustness of the TAM. The study found the relationship between PS and BIU is only significant for the full data and for the UK. Similarly, the study found that there is a negative non-significant effect of PR on the BIU for the full data, Australia, Canada, the USA and India, with however a positive non-significant effect of PR on the BIU for the UK data, and thus confirming the existence of differences in adoption behavior across the countries under study for social media adoption and use within organizations. Future studies may focus on using more advanced techniques to explore the presence of unobserved heterogeneity [26].

Finally, the study found that computer experience has a direct positive significant effect only on the BIU for the full data and data collected from Canada. However, the standardized path coefficient related to the moderating effects is only significant for the full data and data collected from the UK, and therefore reinforcing the existence in adoption behavior across the countries under study for social media adoption and use within organizations. Looking at the impact of these differences on the organizational performance should be included into future research directions.

References

1. Stephen, A.T., Toubia, O.: Deriving value from social commerce networks. J. Mark. Res. **47** (2), 215–222 (2010)
2. Culnan, M.J., McHugh, P.J., Zubillaga, J.I.: How large U.S. companies can use twitter and other social media to gain business value. MIS Q. Executive **9**(4), 243–259 (2010)
3. Burke, W.Q., Fields, D.A., Kafai, Y.B.: Entering the clubhouse: case studies of young programmers joining the online scratch communities. J. Organ. End User Comput. **22**(2), 21 (2010)
4. IBM: Social Commerce Defined, IBM, Editor. IBM (2009)
5. Herhold, K.: How Businesses Use Social Media: 2017 Survey (2017). https://clutch.co/ agencies/social-media-marketing/resources/social-media-survey-2017. Cited 16 Nov 2017
6. STATISTA: Social network advertising revenue from 2014 to 2017 (in billion U.S. dollars) (2017). https://www.statista.com/statistics/271406/advertising-revenue-of-social-networks-worldwide/. Cited 15 Nov 2017
7. Fosso Wamba, S., et al.: Role of intrinsic and extrinsic factors in user social media acceptance within workspace: assessing unobserved heterogeneity. Int. J. Inf. Manage. **37** (2), 1–13 (2017)
8. Fosso Wamba, S., Carter, L.: Social media tools adoption and use by SMES: an empirical study. J. Organ. End User Comput. **26**(2), 1–17 (2014)
9. Fosso Wamba, S., Edwards, A.J.: Factors related to social media adoption and use for emergency services operations: the case of the NSW SES. In: 20th Americas Conference on Information Systems, Savannah, Georgia, USA (2014)
10. Meire, M., Ballings, M., Van den Poel, D.: The added value of social media data in B2B customer acquisition systems: A real-life experiment. Decis. Support Syst. **104**(Supplement C), 26–37 (2017)
11. Dong, J.Q., Wu, W.: Business value of social media technologies: evidence from online user innovation communities. J. Strateg. Inf. Syst. **24**(2), 113–127 (2015)
12. Davis, F.D.: Perceived usefulness, perceived ease of use, and user acceptance of information technology. MIS Q. **13**(3), 319–340 (1989)
13. Venkatesh, V., Davis, F.D.: A theoretical extension of the technology acceptance model: four longitudinal field studies. Manage. Sci. **46**(2), 186–204 (2000)
14. Venkatesh, V., et al.: User acceptance of information technology: toward a unified view. MIS Q. **27**(3), 425–478 (2003)
15. Davis, F.D., Bagozzi, R.P., Warshaw, P.R.: User acceptance of computer technology: a comparison of two theoretical models. Manage. Sci. **35**(8), 982–1003 (1989)
16. Trinchera, L.: Unobserved Heterogeneity in Structural Equation Models: a new approach to latent class detection in PLS Path Modeling, p. 338. Universita degli Studi di Napoli Federico II, Napoli (2007)
17. Teuku, N.: The effects of experience, complexity, and computer self efficacy factors towards the use of human resources information system application. J. Res. Bus. Manage. **2**(11), 26–33 (2014)
18. Thong, J.Y.L., et al.: Consumer acceptance of personal information and communication technology services. IEEE Trans. Eng. Manage. **58**(4), 613–625 (2011)
19. Ringle, C.M., Wende, S., Becker, J.-M.: SmartPLS 3 (2014). www.smartpls.com
20. Leung, J., Cheung, W., Chu, S.-C.: Aligning RFID applications with supply chain strategies. Inf. Manage. **51**(2), 260–269 (2014)

21. Fosso Wamba, S., Ngai, E.W.T.: Importance of issues related to RFID-enabled healthcare transformation projects: results from a delphi study. Prod. Plann. Control **26**(1), 19–33 (2015)
22. Guadagnoli, E., Velicer, W.: Relation of sample size to the stability of component patterns. Psychol. Bull. **103**, 265–275 (1988)
23. Chin, W.W.: The partial least squares approach for structural equation modeling (1998)
24. Chin, W.W.: How to write up and report PLS analyses. In: Handbook of Partial Least Squares, pp. 655–690 (2010)
25. Fornell, C., Larcker, D.F.: Evaluating structural equation models with unobservable variables and measurement error. J. Mark. Res. **18**(1), 39–50 (1981)
26. Becker, J.-M., et al.: Discovering unobserved heterogeneity in structural equation models to avert validity threats. MIS Q. **37**(3), 665–694 (2013)

Leadership, Autonomy and Innovation on a High-Tech Organization

Ana Veloso$^{(\boxtimes)}$ ⓘ and Ana Rita Silva$^{(\boxtimes)}$

School of Psychology, University of Minho, Braga, Portugal
alveloso@psi.uminho.pt, a65536@alumni.uminho.pt

Abstract. The purpose of the case study we are presenting was to explore the relationship between transformational leadership and innovative behaviour and the mediating or moderating role of team-level autonomy in the context of high-tech organizations oriented for innovation. A case study was developed in a High-Tech organization, from the north of Portugal. The study research design involved quantitative and qualitative methods. However, for the purpose of this paper, partial results of the quantitative study are being presented. From 143 employees who fulfil the conditions needed to participate in the study (members of a team and didn't have any leadership position), seventy-six answered, which points out for a response tax of 53%. A questionnaire that included three scales was applied: *Global Transformational Leadership* (GTL); *Team-Level Autonomy* (TLA) and *Innovative Behaviour* (IB). It was found a positive relation between transformational leadership and innovative behaviour and another positive relation between transformational leadership and team level autonomy. Surprising, no relation was found between team level autonomy and innovative behaviour. As discussed, this result may be related to the need of maturing relations and work processes, as this firm is young (8 years since its inception).

Keywords: Creativity · Innovation · Leadership · Team
High-tech organizations

1 Introduction

High-tech organizations have specific characteristics, namely a high pressure to present creative and innovative products/services to their clients in an environment of continuous change and competition. Also, they usually employ high-qualified workers and are focused on developing new products from new technology [1].

In such an organizational context, it is important to develop human resource management (HRM) practices as teamwork and support the development of Creativity and Innovation [2].

HRM is recognized as having an important impact on firms' performance because plays an important role in managing employees' knowledge, skills and behaviours, namely leadership [3, p. 1208]. Leadership defined as the ability of influence employees [4], has an important job supporting employees' innovation [5].

The model of transformational leadership of Carless, Wearing and Mann [6] considers seven behaviours through which leaders influence followers: (a) being a

© Springer International Publishing AG, part of Springer Nature 2018
Á. Rocha et al. (Eds.): WorldCIST'18 2018, AISC 747, pp. 393–398, 2018.
https://doi.org/10.1007/978-3-319-77700-9_39

visionary, (b) being innovative, (c) supporting others, (d) giving empowerment, (e) leadership by example, (f) being charismatic and (g) developing each employee.

HRM also supports employees' creativity and innovation for instance, through the design of a process that increases employees' efficiency: work design, teamwork, education and training, performance assessment and rewards [7].

Teamwork is a common HRM practice in High-tech organizations [8]. When is associated with a transformational style of leadership, better results are reached, considering creativity and innovation, if team autonomy is present. Autonomy is associated with high levels of innovation [9, 10]. For instances, employees who experience organizational support, trust and autonomy from their leaders, describe the organization as supportive of Innovation [11].

The purpose of the case study we are presenting was to explore the relationship between transformational leadership and innovative behaviour and the mediating or moderating role of team-level autonomy in the context of high-tech organizations oriented for innovation (see Fig. 1).

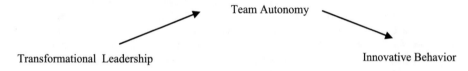

Transformational Leadership   Innovative Behavior

Fig. 1. Model of the study.

2 Method

A case study was developed in a High-Tec organization, from the north of Portugal. This organization was chosen because the research team had an easy and fast access to the Human Resource Management Department.

The main activity of this organization is the development of an online platform to sell luxury brands. The organization was at the time of the study, 8 years old and employed 500 professionals.

The study research design involved quantitative and qualitative methods. First, it was asked to the participants to complete and online survey. The quantitative results of the questionnaire were explored with interviews with special attention to explore employees' perceptions about innovation, leadership and team autonomy. However, for the purpose of this paper, only partial results of the quantitative study are being presented.

2.1 Sample

An invitation to participate in the study was sent to 143 employees who were members of a team and didn't have any leadership position. Seventy-six answered, which points out for a response tax of 53%.

The sample reported an average age between 21 and 30. Fifty five percent of the participants were female and 74% had high level of education (master degree level or higher). All the participants were members of a team and none had any leadership position.

2.2   Instruments

The participants answered a questionnaire that included three scales:

Global Transformational Leadership (GTL). GTL was developed by Carless, Wearing and Mann [6] and adapted to the Portuguese population by Beveren [12] to measure transformational leadership, is a short scale of seven items. The response format of GTL is a 5-point Likert scale, ranging from 1− "rarely or never" to 5− "very frequently, if not always". This scale assesses at what extent a person is visionary, innovative, supportive, participative and worthy of respect. An example of an item is: "My leader communicates a clear and positive vision of the future". GIL validation reliability reached an alpha of Cronbach of 0.93.

Team-Level Autonomy (TLA). TLA was adapted to the Portuguese population by Beveren [12] from the original scale created by Langfred [13]. The 8 item scale was used to evaluate the autonomy of the group. An item example is: "The team is free to decide how to do its own work". The response format of TLA is a 5-point Likert scale, ranging from 1− "rarely it applies" to 5− "it applies almost always". TLA validation reliability reached an alpha of Cronbach of 0.90.

Innovative behaviour (IB). A scale developed by Scott and Bruce [11] and adapted to the Portuguese population by [14] was used to measure innovative behaviour. The 5 items scale measure how often respondents believe they exhibit several innovative behaviours in their workplace. An example of an item is "I promote and support others' ideas". Response options ranged from 1 to 6 (1 = strongly disagree; 6 = strongly agree). Exploratory factor analysis showed one single factor, with an alpha of Cronbach of 0.84.

3   Analysis and Results

A psychometric analysis of the scales was carried on based on the descriptive measures of the answers to the questionnaire: average, standard- deviation (SD) minimum and maxim. See Table 1.

Table 1. Descriptive statistics of the scales.

Scale	Minimum	Maxim	Average	Standard deviation
GTL	1.29	5.00	3.56	0.88
TLA	1.29	5.00	3.21	0.80
IB	3.00	5.80	4.72	0.63

3.1   Correlational Analysis

To explore the relation between the variables of the study a correlation analysis was made (see Table 2).

Table 2. Correlations between scales.

Scale	GTL	TLA	IB
GTL	–	0.53**	0.25*
TLA	0.53**	–	n.s.
IB	0.25*	n.s.	–

* p < .05, **p < .01

A positive correlation was found between transformational leadership and innovative behaviour ($r_s = 0.25$, p < .05) and with team autonomy ($r_s = 0.53$, p < .01). However, there is no significant relation between team level autonomy and innovative behaviour.

4   Discussion of Results

The purpose of this study was to explore the relationship between transformational leadership and innovative behaviour and the mediating or moderating role of team-level autonomy in the context of high-tech organizations oriented for innovation.

First, it was explored the presence of the variables in the point of view of the employees. In this organization, the transformational style of leadership has reached high levels as well team autonomy (although with lower values) and innovative behaviour (see Table 2).

As expected, considering the research in this field, we found a positive relation between transformational leadership and innovative behaviour [6, 15]. In this organization, leaders who, for instance, are supportive, visionary, help team members to develop as individuals, will lead their team to have more innovative behaviours and help the organization to have more success. We also found a relation between transformational leadership and team level autonomy. However, the values this relation reached, points out for a potential possibility of further development, if the leader increases the team level autonomy allowing that the team manage their own work and think creatively [16].

There was not found any statistical relevant relation between team level autonomy and innovative behaviour. This result was not expected as most of the research indicates that innovative behaviour and autonomy are related as the latter, when present, allows team members to have creative and innovative thinking [9, 10, 17]. One possible explanation of this result could be related with the organizational maturity and its fast development. Gratton, Hope-Hailey, Stiles and Truss [18] refer that time is important when groups and organizations are being analysed because some behaviour as, for example, leadership, need time to develop and have an impact: leaders need time to prepare their team to work together and also to mature processes and relations.

It was also our objective to explore the relationship between transformational leadership and innovative behaviour, but also the mediating or moderating role of team-level autonomy. However, as no relation was found between team autonomy and Innovative behaviour, it was not possible to go further with our analysis [19] (Fig. 2).

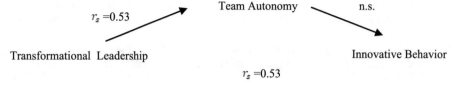

Fig. 2. Correlations between variables

Our results, confirm previous research findings concerning the relations between transformational leadership, team level autonomy and innovative behaviour. However, the absence of a relation between team level autonomy and innovative behaviour should be further explored. Also, the importance of time and the relation of organizational maturity, specifically the degree of development and maturity of processes and relation, and the impact of HRM practices as teamwork and the development of leaders, in such a specific organization as high-tech organizations requires some attention from researchers.

References

1. Jassawalla, A.R., Sashittal, H.C.: An examination of collaboration in high-technology new product development processes. J. Prod. Innov. Manag. **15**(3), 237–254 (1998)
2. Rodrigues, F.C., Veloso, A.: A relevância da Gestão de Recursos Humanos para a inovação organizacional, Special number "Inovação, Criatividade e Empreendedorismo" of Revista Psicologia: Organizações e Trabalho 13(3), 41–56 (2013). ISSN 1984-6657
3. Jiménez-Jiménez, D., Sanz-Valle, R.: Could HRM support organizational innovation? Int. J. Hum. Resour. Manage. **19**(7), 1208–1221 (2008). https://doi.org/10.1080/0958519080210 9952
4. Ferreira, J.C., Neves, J., Caetano, A.: Manual de Psicossociologia das Organizações. McGraw-Hill, Lisboa (2001)
5. Bedani, M.: O Impacto dos Valores Organizacionais na Percepção de Estímulos e Barreiras à Criatividade no Ambiente de Trabalho. Revista Administrativa Mackenzie **13**(3), 150–176 (2012)
6. Carless, S., Wearing, A., Mann, L.A.: Short Measure of Transformational Leadership. J. Bus. Psychol. **14**(3), 389–405 (2000)
7. De Leede, J., Looise, J.K.: Innovation and HRM: Towards an Integrated Framework. Creat. Innov. Manage. **14**, 108–118 (2005). https://doi.org/10.1111/j.1467-8691.2005.00331
8. Veloso, A., Keating, J.: Gestão de recursos humanos em pme's de elevada tecnologia. Psicologia **22**(1), 35–58 (2008)
9. Amabile, T.M., Conti, R., Coon, H., Lazenby, J., Herron, M.: Assessing the Work Environment for Creativity. Acad. Manag. J. **39**(5), 1154–1184 (1996)
10. Eisenbeiß, S.A., Boerner, S.: Transformational Leadership and R&D Innovation: Taking a Curvilinear Approach. Creat. Innov. Manage. **19**(4), 364–372 (2010). https://doi.org/10. 1111/j.1467-8691.2010.00563.x
11. Scott, S., Bruce, R.: Determinants of Innovation Behavior: A Path Model of Individual Innovation in the Workplace. Acad. Manag. J. **37**(3), 580–607 (1994)

12. Beveren, P.V.: Liderança Transformacional e Autonomia Grupal: Adaptação de Instrumentos de Medida (Dissertação de Mestrado) (2015). Retrieved from: http://hdl.handle.net/10316/29078

13. Langfred, C.W.: Autonomy and Performance in Teams: The Multilevel Moderating Effect of Task Interdependence. J. Manag. **31**(4), 513–529 (2005). https://doi.org/10.1177/0149206304272190

14. Gomes, J.F.S., Veloso, A., Roque, H., Ferreira, A.T.: Adaptação e validação de um Questionário de Comportamento Inovador, working paper (2017)

15. Anderson, N., Potočnik, K., Zhou, J.: Innovation and creativity in organizations: A state-of-the-science review, prospective commentary, and guiding framework. J. Manage. **40** (5), 1297–1333 (2014). https://doi.org/10.1177/0149206314527128

16. Piccolo, R.F., Colquitt, J.A.: Transformational leadership and job behaviours: the mediating role of core job characteristics. Acad. Manage. J. **49**(2), 327–340 (2006). https://doi.org/10.5465/AMJ.2006.20786079

17. Amabile, T.M.: A model of creativity and innovation in organizations. Res. Organ. Behav. **10**(1), 123–167 (1988)

18. Gratton, L., Hope-Hailey, V., Stiles, P., Truss, C.: Linking individual performance to business strategy: the people process model. Hum. Resour. Manage. **38**(1), 17–31 (1999)

19. Baron, R., Kenny, D.: The moderator-mediator variable distinction in social psychological research: Conceptual, strategic and statistical considerations. J. Pers. Soc. Psychol. **51**, 1173–1182 (1986)

Performance Appraisal of Higher Education Teachers' in Information Systems and Technology: Models, Practices and Effects

Ana T. Ferreira-Oliveira[1,2(✉)], Sandra Fernandes[3(✉)], and Joana Santos[1(✉)]

[1] Technology and Management School, Viana do Castelo Polytechnic Institute, Viana do Castelo, Portugal
{ateresaoliveira,joana}@estg.ipvc.pt
[2] Faculty of Economics, Porto University, Porto, Portugal
[3] Portucalense Institute for Human Development, Portucalense University, Rua Dr. António Bernardino de Almeida, 541, 4200-072 Porto, Portugal
sandraf@upt.pt

Abstract. Despite the increasing awareness of the relevance of technologies to various life domains and environments, its application to research on Information Systems and Technology Teachers' Evaluation in Higher Education is still very limited. The IT sector has special characteristics, such as the high employability rates that are very relevant to integrate in the discussion about the teachers' performance appraisal. Universities need to reflect about the retention of teachers with good and competitive performance appraisal systems. Also, performance appraisal models had been suffering a change in its theoretical and empirical relevance towards a process-based approach. This short paper presents a work in progress that intends to analyse the state of the art on teacher evaluation in Higher Education, specifically on Information Systems and Technology Teachers', review national and international teacher evaluation frameworks in Higher Education, identify the main principles and assumptions underlying teacher evaluation models in Higher Education and characterize the existing models and practices of teacher evaluation, at national and international level, their main results and conclusions.

Keywords: Teacher evaluation · Quality standards
European Higher Education Area (EHEA)

1 Introduction

Teacher Evaluation in higher education is a complex and controversial issue, which has been subject to great reflection and attention. It is considered an important strategy to improve the quality of higher education institutions. Therefore, it should be analysed and understood, in the European context, according to the recent Standards and Guidelines for Quality Assurance in the European Higher Education Area, defined by the European Association for Quality Assurance in Higher Education [1], which provide a common European framework for the analysis of quality at national and institutional

© Springer International Publishing AG, part of Springer Nature 2018
Á. Rocha et al. (Eds.): WorldCIST'18 2018, AISC 747, pp. 399–404, 2018.
https://doi.org/10.1007/978-3-319-77700-9_40

levels. A key goal of the European Standard Guidelines is to contribute to the common understanding of quality assurance for learning and teaching across borders and among all stakeholders. The ESG have been used by institutions and quality assurance agencies as a reference document for internal and external quality assurance systems in higher education. Teachers and academics play an important role in the development of national and institutional quality assurance systems across the European Higher Education Area (EHEA). These quality standards are an opportunity for Higher Education institutions to reflect upon their internal mechanisms for quality assurance, where teacher evaluation appears not only as a management tool, but as an essential resource to contribute to teachers' professional development, promoting better individual and collective perform-ance of teachers within the university.

This study will aim to contribute to the development of effective evaluation processes that cover both individual and institutional objectives and lead to the quality of Higher Education in Portugal.

The Information Systems and Technology Teachers' constitute a group of teachers considering the high employability rates of the sector. Higher Education Institutions (Heis) need to preserve, retain and attract the best professionals to fulfil their goals in this specific development area. It is crucial to develop studies where we can define models and practices that can be used as best practices, at this particular sector, aiming at retaining talents.

1.1   Teacher Evaluation: Perspectives and Objectives

Teacher Evaluation necessarily involves considering a number of issues that fit and determine a particular definition of what it means to be a teacher and how teachers perceive the roles inherent to the teaching profession.

Assuming that evaluation involves gathering information to judge the merit and/or value of the teacher [2] this means that different definitions of teaching, require different ways of gathering information and making value judgments. This is, the concept of being a teacher necessarily determines the type of data collection and the use of that informa-tion to judge the merit and/or value and, therefore, conditions the evaluation process itself [3]. The diversity of perspectives on what being a teacher is and what it means to teach has serious implications for teacher evaluation and the lack of consensus on this matter is one of the main constraints for the development of an effective teacher evalu-ation system. According to Hadji [4], the "difficulty in evaluating teachers derives more from the uncertainty that outweighs the very essence of teaching and the lack of consensus in this regard, rather than technical problems, always secondary [...] The purpose of teacher evaluation is difficult to establish as it is difficult to define" (p. 32).

Data collection is considered one of the most arguable and least understood compo-nents of the teacher evaluation process. The literature reveals that teachers' work is best described and evaluated when there are different types of evidence collected from multiple sources and multiple methods. The most common practice in any evaluation system is the use of multiple kinds of evidence or data sources (self-evaluations, peer observations, student ratings). The use of a multiplicity of sources is related to the fact that the majority of tasks are themselves multifaceted and that the use of multiple sources increases the validity and reliability of evaluation and reduces the subjectivity, providing more opportunities for evaluating both the quality and quantity of activities. According to Braskamp and Ory [5] an effective evaluation should cover both individual and institutional objectives, reflect the complexity of teachers' work, promote the identity and uniqueness of each teacher and their professional development, communicate the objectives and institutional expectations in a clear way and, finally, should promote collegiality among faculty.

Teacher evaluation should be considered as an important strategy to develop quality assurance in Higher Education and teachers' professional development and not seen as an attack to teachers' professionalism [6]. Also of great interest in this discussion is the link between research and teaching in teacher evaluation models [7, 8]. Is it possible to link formative purposes (targeted for professional development) and summative (oriented to accountability) in the same framework? The idea of an evaluation that promotes learning and professional development is still underdeveloped in many higher education institutions and evaluation is still not viewed as a way of thinking differently and adjusting work while it is ongoing, instead of doing this after at the end.

2 Methodology

This study is part of a broader ongoing research project which aims to analyze the state of the art on teacher evaluation in Higher Education, specifically on Information Systems and Technology Teachers', review national and international teacher evaluation frameworks in Higher Education, identify the main principles and assumptions underlying teacher evaluation models in Higher Education and characterize the existing models and practices of teacher evaluation, at national and international level, their main results and conclusions. It seeks to identify the key dimensions considered in national policy documents and frameworks developed amongst higher education institutions in Portugal. Based on a review of existing literature on performance appraisal in higher education and on several national policy documents, this study seeks to explore national case studies, looking at different existing frameworks and guidelines used for the appraisal of faculty, identifying the overall dimensions, indicators and weights that have been set out to evaluate faculty performance. To achieve this goal, the analysis of stakeholders' perspectives regarding the design, development and implementation of performance appraisal frameworks in Portuguese higher education institutions will be of crucial importance.

The study seeks to find answers to this question and other research questions, such as:

- What are the current teacher evaluation policies in Higher Education?
- What are the main principals and assumptions underlying the legal documents?
- What are the main goals and purposes defined for teacher evaluation?
- What are the key dimensions included in the (existing) teacher evaluation frameworks?
- What tools and sources of information are used?
- What are academic staff perceptions regarding teacher evaluation policies and practices?

– How are the research and teaching dimensions considered in the process?

The lack of scientific studies and existing literature on teacher evaluation in the context of Higher Education in Portugal are a significant opportunity for the development of this study, its objectives and outcomes. In general, this study aims to develop three main objectives and their specific objectives:

- Analyze the state of the art on teacher evaluation in Higher Education, specifically on Information Systems and Technology Teachers'.
    - Review national and international teacher evaluation frameworks in Higher Education, specifically on Information Systems and Technology Teachers'.
    - Identify the main principles and assumptions underlying teacher evaluation models in Higher Education.
    - Characterize the existing models and practices of teacher evaluation, at national and international level, their main results and conclusions.
- Identify teachers and academic leaders' perspectives regarding the implementation of Information Systems and Technology Teachers evaluation frameworks in higher education institutions
    - Compare the results and effects of teacher evaluation in different Portuguese Higher Education institutions
    - Identify the main opportunities and challenges that the teacher evaluation process raises for Higher Education institutions

Discuss the roles of the university professor (research, teaching, management and extension to the community) considering the current policies and practices of teacher evaluation in Higher Education institutions in Portugal.

- To support national higher education institutions and their stakeholders to address the quality standards defined by the European Higher Education Area (EHEA).
    - Implement an action-research plan in two public higher education institutions, aimed at involving teachers and stakeholders in a collaborative process of finding solutions to their own problems, leading to a change process;
    - Develop open access resources based on best practices of successful national case studies of teacher evaluation;
    - Define a set of guidelines, recommendations and suggestions for improving quality in higher education, published in a practical guide (handbook);
    - Draw inputs to support the link between research and teaching in teacher evaluation frameworks used in higher education institutions;
    - Publish an ebook with Best Practices of Teacher Evaluation in Higher Education in Portugal

To carry out the study, international studies with similar goals to this study will be carefully analyzed, such as the ones following [9, 10] in Australia, [11] in North America and in the United Kingdom [12]. The methodology used in these international studies comprised a survey of institutions focusing on existing policies and practices, a literature review, consultation with academic staff (via survey and case studies) on tentative proposals made for recognizing and rewarding good teaching across the university system, a comparison of the findings from the case studies, staff and institutional surveys, amongst other methods.

The study will employ both a qualitative and a quantitative methodology. The research will be developed in two different phases. The first phase will include an update of the review of the state of the art and the development of a critical analysis of the existing performance appraisal frameworks implemented amongst Portuguese higher education institutions. This includes the review and analysis of articles in recognized international peer-reviewed journals, as well as research reports and publications from assessment agencies such as the national Agency for Assessment and Accreditation of Higher Education (A3ES) and the European Association for Quality Assurance in Higher Education (ENQA).

For data collection, data will be selected and compared from different public higher education institutions in Portugal. In the first phase, data collection will be based on the application of an online survey addressed to all Information Systems and Technology Teachers from selected higher education institutions. The questionnaire will be designed and validated before application, with the support of an international consultant from ENQA and/or A3Es. The purpose is to discover how far Information Systems and Technology Teachers agree with the principles of the performance appraisal frameworks being used and their perceptions of how well the various indicators have been implemented in their organic unit and/or higher education institution. Also in this phase, individual semi-structured interviews to stakeholders and academic leaders will be carried out, inquiring about positive/negative outcomes of the current teacher evaluation frameworks used in the institutions, the problems and constraints, the unforeseen effects, the implications for teachers' career progression and professional development, amongst other important issues that might emerge from the literature review update.

A set of guidelines, recommendations and suggestions for improving teacher evaluation practices in higher education will be developed at the end of the study, based on findings from the implementation of an action research plan. The publication of an ebook with Best Practices of Information Systems and Technology Teachers Evaluation in Higher Education in Portugal and the development of open access resources included in a Handbook for Information Systems and Technology Teachers Evaluation are some one of the expected outcomes. Finally, the study intends to develop a culture of continuous improvement of Higher Education, through the systematic discussion and reflection, with academic leaders and teachers, about quality in higher education.

3   Final Remarks

Despite the increasing awareness of the relevance of technologies to various life domains and environments, its application to research on Information Systems and Technology Teachers' Evaluation in Higher Education is still very limited. The IT sector has special characteristics, such as the high employability rates that are very relevant to integrate in the discussion about the teachers' performance appraisal. Universities need to reflect about the retention of teachers with good and competitive performance appraisal systems. Also, performance appraisal models had been suffering a change in its theoretical and empirical relevance towards a process-based approach. It is relevant to continue developing more efficient human resource management systems that integrate adequate performance appraisal models considering the high competitiveness of the sector and the need to respond to the personal and professional expectations of teachers, towards their professional development and subsequently higher competitiveness in higher education institutions.

References

1. ENQA, ESU, EUA & EURASHE.: Standards and Guidelines for Quality Assurance in the European Higher Education Area (ESG), Brussels, Belgium (2015)
2. Flores, M.A.: Da Avaliação de Professores: Reflexões sobre o caso Português. Revista Iberoamericana de Evaluacion Educativa 2(1), 240–256 (2009)
3. Flores, M.A.: A Avaliação de Professores numa Perspectiva Internacional: sentidos e implicações. Areal Editores, Porto (2010)
4. Flores, M.A., Veiga Simão, A.M.: Competências desenvolvidas no contexto do Ensino Superior: a perspectiva dos diplomados. In Actas das V Jornadas de Redes de Investigación en Docencia Universitaria, Universidade de Alicante, Espanha (2007)
5. Braskamp, L.A., Ory, J.C.: Assessing faculty work: Enhancing individual and institutional performance. Jossey-Bass Publishers, San Francisco (1994)
6. Harley, S.: The impact of research selectivity on academic work and identity in UK universities. Stud. High. Educ. 27(2), 187–205 (2002)
7. Sikes, P.: Working in a 'new' university in the shadow of the Research Assessment Exercise. Stud. High. Educ. 31(5), 555–568 (2006)
8. Young, P.: Out of balance: lecturers' perceptions of differential status and rewards in relation to teaching and research. Teach. High. Educ. 11(2), 191–202 (2006)
9. Ramsden, P., Margetson, D., Martin, E., Clarke, E.: Recognising and Rewarding Good Teaching in Australia Higher Education. Australian Government Publishing Service, Canberra (1995)
10. Ramsden, P., Martin, E.: Recognition of good university teaching: policies from an Australian study. Stud. High. Educ. 21(3), 299–315 (1996)
11. Fairweather, J.: Faculty Work and Academic Trust: Restoring the Value of Teaching and Public Service in American Academic Life. Allyn and Bacon, Boston (1996)
12. Gibbs, G., Habeshaw, T.: Recognising and Rewarding Excellent Teaching. Open University, Milton Keynes (2003)

Author Index

© Springer International Publishing AG, part of Springer Nature 2018   405
Á. Rocha et al. (Eds.): WorldCIST'18 2018, AISC 747, pp. 405–406, 2018.
https://doi.org/10.1007/978-3-319-77700-9

Printed in the United States
By Bookmasters